Doing Feminist Urban Research

Doing Feminist Urban Research introduces the reader to the newly emerging 21st-century global landscape of feminist urban research. It showcases decolonising practices, partnerships and teamwork, new standards such as EDI, geo-ethnographic methodologies, software-enhanced qualitative data analysis, and knowledge mobilisation.

This book delves into both the institutional and lived realities of the practice of feminist urban research for the 21st century via the insights of the GenUrb transnational research project. Through reflection exercises based on real-life examples, it covers feminist methodologies and research techniques, critically examining the 'field' through comparison and feminist geo-ethnographies. It guides readers through navigating the politics of decolonising research, working across differences, and embracing feminist ethics and activism. The book also explores data through the practices of translation, data management, data analysis, and the use of NVivo. And it further introduces professional standards, including EDI, collaboration with partners, engagement in teamwork, the handling of crises, such as pandemics, and knowledge mobilisation, including utilising social media. Accompanying web resources will assist scholars and students with additional audio files and documents.

This book's practical guidance will help those starting to contemplate and engage in qualitative feminist urban research as well as those teaching the practice and politics of research. It will appeal to practitioners in urban studies, geography, gender and women's studies, sociology, anthropology, global studies, and development studies.

Linda Peake, FRSC, is a professor in the Faculty of Environmental and Urban Change at York University, Toronto, Canada where she was also Director of the City Institute (2013–2023). She is PI on the SSHRC Partnership Grant, Urbanisation, gender and the global south: a transformative knowledge network (GenUrb), a Trustee of the Urban Studies Foundation, and an Associate Editor on the AAG *International Encyclopedia of Geography*. Her latest publications include the books *Urbanisation in a Global Context* (2nd edition, edited with Alison Bain, 2022), *A Feminist Urban Theory for Our Time: Rethinking Social Reproduction and the Urban* (edited with Elsa Koleth, Gökbörü Tanyildiz, Raj Narayanareddy, and darren patrick, 2021), and the forthcoming *Elgar Handbook on Gender and Cities* (edited with Grace Adeniyi-Ogunyankin and Anindita Datta).

Nasya S. Razavi was a postdoctoral fellow with GenUrb (2019–2024) and is lead researcher on the Cochabamba City Research Team (CRT). She is currently the Latin America Program Manager at Inter Pares, a feminist social justice organisation based in Ottawa. Nasya completed her PhD at the Department of Geography and Planning at Queen's University, Kingston, Canada, in 2019, which she has published with Routledge as *Water Governance in Bolivia: Cochabamba since the Water War*. Nasya adopts a feminist decolonial approach to her work in international development, gender, and environmental and social justice.

Araby Smyth was a postdoctoral fellow with GenUrb (2021–2024), researching place ecologies of finance and debt. She is currently an Assistant Professor in the Department of Geography and Environment at Mount Allison University. Her research has been funded by the Antipode Foundation,National Science Foundation (USA), and Society of Woman Geographers. She has published in geography journals such as *Antipode*, *Environment and Planning A: Economy and Space*, and *Transactions of the Institute of British Geographers*. She is an editor on the Editorial Collective of the journal *ACME: An International Journal for Critical Geographies*.

Doing Feminist Urban Research

Insights from the GenUrb Project

Edited by Linda Peake, Nasya S. Razavi, and Araby Smyth

Routledge
Taylor & Francis Group
LONDON AND NEW YORK

Designed cover image: gettyimages.ca/Jackyenjoyphotography

First published 2025
by Routledge
4 Park Square, Milton Park, Abingdon, Oxon OX14 4RN

and by Routledge
605 Third Avenue, New York, NY 10158

Routledge is an imprint of the Taylor & Francis Group, an informa business

British Library Cataloguing-in-Publication Data
A catalogue record for this book is available from the British Library

Library of Congress Cataloging-in-Publication Data
Names: Peake, Linda, 1956– editor. | Razavi, Nasya Sara, editor. | Smyth, Araby, editor.
Title: Doing feminist urban research / edited by Linda Peake, Nasya S. Razavi, and Araby Smyth.
Description: Abingdon, Oxon ; New York, NY : Routledge, 2025. | Includes bibliographical references and index.
Identifiers: LCCN 2024016894 (print) | LCCN 2024016895 (ebook) | ISBN 9781032668673 (hardback) | ISBN 9781032668680 (paperback) | ISBN 9781032668727 (ebook)
Subjects: LCSH: Cities and towns—Research. | Feminism. | Feminist anthropology.
Classification: LCC HT110 .D64 2025 (print) | LCC HT110 (ebook) | DDC 307.76072—dc23/eng/20240527
LC record available at https://lccn.loc.gov/2024016894
LC ebook record available at https://lccn.loc.gov/2024016895

ISBN: 9781032668673 (hbk)
ISBN: 9781032668680 (pbk)
ISBN: 9781032668727 (ebk)

DOI: 10.4324/9781032668727

Typeset in Sabon
by Apex CoVantage, LLC

Doing Feminist Urban Research: Insights from the GenUrb Project
Edited by Linda Peake, Nasya S. Razavi, and Araby Smyth
With City Research Team contributions from

Cochabamba
Nasya S. Razavi
Sonia Pardo Burgoa

Delhi
Anindita Datta
Swagata Basu

Georgetown
Karen de Souza
Joy Marcus
Linda Peake

Ibadan
Grace Adeniyi-Ogunyankin
Sylvester Nmormah
Bukola Omolara Odunola

Ramallah
Natasha Aruri
Mai Al-Battat

Shanghai
Penn Tsz Ting Ip
Tsung-Yi Michelle Huang

Dedication

We dedicate this book:

To Kemi (March 8, 1977, to July 31, 2023)

To the participants who engaged in the research, gave their time, and shared their stories.

To the readers, the next generation of researchers. We hope this book sparks a desire to commit to social justice and transformative knowledge production in the hope of more just urban futures.

Brief contents

Detailed contents

Figures and tables

Figures

Tables

Companion website materials

Audio clip

Documents

Editor, author, and contributor biographies

Editors

Linda Peake, FRSC, is a professor in the Faculty of Environmental and Urban Change at York University, Toronto, Canada, where she was also director of the City Institute (2013–2023). She is principal investigator on the SSHRC Partnership Grant, *Urbanisation, gender, and the global South: a transformative knowledge network* (GenUrb), a trustee of the Urban Studies Foundation, and an associate editor of the *AAG international encyclopedia of geography*. Her latest publications include *Urbanisation in a global context* (2nd edition, edited with Alison Bain, 2022); *A feminist urban theory for our time: rethinking social reproduction and the urban* (edited with Elsa Koleth, Gökbörü Tanyildiz, Raj Narayanareddy, and darren patrick, 2021); and the forthcoming *Elgar handbook on gender and cities* (edited with Grace Adeniyi-Ogunyankin and Anindita Datta).

Nasya S. Razavi was a postdoctoral fellow with GenUrb (2019–2024) and lead researcher on the Cochabamba City Research Team (CRT). She is currently the Latin America program manager at Inter Pares, a feminist social justice organisation based in Ottawa. Nasya completed her PhD at the Department of Geography and Planning at Queen's University, Kingston, Canada, in 2019, and has since published it with Routledge as *Water governance in Bolivia: Cochabamba since the Water War* (2022). Nasya adopts a feminist decolonial approach to her work in international development, gender, and environmental and social justice.

Araby Smyth was a postdoctoral fellow with GenUrb (2021–2024), researching place ecologies of finance and debt. She is currently an Assistant Professor in the Department of Geography and Environment at Mount Allison University. Her research has been funded by the Antipode Foundation, National Science Foundation (USA), and Society of Woman Geographers. She has published in journals such as *Antipode*, *Environment and Planning A: Economy and Space*, and *Transactions of the Institute of British Geographers*. She is on the Editorial Collective of the journal *ACME: An International Journal for Critical Geographies*.

Authors

Jenna Blower worked with GenUrb as a graduate assistant and MITACS intern in 2020. She is a PhD student in anthropology in the Faculty of Liberal Arts and Professional

Studies at York University, Toronto, examining transnational labour migration and care work.

Mantha Katsikana was a graduate assistant with GenUrb (2020–2023). She is a PhD student in geography in the Faculty of Environmental and Urban Change, York University, Toronto, working on feminist narratives of the austerity crisis in Athens, Greece. Mantha's work is informed by her long-term engagement with feminist, queer, and solidarity movements in Athens.

Elsa Koleth was a postdoctoral fellow with GenUrb (2017–2021), working on various publications and, with Biftu Yousuf, on the construction of the project's NVivo database. Elsa completed her PhD at the University of Sydney in the field of migration studies with a thesis entitled *Haunted borders: temporary migration and the recalibration of racialised belonging in Australia.*

Mel Mikhail was a graduate assistant with GenUrb (2020–2023). She is a PhD candidate in gender, feminist, and women's studies at York University, Toronto, whose work brings together anti-colonial theories of space and urban political economies of finance.

Carmen Ponce was a postdoctoral fellow with GenUrb (2022–2023). Her research centres on spatialities of violence against women in the GenUrb cities. She completed her PhD in economics at the Pontificia Universidad Católica del Perú. Her dissertation focused on the role of increasing climate variability on household livelihood diversification in the Peruvian Andes.

Wiley Sharp was a graduate assistant with GenUrb (2022–2023). They are a PhD student in the Department of Geography and Planning at the University of Toronto, and they completed their MA in geography in the Faculty of Environmental and Urban Change at York University, Toronto. Wiley's work examines the place-making practices of queer and trans* youth, and their MA thesis explored constellations of queer and trans* places in suburban Toronto.

Biftu Yousuf was GenUrb's in-house NVivo specialist (2021–2023) and a PhD student in geography in the Faculty of Environmental and Urban Change at York University, Toronto. She is a QSR-certified NVivo expert and has worked on several research projects to facilitate the application of grounded theory and feminist methodologies using NVivo. Biftu's dissertation uses feminist political geography/geopolitics to probe the role of racialised diasporas in Canada's refugee protection project and is entitled *Invisibilised providers: the role of racialised diasporas in refugee sponsorship.*

Contributors

Grace Adeniyi-Ogunyankin is a co-investigator with GenUrb and the Ibadan CRT's lead researcher. She is associate professor in both the Departments of Geography and Planning and Gender Studies at Queen's University, Kingston, Canada, and holds a Canada Research Chair. Her current research examines the impact of contemporary urban transformations on youth identity, labour practices, psychosocial well-being, and future orientation in Lagos and Ibadan, Nigeria.

Damilola Agbalajobi is a member of the Ibadan CRT. She is a senior lecturer in the Political Science Department, Obafemi Awolowo University, Ife-Ife, Nigeria. She is also a fellow of the African Studies Centre, Leiden, and a research fellow of the Conflict Research Network – West Africa.

Mai Al-Battat is a researcher with GenUrb's Ramallah CRT. Mai holds a BA in architectural engineering from Birzeit University, Palestine, and she is currently completing her MA in integrated design research at the Cologne University of Applied Sciences, Germany.

Natasha Aruri is the co-lead of GenUrb's Ramallah CRT. She is the co-founder and director of UR°BANA, an interdisciplinary research and design studio, which is a partner in the GenUrb research. Her work focuses on cities of exasperated insecurities, the spatial politics of and resistance to (neo)colonialism, and dynamic people-based spatial design strategies. Between 2019 and 2023, she was a postdoctoral researcher and coordinator at K-LAB, Technical University Berlin, working on critical mapping projects for socio-political change.

Swagata Basu is a member of the Delhi CRT. She is an associate professor and head of the Department of Geography at SSV College, Hapur, India. Swagata's research interests are in violence against women and intimate partner violence as well as issues related to women's right to the city. She works closely with women's organisations that address the livelihood issues of women in low-income urban neighbourhoods.

Kehinde Mary Bello is a member of the Ibadan CRT. She is an economist with a PhD in economics from Obafemi Awolowo University, Nigeria. She has taught at various universities in Nigeria, including Kings University, Odeomu, Redeemer's University, Ede, and the University of Lagos, Akoka.

Leeann Bennett is the project manager of GenUrb (2017–2024). She has an MA (and ABD PhD candidacy) in women's studies from York University. Her research interests include sexuality studies, particularly sex work and LGBTQ activism, Caribbean studies, transnational feminism, feminist methods and methodologies, and popular culture.

Susan Collymore, Halima Khan, Joy Marcus, Vanessa Ross, and Wintress White are all grassroots women, researchers on the Georgetown CRT, and long-time members of Red Thread. Red Thread is a women's organisation located in Georgetown, Guyana, committed to organising with women and communities, always starting from the interests of grassroots women, to cross divides. It works on issues of domestic violence, sexual offenses, poverty reduction, and the environment, among others, and has an in-house research team. It is their preference to be listed collectively.

Anindita Datta is the lead researcher on the Delhi CRT in GenUrb. She is professor and head of the Department of Geography, Delhi School of Economics, University of Delhi. Her research interests are in feminist geography, conceptual traditions in geography, and the social geography of India, with a particular interest in geographies of care, Indigenous feminisms, spaces of resistance, everyday geographies, and issues of gendered and epistemic violence.

Karen de Souza is co-lead on the Georgetown CRT, a life-long political activist, and co-ordinator of Red Thread, a partner in the GenUrb project. Since 1992, she has worked on numerous Red Thread research projects with Linda Peake.

Tsung-Yi Michelle Huang is a member of the Shanghai CRT. She is a professor in the Department of Geography at National Taiwan University. Her research specialisation lies in contemporary urban culture in East Asia. Her latest work (2020) is *New sensibilities in mainland China and Hong Kong: Politics of emotion in the dreams of development* (in Chinese).

Penn Tsz Ting Ip is lead researcher with the Shanghai CRT. She is a senior lecturer at the Department of Humanities, Language and Translation, the School of Arts and Social Sciences at Hong Kong Metropolitan University. She was previously assistant professor at the Department of Cultural Industry and Management in the School of Media and Communication at Shanghai Jiao Tong University (2019–2023). Her research interests include gender and women's studies, urban studies, migration, globalisation, and affect theory. She is currently editing *Feminist explorations of urban China*, under review with Routledge, as well as an English translation of her book *Ordinary women*.

Sylvester Nmormah has an MA in Economics from Obafemi Awolowo University in Ife-Ife, Nigeria. He is a research assistant with the Ibadan CRT and a program officer for the Centre for Human Development, also a GenUrb partner.

Bukola Omolara Odunola is a research assistant with the Ibadan CRT, employed at the Centre for Human Development in Nigeria, a GenUrb partner. She is also currently a PhD student in the Department of Sociology in the Faculty of Social Sciences, University of Ibadan, Nigeria.

Monica Orisadare is a research assistant with the Ibadan CRT. She is an associate professor in the Department of Economics, Obafemi Awolowo University, Ife-Ife, Nigeria.

Sonia Pardo Burgoa is a researcher with the Cochabamba CRT. She is the director of the Centro de Estudio y Trabajo de la Mujer (CETM), a non-governmental organisation dedicated to women's rights activism and research with links to San Simone University, in Cochabamba, Bolivia. CETM is also a partner in the GenUrb project. Sonia works primarily with Quechua women from the Cochabamba Valley, supporting processes of social transformation and gender equality.

Acknowledgements

The GenUrb project has been a labour of love for many people. Some GenUrb members have been with us from start to finish; others have joined us for periods along the way. Many have contributed their time and labour to the collective endeavour in ways that have gone beyond expectations and requirements, for which no level of thanks is enough. In particular, we thank the City Research Team (CRT) leads and the postdoctoral fellows who have been committed to the collective in ways that could not have been foreseen. Let's dive into who they are.

The CRTs formed the backbone and beating heart of the project. Without them there would be no GenUrb. Not only were they responsible for the research – collecting data, interpreting, translation, coding data in NVivo, and analysis – but also organising a wide range of public-education and knowledge mobilisation activities, from publications, conferences, and workshops to art exhibitions and radio shows. They also participated in meetings too numerous to mention, face to face and on Zoom. And they also cared deeply about the women who participated in this research.

Nasya S. Razavi, PhD, is the Cochabamba team lead and a SSHRC postdoctoral fellow while also Latin America program manager at Inter Pares, Ottawa, Canada. She has worked alongside Sonia Pardo Burgoa, the director of the women's organisation Centro de Estudios y Trabajo de la Mujer (one of our partners) in Cochabamba. At the start of the project, Nina Laurie, professor at the University of St. Andrews, Scotland, and Maria Esther Pozo, professor and (then) vice chancellor at the Universidad Mayor de San Simón, Bolivia, served as joint team leads of the Cochabamba CRT. For extra research support, the CRT employed Ingrid Baldivieso, an undergraduate student at Universidad Mayor de San Simón, Ida Peñaranda, an activist-researcher, and PhD student, Victor Hugo Mamani Yapura.

In the Delhi CRT, Anindita Datta, professor and head of the Department of Geography at the Delhi School of Economics, University of Delhi, worked with Swagata Basu, professor at the Department of Geography, SSV College, Hapur, alongside three field assistants, three research associates, and twelve graduate students.

In Georgetown, the CRT comprised members of the women's organisation Red Thread (also a partner in the research) – Susan Collymore, Halima Khan, Joy Marcus, Vanessa Ross, and Wintress White – working with the team co-leads Karen de Souza, coordinator of Red Thread, and Linda Peake, professor at the Faculty of Environmental and Urban Change, York University.

The Ibadan CRT was led by Grace Adeniyi-Ogunyankin, associate professor and Canada Research Chair at the Departments of Geography and Planning and Gender Studies, Queen's University. Initially, Grace was team co-lead with Similola Afonja, professor emerita and executive director of the Centre for Human Development (CHD), a civil society organisation based in Ibadan (and a GenUrb partner). Sylvia Bawa, associate professor at the Department of Sociology, York University, also joined the CRT. Research assistance in Ibadan was provided by Damilola Agbalajobi, senior lecturer and head of the Department of Political Science at Obafemi Awolowo University, Ife-Ife; Tunrayo Abimbola Adeyemi, PhD student, Kehinde Bello PhD, Sylvester Nmormah, CHD program officer, and Monica Orisadare, associate professor, all at the Department of Economics, Obafemi Awolowo University, Ife-Ife; and Bukola Omolara Odunola, PhD student at the Department of Sociology, Faculty of Social Sciences, University of Ibadan. In Toronto, further research assistance was provided by Wumi Asubiaro-Dada, PhD student at the Centre for Criminology and Sociolegal Studies, University of Toronto.

The Ramallah CRT was co-led by Dr. Natasha Aruri, postdoctoral researcher at K LAB, Technische Universität Berlin, and co-founder of the thinktank UR°BANA (another GenUrb research partner), and Dr. Andreas Brück, managing director of K LAB, senior lecturer, Technische Universität Berlin, and co-founder of UR°BANA. They worked closely with Mai Al-Battat, MA student, Integrated Design Research, Cologne University of Applied Sciences. Further research assistance was provided by Wala' Lmlah, office manager, Pal Delta, Ramallah; Alaa' Fayyaleh, MA student, Birzeit University; and Basil Farraj, lecturer, Birzeit University. Two Ramallah-based civic-mobilisers and consultants also joined their team: Amal Juma', journalist, trainer, and communications director of the Women's Affairs Technical Committee; and Dr. Sahar Qawasami, co-founder and director of Sakiya.

The Shanghai CRT was led by Penn Tsz Ting Ip, senior lecturer at the Department of Humanities, Language and Translation, School of Arts and Social Sciences, Hong Kong Metropolitan University, aided by Tsung-Yi Michelle Huang, professor at the Department of Geography, National Taiwan University. Research assistance was provided by Yuk Ting Ho and Tamires Lietti, both MA students, Department of Cultural Industry Management, School of Media and Communication, Shanghai Jiao Tong University; Jie Xu, MA student, Centre for Gender Studies, SOAS, University of London; and Haiyan Zhou, PhD student, Department of Sociology, Shanghai University. The following members of the CRT played active roles in networking: Jing Wang, associate professor, Department of Cultural Industry Management, School of Media and Communication, Shanghai Jiao Tong University; Liu Xi, associate professor, Department of China Studies, Xi'an Jiaotong-Liverpool University, Suzhou; and Zhang Yu, lecturer and executive director of the Gender and Culture Research Center, University of Shanghai for Science and Technology, Shanghai.

In addition to the research partners mentioned above, we also benefitted from the presence of institutional partners. These included the UN-Habitat Gender Hub, represented by Angela Mwai, and (the now defunct) Women in Cities International, represented by Kathryn Travers.

Alongside the CRTs, and based at the City Institute, York University, was the management team, comprising Linda Peake, the project's principal investigator, and Leeann Bennett, project manager, who were often joined by the postdoctoral fellows. Also active at specific periods of time were the comparative team of Beverley Mullings, professor at the Department of Geography and Planning, University of Toronto, Canada; Faranak Miraftab, professor at the Department of Urban and Regional Planning, University of Illinois, Urbana-Champaign, USA; and Cristina Temenos, reader, at the Department of Geography, Manchester University, UK. The knowledge mobilisation team comprised Jack Gieseking, PhD, Research Fellow at the Five College Women's Studies Research

Center, Hampshire College, USA, and Sara Koopman, associate professor, at the School of Peace and Conflict Studies, Kent State University, USA.

At York University, GenUrb benefitted enormously from postdoctoral fellows Elsa Koleth PhD, also a research associate, Carmen Ponce, PhD, Nasya S. Razavi, PhD, and Araby Smyth, PhD – who have each spent at least a year, mostly more, working closely with the principal investigator and project manager. They have participated in research, analysis, public education, and knowledge mobilisation, including writing for publication, sourcing extra funding, and often picking up the pieces when multiple unforeseens landed in our laps.

We also employed York University students to work in specific roles. Darren Patrick, PhD, operated for a short time as our editor and publications manager. Esteban Sabatasso, MA student, was our website developer. For a couple of years, Biftu Yousuf, PhD student, worked tirelessly as our NVivo specialist, going above and beyond to make our data make sense, across languages, across Macs and PCs, and across the limitations of NVivo itself. Waking up to start work at 4 a.m. every day for a year to incorporate all the data inputted the day before so that it was ready for all the CRTs to work with was not part of her job description!

Students at York University also took on roles as transcribers and translators. Javier Garate Alfaro, MA student and member of York's Centre for Research on Latin America and the Caribbean (CERLAC), and Eleonore Douglas and Carmen Ramirez, also CERLAC members, worked as transcribers and translators for the Cochabamba CRT. Saaraa Esau, master's student, and PhD students Karen Naidoo, Serene Paul, and Angela Stanley all worked alongside members of Red Thread, working as transcribers for the Georgetown CRT.

Several York students also worked as research assistants, supervised by the project manager. These included:

- undergraduates: Dara Dillon (a course intern), Bri Gardner, Melissa Hernández Jasso, Javeria Mirza, Rawan Mostafa, Anjli Saddi, and Siya Zhang;
- master's students: Eyram Agbe, Vanessa Bart-Plange, Jamilla Mohammud, Wiley Sharp, and two other students who chose not to be named;
- PhD students: Jenna Blower, Mantha Katsikana, Sehrish Malik, Mel Mikhail, and Zhi Ming. Nehal El-Hadi, a PhD graduate from the University of Toronto, also worked as a research assistant for a short period while a visiting scholar at the City Institute, York University.

Our external advisory board provided guidance when needed and comprised: Cindi Katz, professor, Earth and Environmental Sciences, CUNY Graduate Center, USA; Richa Nagar, professor, Gender, Women and Sexuality Studies, University of Minnesota, USA; Susan Parnell, professor and chair in Human Geography, University of Bristol, UK; and Geraldine Pratt, professor, Canada Research Chair in Care Economies and Global Labour and head of the Department of Geography, University of British Columbia, Canada.

We want to recognise and thank the Social Sciences and Humanities Research Council (SSHRC) of Canada (SSHRC Partnership Grant File # 895–2017–1011) and York University not only for funding the research but also for helping to resolve a range of financial, technical, and practical issues to which partnership grants give rise.

At SSHRC, we received help from Eric Bergeron, Marou Mahjoub, Lauren Mathieson, Nadine Miville, Chantelle Perreault, and Caroline Vandergoten, program officers in the Research Grants and Partnership Division, SSHRC.

At York University, our host partner, we received help from:

- Finance Services, Research Accounting, and General Accounting: including Francesca Abrusezze, assistant manager, Research Accounting; Xi (Cindy) Chen, accounts administrator, Research Accounting; Firas Hago, manager, General Accounting; Grace Law, accounts payable clerk, General Accounting; Jeannie Montgomery, general accounting clerk, Research Accounting; Sonia Fermin-Peet, expense compliance administrator, General Accounting; Paula Perri, manager, Finance Services Products; Domenica Semplicio, accounts payable clerk, General Accounting; and Angela Zeno, associate director, Research Accounting;
- Student Financial Services: Adrienne Beamer, student financial liaison officer; and Karen Warner, manager, Scholarships and Bursaries;
- Office of Research Ethics: Wendy Jokhoo, manager; and Michael Xavier, ACC coordinator;
- Office of Research Services: Joan Broussard, director, Research Grants; Emma Doran, strategic and institutional research specialist; Yewn Forrester, strategic institutional research initiatives coordinator; Diana Frasca, strategic and institutional research specialist; Denise Jagdeo-Singh Martinez, research grants administrator; and Lisa Rumiel, strategic and institutional research specialist;
- Faculty of Liberal Arts and Professional Studies: David Cuff, director, Strategic Research and Partnerships; and Helen Papacharalambous, faculty research administrator;
- Faculty of Environmental and Urban Change: Paul Elliott, executive officer; Michael Faranda, coordinator, Technical Support & Applications; Alice Hovorka, dean; Philip Kelly, associate dean research; Rhoda Reyes, research officer; and Barb Wells, executive assistant;
- Faculty of Graduate Studies: Kim McIntyre, postdoctoral services and professional skills coordinator, Graduate Academic Affairs; and Lindsay Gonder, manager, Graduate Studies, Office of the Associate Dean, Research and Graduate Studies;
- Office of the Vice-President of Research and Innovation: Anna Gordon, former senior human resources officer; Celia Haig Brown, former vice-president Research; Jennifer Hyndman, vice-president Research; Felix Moses, senior executive officer; Chandra Persaud, senior financial officer; and Emma Yuen, senior human resources officer;
- Innovation York: Michael Johnny, manager, Knowledge Mobilisation.

We would also like to thank the Payroll Department at York University, the York University Library, York University Information Technology, and anyone we might have accidentally omitted, for all their assistance.

Putting a book together was not an initial aim of the GenUrb project. We originally set out to produce a much more modest set of research training modules, hosted on our own website, but our ambitions grew as we realised that the material we wanted to cover comprised a book-length project. Moreover, we wanted to ensure that the material we produced would continue to exist and be accessible beyond the life of the project. Hence, we started looking for a publisher who would be willing to produce an open access book, providing both web-based and print copies, the latter to ensure that those with little or unreliable internet access were able to utilise it. Routledge was our first choice, and our experience with them, and specifically with Faye Leerink, editor, Environmental Studies, Geography, Social Sciences, Tourism, Hospitality and Events, Urban Studies, and Prachi Priyanka, editorial assistant, has been extremely positive. Their belief in this book gave us the final push to complete it.

Writing and editing the book, with so many contributors, felt at times like a never-ending task, but finish we did! The chapter authors comprise the principal investigator, postdoctoral fellows (Elsa Koleth, Carmen Ponce, Nasya S. Razavi, and Araby Smyth), and graduate students (Jennifer Blower, Mantha Katsikana, Mel Mikhail, and Wiley Sharp) at York University. There was much sending back and forth of drafts, smoothed along by the goodwill of all those involved. The many contributors – members of the CRTs and the project manager – replied graciously to the many requests via email and Zoom meetings. We were helped in the final stages by the copyediting skills of Leeann Bennett, Mel Mikhail, and Wiley Sharp. External reviewers also provided us with feedback on specific chapters: Christian Anderson, Sara Koopman, and Gökbörü Tanyildiz. The book was given a final polish with the excellent professional editing skills of Ann Varley, emeritus professor at the Department of Geography, UCL, UK. Thanks and gratitude abound.

The work of all these people would not have been possible without the participants in our project, who, for the sake of privacy and confidentiality, obviously cannot be named. Their spirited participation in the face of economic hardship was not taken for granted, and we hope we have been able to give back some of what we took. Some of us are still in touch and continue to provide resources. In a few cases, deeper contacts have been made and flourish.

Finally, we have to single out Leeann Bennett, our project manager, based at the City Institute, with whom we have worked closely. We have only immense appreciation for the professionalism, camaraderie, support, and kindness everyone in GenUrb has experienced with her.

Linda Peake, Nasya S. Razavi, and Araby Smyth

GenUrb

Urbanisation, gender, and the global south:
A transformative knowledge network

Introducing GenUrb

Linda Peake, Araby Smyth, and Nasya S. Razavi

Through the experiences of the feminist urban research project *Urbanisation, gender, and the global South: a transformative knowledge network* (GenUrb), this book aims to enrich your understanding of what a feminist approach to urban research involves in the context of transnationalism. This chapter outlines how a global network of feminist urban researchers evolved and introduces the cities in which research was conducted – Cochabamba, Bolivia; Delhi, India; Georgetown, Guyana; Ibadan, Nigeria; Ramallah, Palestine; and Shanghai, China. It also presents those engaged in the GenUrb partnership, based in the research cities and at the City Institute at York University in Toronto, Canada. The chapter concludes with an outline of the six sections of the book – the building blocks of feminist decolonial urban research, the context of 21st-century urban feminist research and policy, feminist research standards, feminist methodologies and research methods, feminist data analysis, and feminist approaches to knowledge mobilisation.

Section 1. Introducing *Doing feminist urban research: insights from the GenUrb project*

This book is designed to do what many research textbooks do. It will introduce you to feminist methodologies, research methods, data analysis, and knowledge mobilisation practices, including using social media. Unlike many conventional research textbooks, however, it addresses the grounded, emotional, and relational nature of feminist research, and it provides examples of how embodiment, positionality, and critical reflexivity inform the research process. It sets out the professional standards, procedures, and protocols, beyond feminist ethics, to which feminist research must adhere. It equips you with an understanding of 21st-century global political landscapes, including those of COVID-19 and the Sustainable Development Goals (SDGs), as well as the ethical contexts of knowledge production within which research is situated. It addresses the need to decolonise feminist knowledge production, how comparative research contributes to this goal, and the role that activism plays in feminist research. It delves into questions around data that go beyond analysis, including the use of software, particularly NVivo, as well as issues of translation and interpreting, data management, and ownership. Finally, the text engages with how feminist researchers often eschew the model of the lone researcher to work together in teams and partnerships.

For all that research is a process bound by professional standards, ethics, and a rigorous approach to the application of methods and data analysis, it is also entangled in power and personal relations and subject to the vagaries of everyday life. This can mean that what is written on paper – research proposals, grant applications, ethics reviews –

DOI: 10.4324/9781032668727-1

may not easily be transferred into the 'field', and what is assumed to be under the control of the researcher may take on a life of its own. So we also take you through the experiences of the GenUrb project, exploring the messy realities of research practice. We provide an insight into the lived realities of research, which is prone to jolts, stoppages, power plays, and unwelcome whirlwinds of emotions, as well as unexpected moments of joy, satisfaction at things going well, and the possibilities opened by the growth of trust and solidarity.

GenUrb members wrote this book to share our experiences and make a pedagogical contribution to the training of future generations of feminist urban scholars. It is authored and edited collaboratively by various members of the GenUrb research team at different stages of their careers, ranging from graduate students to full professors. What emerges are sometimes collective, sometimes discordant GenUrb voices that share the challenges of engaging in feminist research and the limits of the ability to work through tensions and contradictions, as well as what it means to collaborate through teamwork, working across affinities and difference, in an ongoing, dialogic, and collaborative process.

The following section introduces you to the GenUrb project, its goals, research focus, and geographical scope.

Section 2. Introducing GenUrb

This book is informed by experiences from the transnational feminist urban research project GenUrb (2017–2024), which works across six cities and involves an inner core of over 20 researchers set amidst a broad partnership that has involved approximately 60 more students, scholars and activists, members of women's grassroots and civil-society organisations, and policy-shaping institutions in the global North and South. In what follows we outline the evolution of GenUrb's network of feminist researchers, introduce the cities where the research took place, and present its members to you.

(i) The GenUrb partnership

GenUrb is a multi-sited, longitudinal research project funded by the Social Sciences and Humanities Research Council of Canada (SSHRC) since 2017 and has its administrative base at the City Institute at York University in Toronto, Canada.

GenUrb functions through networks of grounded solidarities between scholars, activists, and local communities, employing participatory and collaborative methods of research and knowledge creation. We are engaged in research in six cities in the global South on the everyday place- and life-making practices of women living in urban neighbourhoods marginalised by underdevelopment and economic insecurity. We explore the creation of place ecologies of economic precarity and violence, analysing the spatialities through which they are constituted. Attentive to historical specificities of colonialism, capitalism, patriarchy, and everyday micropolitics, we explore the changing nature of social reproduction and production in these women's lives alongside the spaces of hope, the cracks, and the slippages that reveal possibilities for different ways to imagine the urban. In so doing, we also engage with 'policy shapers': those whose knowledge of gender and urban planning and policy enables us to explore the extent to which the SDGs (specifically SDG 5 on gender equality and SDG 11 on sustainable cities) have affected the lives of women living in these cities.

The project involves research undertaken by each of six City Research Teams (CRTs), employing a mixture of geo-ethnographic strategies, including semi-structured interviews with policy shapers, as well as in-depth interviews and life-history interviews with women living in marginalised neighbourhoods. During the COVID-19 pandemic, these methods were sometimes supplemented by digitally mediated methods, allowing links to be maintained with participants through both in-person and virtual communications. We have also undertaken a range of knowledge-mobilisation and public-education activities, such as academic, policy, and creative outputs, workshops, and public events, including an international conference to mark its mid-term point held at York University in Toronto in September 2019. In addition, we have incorporated objectives related to research training and pedagogy through, for example, employing research assistants, facilitating training workshops and reading groups, creating undergraduate and graduate courses, and authoring this book, among other publications.

GenUrb example 0.1 outlines how the GenUrb partnership came into being and developed.

GenUrb example 0.1: The evolution of the GenUrb partnership

The GenUrb partnership was built in the decade before the project began. The starting point for the partnership was the research that I, Linda Peake, have been conducting since the early 1990s with the Guyanese grassroots women's organisation Red Thread. I asked: can this partnership that enabled the production of knowledge outside the academy be replicated on a global scale in the service of transnational feminist praxis? The various global networks of feminist geography scholars of which I was a part were the primary factor facilitating this potential replication. They included both scholarly and friendship groupings, including CWAG (the Canadian Women and Geography Group of the Canadian Association of Geographers, now Feminist Intersectional Solidarity Group, FIGS), GPOW (Geographical Perspectives on Women Specialty Group of the American Association of Geographers, now the Feminist Geographies Specialty Group, FGSG), the WGSG (Women and Geography Study Group of the Royal Geographical Society/Institute of British Geographers, now the Gender and Feminist Research Group, GFRG), and the Commission on Gender and Geography of the International Geographical Union. While CWAG and GPOW consolidated a North American network, the Commission on Gender and Geography enabled a network that, over time, has become global in reach. Beyond these networks, which in places harked back over 50 years, I also drew upon individual feminist urban scholars around the globe.

Red Thread, in conjunction with these groups, formed the foundation of a global partnership. This network was utilsed to organise a workshop at York University in 2010 on 'Feminist interventions into the urban' (see Peake and Rieker, 2013). The workshop established the clear need for a feminist comparative project that could transcend national and regional boundaries to explore globally women's engagement in urban placemaking and that could also extend beyond the academy and connect to women's grassroots networks and organisations. The broad and deep range of knowledge required for understanding the dynamics of cities and working across cultural, urban, and regional differences necessitated a coordinated and collaborative approach that only a partnership could provide. Although considered a success, this three-day meeting also brought to the surface differing, seemingly incompatible strands of feminism, personality clashes, and, most importantly, racialised discord. The fallout from these interactions, although not visible to all participants, left me with a feeling that bringing together a global team over a long-term period could be too fraught with political differences to succeed.

Over the following two years, however, the lure of the research, of being able to engage in a globally organised project that could make a substantive intervention into urban theory through a feminist methodology and that could also contribute to understandings of how women's everyday lives were being lived in cities of the global South, did not go away. I decided, once again, to take forward the idea of a comparative feminist urban research project. In 2013, I contacted several feminist urban scholars who had been at the workshop about forming a research team and submitting a grant application. There was a resounding 'yes' to this request.

An opportunity to formulate a research team and to apply for monies came from a 2014 call from the International Social Science Council (ISSC) for research on 'Transformations to sustainability', with our application successfully reaching the shortlist for funding. The ISSC seed monies enabled the investigation of research field sites in 2014 in Cairo, in 2015 in Georgetown, Mumbai, Ramallah, and Shanghai, and in 2016 in Cochabamba and Ibadan. Our partnership met again in January 2015 at York University, and CRTs for Cairo, Cochabamba, Georgetown, Ibadan, Mumbai, Shanghai, and Tehran were formed.

While we were unsuccessful in securing an ISSC grant, the team members unanimously agreed to apply for a SSHRC Partnership Grant. A renewed urgency to conduct this research arose from the inauguration of the SDGs (2015) and UN-Habitat III's New Urban Agenda (October 2016). These global developments concerning the urban, with gender and poverty at their core, demanded our critical attention.

In late 2016, the project had to make the difficult decision to cancel plans for a CRT in Tehran. This team was to have been led by a prominent Iranian Canadian feminist scholar who, on a journey to Iran in June 2016, was imprisoned, ostensibly for using feminism to subvert Islamic beliefs. After her release in September and extensive discussion, we decided that continuing to pursue research and establishing a CRT in Tehran, given the climate of hostility towards feminism and research on women's issues (among other topics), would pose a potentially significant security risk to other members of the Tehran team. With security outweighing all other considerations, there was agreement not to continue.

With the CRTs established, we then brought on board other partners. When the project started, we had twelve partners: six universities (the American University in Cairo, New York University Shanghai, Queen's University in Kingston, Ontario, University of St. Andrews in Scotland, the Tata Institute of Social Sciences in Mumbai, and Trinity College in Hartford); three women's groups (Red Thread in Georgetown, Centro de Estudios y Trabajo de la Mujer in Cochabamba, and a lesbian organisation in Shanghai, who do not wish to be identified); one civil society organisation (Centre for Human Development in Ibadan); one private research and design organisation (UR°BANA in Ramallah/Berlin); and two policy organisations (Women in Cities International/Femmes et Villes and the UN-Habitat Gender Hub). The partners – universities, grassroots women's organisations, civil society organisations, and policy institutions – set the parameters of the research, public education, and knowledge mobilisation activities of the project. In addition to the six CRTs, we established two other teams: the Comparative Research Team and the Knowledge Mobilisation Team, all of whom work with one or more of these partners. In addition, we had a projectwide steering committee, a small management committee, and an external international advisory board.

Our SSHRC partnership grant application was successful, and we received our notice of award in March 2017. Work began on organising the project that summer. Our first tasks were the employment of a project manager and securing clearance from the ethics board at York University. Given our engagement with Indigenous women in Cochabamba, ethics clearance took months longer than anticipated, delaying the start of fieldwork.

Over time, membership of the project (as is common with projects of this duration and large size) has changed. People and partners have been added, and others have left. In the cases of the comparative research and knowledge mobilisation teams, for example, their centrality to the work of GenUrb has waxed and waned depending upon the stage of the project. And others, such as postdoctoral fellows and graduate students, have left as their contracts have ended, while other members have moved in and out of the project as their time has allowed. In some cases, change has gone beyond individual members to whole teams leaving and joining. CRTs established in Cairo and Mumbai were to leave, while the Delhi CRT was created in the fourth year of the project. The contours of the GenUrb partnership are dynamic, flexible, and responsive, with the glue holding the partnership together being a broad agreement on a feminism concerned with the factors underlying the changing material conditions of women experiencing precarity in cities of the global South.

By Linda Peake

(ii) The cities in GenUrb

Although GenUrb comprises a global network our focus is on cities located in the regions geopolitically demarcated as the global South: East and South Asia, Latin America and the Caribbean, Middle East and North Africa, and Sub-Saharan Africa. Our collective decision to conduct research in the global South was politically charged by the desire to take our existing analyses and knowledge of these individual cities and to bring them into collaboration, without the need to position them in relation to cities in the global North. The six cities in which we worked varied dramatically in size and density and as regards the national rate of urbanisation. The cities involved in the GenUrb research project include two megacities, Delhi, India, and Shanghai, China, one medium-sized city, Ibadan, Nigeria, and three small-sized cities, Cochabamba, Bolivia, Georgetown, Guyana, and Ramallah, Palestine. There was an intentional choice to work in different-sized cities in countries with different rates of urbanisation (the projected annual average rate of change of the size of the urban population, which, as census data showed, ranged from 1.01 percent in Guyana to 3.92 percent in Nigeria). Despite an academic focus on megacities (with populations of over ten million) in studies of the global South, globally, nearly half of the urban dwellers reside in smaller, non-capital cities. We wanted to ensure that our comparative analysis paid attention to both the large and smaller cities. In addition, China, India, and Nigeria are together expected to account for over one-third of projected growth of the world urban population by 2050 (United Nations, 2019), hence our decision to include cities from these countries. Amid this diversity there was also a decision to add a modicum of 'control' by engaging primarily with cities in 'lower middle-income countries' (United Nations, 2019), although China, classified as an upper-middle-income country, was also included given its significant impact on global urbanisation rates.

Many factors were considered when discussing how many and which cities might be included in the GenUrb project. For instance, project members believed six cities to be the largest manageable number for working across languages and time zones and still ensuring the production of credible comparative research. As outlined in GenUrb example 0.1, the choice of cities was determined primarily by the need for the presence within it of feminist urban scholars aligned with the goals of GenUrb.

The 'GenUrb cities' are diverse in their economic, social, environmental, and historical contexts – yet they share several commonalities. Despite different rates of urbanisation, they are all experiencing population growth, speared by natural increase and rural-to-urban migration, the latter particularly relevant in Shanghai, Cochabamba, and Delhi. All have colonial histories (and presents) and are capital cities or important regional centres, characterised by their residents' experiences of inequality, insecurity, class disparities, and differentiated access to basic services. Unequal economic growth in each city is evidenced by stark contrasts between wealthy and underdeveloped districts. The cities also share urban challenges such as an absence of formal employment in many sectors of the economy, which – alongside the rising costs of inadequate housing and education, the crumbling of electricity, water, and sanitation infrastructure, and traffic congestion – increases the pressures on women's everyday use of time and mobility as they seek to secure their livelihoods and navigate the city.

The choice of which neighbourhood to work in was determined by each CRT with the simple stipulation that it be an area characterised by underdevelopment and economic insecurity. State withdrawal is evident in most of the urban neighbourhoods studied in the GenUrb project, although the neighbourhood in Shanghai is marked by heavy state governance and surveillance, and much of the neighbourhood in Ramallah is under Israeli security control.

Several neighbourhoods in the study were initially housing schemes, such as the Workers' New Village in Shanghai and the resettlement colonies in northern Delhi, intended to decongest the inner city. The communities of Challenge in Ibadan and the Sacaba zone of Cochabamba are situated in areas of peak urban renewal, which are exacerbating spatial inequalities, driving the displacement of the urban poor. Sophia, initially on the outskirts of Georgetown though now surrounded by built-up areas, was built on the site of a sugar plantation that was occupied by squatters and eventually regularised. The agricultural village of Ein Qiniya, however, still lies at the limits of the municipality of Ramallah. Different forms and high rates of gendered violence are present across all cities, experienced through what Anindita Datta (2016, p. 178) has called 'genderscapes of hate'. Yet, in the light of our engagement with grassroots women in these cities and with researching 'the everyday, the ordinary, the mundane' of these women's lived experiences (Butcher and Maclean, 2018, p. 688), we maintain that these cities contain vibrant and dynamic multiplicities, revealing stories of adversity and sorrow but also of hope and resistance.

(iii) The people in GenUrb

The core members of the CRTs were academics and in two of the CRTs they worked directly with grassroots women, in Georgetown (Red Thread) and Cochabamba (Centro de Estudios y Trabajo de la Mujer, CETM), while the Ibadan CRT worked with the research-based civil society organisation the Centre for Human Development (CHD). Given that all three of these partners have histories of engaging in research, no distinctions were made between their CRTs, and the CRTs composed entirely of academics in terms of their incorporation into research activities, including designing interview guides, conducting interviews, devising and hosting public-education and knowledge-mobilisation events, and engaging in written work. In what follows, we identify each of the CRT members. In some cases, we have omitted names to follow the wishes of their respective CRT, primarily to safeguard identities and avoid the researchers being identified with a feminist project. Given that the GenUrb project operates in countries in which government tolerance for and reactions to feminism in general and to feminist research specifically have deteriorated during the period of the project, we have put the safety of project members and participants above all other considerations.

The Cochabamba CRT was initially co-led by Dr. Nina Laurie, professor of geography at the University of St. Andrews in Scotland, and Dr. María Esther Pozo, Professor at the Centre for Planning and Management of the Universidad Mayor de San Simón in Bolivia. After Nina's departure in 2018 (for reasons unconnected to the project), Canadian postdoctoral fellow Dr. Nasya S. Razavi stepped in as co-lead. At the same time, Dr. Pozo also took a more distanced role after being appointed vice chancellor. The Cochabamba CRT also includes Sonia Pardo Burgoa, the director of CETM. CETM was established in the 1980s by socialist feminists at the Universidad Mayor de San Simón. It is a women's activist and non-governmental organisation that works with Indigenous Quechua women from the Cochabamba Valley, supporting social transformation and gender equality. Nasya also employed various research assistants in Bolivia, including Ingrid Baldivieso, Victor Hugo Mamani Yapura, and Ida Peñaranda, and, at York University, Eleanor Douglas, Carmen Ponce, Carmen Ramirez, and Javier Garate Alfaro, to help with a variety of research activities.

The Delhi CRT was the latest addition to the project, joining in 2020. It is led by Dr. Anindita Datta, professor of geography and head of department at the Delhi School of Economics, University of Delhi, who worked with Dr. Swagata Basu, associate professor of geography and head of department at SSV College in Hapur. The Delhi CRT is not partnered with any non-academic organisation, but it conducted its fieldwork with a team of three field assistants who were associated with an organisation (that has to remain anonymous) of trained paralegal workers and development practitioners. Support was also received from twelve graduate students and three research associates hired locally to assist with translation. One of these research associates also helped with the coding of the translated data and conference presentations.

The Georgetown CRT is led by the grassroots women's organisation Red Thread. Red Thread's goal is to organise with women, beginning with grassroots women, to cross divides and transform their conditions (Peake and Trotz, 1999; Andaiye, 2000).

They provide services to women and children caught up in unequal power relations and simultaneously work to change those relations. Karen de Souza, one of Red Thread's co-founders and its administrator, is co-lead with the GenUrb principal investigator, Dr. Linda Peake. Red Thread members include Susan Collymore, Halima Khan, Joy Marcus, Vanessa Ross, and Wintress White. They form an experienced in-house research team, having all been engaged in research for over three decades (see redthreadguyana.org). In addition, the CRT employed three PhD students to help with transcription – Karen Naidoo, Serene Paul, and Saaraa Esau – as well as the project manager Leeann Bennett.

The Ibadan CRT is co-led by Dr. Grace Adeniyi-Ogunyankin, associate professor of geography at Queen's University in Kingston, Ontario, Canada, Dr. Sylvia Bawa, associate professor of sociology at York University in Canada, and Dr. Similola Afonja, professor emeritus of sociology at Obafemi Awolowo University in Ife-Ife, Nigeria, and executive director of the Centre for Human Development (CHD). CHD is a Nigerian research organisation based in Ibadan whose prime mission is to partner with governments, donor agencies, non-government agencies, and local community organisations to develop sustainable development programmes and policies. Other members of the Ibadan CRT include Kehinde Bello, who completed her PhD at Obafemi Awolowo University, Bukola Omolara Odunola, a PhD student at the University of Ibadan, Dr. Damilola Agbalajobi, professor and head of the department of Political Science at Obafemi Awolowo University, two members of CHD (Sylvester Nmormah and David Olagunju), and Dr. Monica Orisadare, associate professor at Obafemi Awolowo University, Ife-Ife. Extra research assistance was provided by Wumi Asubiaro-Dada, PhD student at the University of Toronto and Tunrayo Abimbola Adeyemi, PhD student at Obafemi Awolowo University, Ife-Ife.

The Ramallah CRT is led by Dr. Natasha Aruri, a Palestinian scholar and postdoctoral fellow at Technische Universität Berlin, working with master's student Mai Al-Battat. The team partners with UR°BANA, particularly Dr. Andreas Brück, professor of urban planning at Technische Universität Berlin. UR°BANA is a 'think-do tank' that aims to co-create improved urban environments, spaces, and places. It has the ear of policy shapers in Palestine who are listening to UR°BANA's suggestions about how to deliver a practical programme for sustainable development. The Ramallah CRT also works with Sakiya, a progressive academy for experimental knowledge production and sharing, based in Ein Qiniya, bringing together local agrarian traditions of self-sufficiency with contemporary art and ecological practices.

The Shanghai CRT is led by Dr. Penn Tsz Ting Ip, previously assistant professor in the Department of Cultural Industry and Management in the School of Media and Communication at Shanghai Jiao Tong University and currently a senior lecturer at the School of Arts and Social Sciences at Hong Kong Metropolitan University. Dr. Tsung-Yi Michelle Huang, professor of geography at National Taiwan University, was also initially co-CRT lead, and she has continued to play a supportive role alongside several other academics and graduate students in Shanghai. Networking activities were undertaken by Dr. Jing Wang, associate professor in the Department of Cultural Industry Management, School of Media and Communication, Shanghai Jiao Tong University; Dr. Liu Xi, associate professor in the Department of China Studies, Xi'an Jiaotong-Liverpool University, Suzhou; and Dr. Zhang Yu, lecturer and executive director of the Gender and Culture Research Center at the University of Shanghai for Science and Technology, Shanghai. Research assistance was provided by Yuk Ting Ho and Tamires Lietti, both MA students in the Department of Cultural Industry Management, School of Media and Communication, Shanghai Jiao Tong University; Jie Xu, an MA student in the Centre for Gender Studies, SOAS, University of London; and Haiyan Zhou, a PhD student in the Department of Sociology, Shanghai University. The CRT has no permanent partner organisation, but they have worked with a wide variety of individuals and organisations.

The core members of the CRTs introduce themselves in Audio clip 0.1, while in GenUrb example 0.2, the Delhi CRT members provide an account of the ways in which feminist relations of friendship and mentorship have influenced their ability to engage in GenUrb.

Audio clip 0.1: Introducing the City Research Team members

Audio clips of the Cochabamba, Delhi, Georgetown, Ibadan, Ramallah, and Shanghai CRT members introducing themselves is available on the book's website www.routledge.com/9781032668680.

By Grace Adeniyi-Ogunyankin, Similola Afonja, Natasha Aruri, Anindita Datta, Swagata Basu, Susan Collymore, Karen de Souza, Tsung-Yi Michelle Huang, Penn Tsz Ting Ip, Joy Marcus, Sylvester Nmormah, Bukola Omolara Odunola, Monica Orisadare, Linda Peake, Nasya S. Razavi, Vanessa Ross, Wintress White, Zhang Yu, and Liu Xi

GenUrb example 0.2: Feminist friendships and mentoring in the Delhi City Research Team

The Delhi CRT represents a case of feminist associations of friendship and mentoring existing prior to joining the GenUrb project having been strengthened over the years of the research. Indeed, these relations of friendship made it possible for the members of the Delhi CRT to participate in the GenUrb project in the first place. Here, we trace three levels of feminist friendship and mentoring that played out simultaneously in transnational and local contexts.

Transnationally, principal investigator Linda Peake's work had influenced both Anindita and Swagata during their early career phases. Anindita first met Linda in person in 2010 when Linda came to Delhi to participate in a conference Anindita had co-organised. This was the first major conference on geography and gender to be organised in India. The three-day meeting made a great impact, and Linda was very encouraging of Anindita's efforts to create space for feminist geographies within mainstream human geography. She also offered valuable advice on post-conference publications. Later, Linda also connected with Anindita in the IGU Commission for Gender and Geography, where they briefly overlapped as steering committee members and met at some of the commission's conferences. Based on these earlier interactions, Linda invited Anindita to GenUrb's first workshop, in 2018 in Mumbai. Two years later, Linda invited Anindita to form a CRT for Delhi and to join the GenUrb project.

Linda's important work and a feminist association through the IGU Commission for Gender and Geography were therefore instrumental in forging this collaboration.

At another level there is also a deep association and friendship between the two Delhi CRT members, Anindita and Swagata. We both studied at the Centre for the Study of Regional Development (CSRD) at Jawaharlal Nehru University for our MAs and PhDs, although we were not contemporaries and are almost a decade apart in age. Swagata was aware that Anindita had written the first PhD on gender and geography at CSRD and had met Anindita fleetingly at several academic events. A common interest in geographies of gender led to several insightful exchanges. We began working closely together in 2006 as part of a textbook-writing committee for the National Council of Educational Research and Training (NCERT). As an offshoot of this large national project, we both continued to be involved in several other workshops and curriculum-development initiatives led by NCERT, while our respective positions in the University of Delhi and SSV College Hapur made it ideal for us to represent geography educators from central and state universities. Thrown together in various programmes, we developed a trust, friendship, and ease of working together. Swagata also participated in several conferences and seminars that Anindita organised and contributed a chapter to Anindita's first book. Knowing Swagata's connections to the field site and interest in feminist research, Anindita invited Swagata to work with her as part of the Delhi CRT. During data collection and all through lockdown we remained connected, noting the similarities in our situations and turning to each other for advice on professional and family matters – enacting a form of care-centred feminist peer mentoring. Our skills as researchers are complementary, and we are able to work and write together seamlessly.

Finally, Swagata's existing standing in the community where the fieldwork was conducted enabled her to appoint CRT members and was the third crucial layer of feminist friendship that underpinned the project. Her friendships with women from the community and her sensitivity to their needs and care-filled efforts to cross the class barrier forged another bond of trust. The field assistants in turn deployed care and kindness with the participants and together these multiple layers of feminist friendship and trust yielded rich interviews for the Delhi CRT.

By Anindita Datta and Swagata Basu

In addition to the CRTs, there are other teams within GenUrb. What we refer to as the 'York team' is based at the City Institute, York University, Toronto, and comprises the management team of the grant's principal investigator, Linda Peake, the project manager, Leeann Bennett, postdoctoral fellows, and a changing array of graduate and undergraduate students (all of whom are listed in GenUrb example 0.3).

GenUrb example 0.3: Members of GenUrb's York University team

Status	*York Team members*
Undergraduates	Dara Dillon, Bri Gardinder, Melissa Hernández Jasso, Javeria Mirza, Rawan Mostafa, and Siya Zhang (and two others who do not wish to be named)
Master's students	Eyram Agbe, Javier Garate Alfaro, Vanessa Bart-Plange, Saaraa Esau, Jamilla Mohammud, Esteban Sabbatasso, and Wiley Sharp (and one other who does not wish to be named)
PhD students	Jenna Blower, Mantha Katsikana, Sehrish Malik, Mel Mikhail, Zhi Ming, Karen Naidoo, Darren Patrick, Serene Paul, Angela Stanley, and Biftu Yousuf
Postdoctoral fellows	Elsa Koleth, Carmen Ponce, Nasya S. Razavi, and Araby Smyth
Staff	Leeann Bennett
Faculty members	Linda Peake

By Leeann Bennett

This book foregrounds the research experiences of GenUrb in chapters authored by members of the GenUrb team. The principal investigator and the postdoctoral fellows in GenUrb, all based in Toronto while writing their sections of the book, hail from Australia via India, Canada, Peru, the USA, and the UK. Other chapter authors, all graduate students based in Toronto, either temporarily or permanently, come from Greece, Ethiopia, the USA, and the greater Toronto area. Together with the CRTs, the members of GenUrb are situated across many borders, coming from and living in many places as well as from both working- and middle-class positionings, identifying as women, as straight, queer or gender fluid (she and they), from first-generation and long-established family backgrounds, and crossing diasporic, settler and global South geographic and political locations. Our feminisms are multiple, and, for those of us based in Canada, are bound up in Canadian settler colonialism and the dominance of the northern academy in circuits of material and discursive power.

Transnational feminist research requires a large commitment of time and emotional engagement. It necessitates developing trust and reciprocity building and taking ethical responsibilities seriously. It also requires learning not only from analysis of data but also from relationships, and this means listening to others and respecting the legitimacy of others' knowledge, and, inevitably, this will mean living through difficult and challenging engagements. Recognising and respecting difference requires not only flexibility in research design but constant reflexivity and care towards others. These commitments are not for the frivolous or for those who prioritise quick publications or single-authored publications. The rewards, though, can be huge including the development of new knowledge, different ways of doing research, and lifelong connections, friendships, and solidarities (see also Howitt and Stevens, 2016).

In the following section, we provide an overview of the material covered in the book and how it is organised.

Section 3. How this book is organised

This book, organised into six parts, provides a specific engagement with feminist urban research drawing on insights from the GenUrb project.

(i) Part I. The building blocks for decolonising feminist urban research

Part I sets out the components that shape the intellectual and practice-based commitments of a feminist approach to decolonising research, addressing who we produce knowledge with (within and beyond the academy), how we can produce knowledge together (employing a comparative analytical lens and engaging with translation), and why and how we do this work (speaking to a feminist politics of activism). Chapter 1 discusses the potential of a comparative methodology to engage the urban through a decolonial lens. It introduces GenUrb's research focus on cities in the global South and discusses some key methodological strategies developed by post-colonial and feminist urban scholars to study the urban in a comparative vein. Chapter 2 shifts the focus from analytical framings to people and speaks to the practice of decolonising feminist knowledge production, turning not to the field but inwards to the members of the GenUrb project itself. It troubles the presumed affinities implied in the feminist transnational collective by tracing the ways in which GenUrb is ruptured by differences manifested along spatial and temporal axes. Chapter 3 addresses how our transnational research, conducted in multiple languages, necessitates engagement with decolonial feminist practices of translation and interpreting. In doing so, it unpacks the colonial roots of Anglophone hegemony in urban studies and its reproduction of epistemic injustice. It explores translation as decolonial feminist praxis, key to dismantling colonial hierarchies and addressing the material legacies of colonisation. Finally, Chapter 4 addresses the feminist politics of scholar-activism within and beyond the neoliberal academy. This chapter highlights how feminist scholar-activists forge ethical relations with communities and engage a feminist ethics of care as well as illuminating the challenges of doing feminist alliance work in antiracist and decolonial struggles.

(ii) Part II. The context of 21st-century urban feminist research

Part II addresses key 21st-century societal and institutional contexts, namely those of COVID-19 and the global agenda of the SDGs, in which feminist urban research and policy take place. In an era of increasing ecological crisis, in which pandemics are becoming global, with devastating consequences for everyday life, we need to reassess how research is conducted. Chapter 5 engages this concern, considering how feminist research and policy processes are affected by and can be creatively adapted to meet pandemic geographies. It provides an overview of how, mid-way through the project, GenUrb pivoted its focus towards the impact of COVID-19 on urban women, adapted research strategies in the context of restricted in-person access, and supported participants beyond the research. Chapter 6 addresses the current global institutional context of the 2030 Agenda and the global SDGs, with particular reference to the synergies between SDG 5, 'Gender equality', and SDG 11, 'Sustainable cities and communities', setting the SDGs within the broader context of feminist engagement with urban policy and global frameworks for urban governance. It explores the gap between the SDGs as global development goals and their local implementation. Further, it provides a roadmap to think through how feminist research can connect to policy outcomes.

(iii) Part III. Feminist research standards

Part III addresses the research standards that feminist scholars must engage with. While the topic of the first chapter on research ethics is commonly addressed in research textbooks, the others (of professional standards, partnerships, and data management) may be less familiar – albeit of increasing importance – in a context in which research is more commonly taking place across national borders and based on teamwork. Chapter 7 establishes feminist ethics as the basis for making decisions throughout the research process. It articulates a feminist ethics of care grounded in intersectionality, positionality, and decolonisation, and it explores how a feminist care ethics can be applied to research. Chapter 8 outlines how professional standards function as a set of expected behaviours and a form of assessment in research. These include standards of institutional research ethics; new policies related to equity, diversity, and inclusion; and the norms about research monitoring, evaluation, and knowledge mobilisation. We consider how these standards – which can be difficult enough for the individual researcher to follow – play out in research that involves teamwork and partnerships. Given the importance of feminist research as a joint endeavour, Chapter 9 discusses the value of research partnerships. It describes how to establish, manage, and lead research partnerships while also addressing the challenges that they pose. Chapter 10 focuses on the importance of data management, opening the data management process to questions of ethics and politics and especially to questions of data justice. It explains the increasingly formalised process of research data management, including the data management plan, in university research contexts and questions of data ownership for feminist researchers.

(iv) Part IV. Feminist methodologies and methods

The fourth group of chapters addresses the methodological approaches and methods employed by the GenUrb project, establishing, first, the philosophical roots of feminist methodologies before turning to how the project engaged with feminist approaches to fieldwork, specifically feminist geo-ethnographies and interviews. First, Chapter 11 introduces the connections between methodology, ontology, and epistemology – the core dimensions of a philosophy of research – before turning to how feminist methodologies engage knowledge production through reflexivity, positionality, the relational nature of research, research ethics, and social change. Chapter 12 introduces fieldwork and the 'field'. It makes practical suggestions for disrupting colonial practices in fieldwork as well as outlining how to prepare for fieldwork and how to use methods such as observation, participant observation, and the taking of field notes, emphasising the importance of cultivating a feminist ethics of care while in the field. Chapter 13 turns to the field-based approach of geo-ethnography, an emerging approach that produces place-based, embodied research findings attuned to scale and to power relations. Chapter 14 focuses on perhaps the most widely used method in feminist research, the interview. It discusses types and modes of interviews while also outlining the need to consider issues such as sample size, participant outreach, consent, the interview guide, and compensation, as well as the significance of location when interviewing, and of maintaining contact with participants.

(v) Part V. Feminist qualitative data analysis

The fifth group of chapters addresses the topics of qualitative data and analysis. Prior to analysis, GenUrb engaged in translation and preparation (coding) of data using the qualitative analysis software NVivo. For those conducting research in a language other than English, Chapter 15 introduces the practicalities of translation and interpreting in distinct stages of the research process, including in literature reviews, in the field, and in knowledge-mobilisation activities. Chapter 16 turns to the nuts and bolts of qualitative data analysis, outlining both deductive and inductive approaches, followed by accounts of the most commonly used forms of qualitative data analysis (content analysis, discourse analysis, narrative analysis, visual analysis, and grounded theory). The chapter also addresses preparing, annotating, coding, and exploring field data as well as the need for reflexive data analysis. The remaining two chapters introduce software-aided analysis for feminist research. Chapter 17 presents the arguments for and against the use of computer software for qualitative research. It also introduces NVivo as a research tool for feminist research and foregrounds a discussion of why, when, and how the GenUrb project used this software. Chapter 18 provides a linear framework for conducting qualitative data analysis with NVivo. It starts by discussing the creation of a design framework and goes on to discuss the preparation of research material, coding, memos, annotations, and data exploration. It concludes by discussing some of the complexities NVivo users often encounter.

(vi) Part VI. Feminist approaches to knowledge mobilisation

The final chapters take up the topic of knowledge mobilisation with a specific focus on the use of social media. Chapter 19 examines the role that effective knowledge mobilisation plays in feminist research projects. It expounds on the role of knowledge mobilisation in various stages of the research process, including prior to, during and after the research. Chapter 20 zooms in on the uses of social media for feminist scholar activism, along with the challenges and possibilities their use entails. The chapter offers an overview of social media tools and platforms that have been used within feminist research, considering social media as spaces for political engagement, solidarity, and knowledge mobilisation, as well as discussing the ethical and political challenges associated with the use of social media.

Section 4. Chapter structure

Within each chapter, you will find a number of defining features. Each begins with a list of learning objectives, stating what you should be able to understand about the chapter topic and what you should be able to do in your own research after reading the chapter. The diverse voices of GenUrb are highlighted in 'GenUrb examples' that provide you with an opportunity to process the material you are learning about through real-world experiences narrated by the people in GenUrb. Reflection exercises also provide an opportunity to engage with the chapter content, either by thinking deeply about aspects of your own research or thinking through hypothetical examples. Key concepts are also identified by the use of bold type in each chapter and defined in the glossary. In some chapters, you will also find references to the book's companion website www.routledge.com/9781032668680, where you will find extra material relating to various chapters. This includes links to the websites of organisations, universities, research tools, and more, audio files featuring the voices of GenUrb members, and short documents that provide extra information on the practicalities of conducting research.

Section 5. Summary

This chapter has introduced you to feminist urban research via the transnational feminist urban research project *Urbanisation, gender, and the global South: A transformative knowledge network* (GenUrb). It has outlined how the GenUrb global network evolved and introduced the cities in which the research was conducted as well as those engaged in the research. The chapter concluded with a discussion of how to use the material included in each chapter and an outline of the six parts of the book. We hope that you engage with this book critically and that it can help you contribute to larger, ongoing conversations about feminist urban research to keep it a critical and reflexive practice.

Part I

The building blocks for decolonising feminist urban research

1 Feminist comparative urban research

Linda Peake, Mel Mikhail, and Elsa Koleth

Learning objectives
On completing this chapter, readers should understand

- approaches to understanding the Global South;
- post-colonial attempts to decolonise urban research;
- post-colonial and feminist scholars' role in producing comparative urban knowledge;
- and three key lineages of feminist comparative urban scholarship.

Further, readers should be able to

- recognise the purpose of comparison in urban studies;
- and determine whether or not comparison is relevant for their study.

This chapter engages with the comparative as an analytical approach to decolonising the study of the urban, drawing on the experiences of GenUrb's transnational project and its focus on cities in the global South. It turns first to a discussion of the global South, of how the term evolved, and how it has been analysed. Various approaches to theorising the urban global South in the 21st century are outlined before reviewing the contributions of post-colonial scholarship to approaches to urban comparison, highlighting strands of post-colonial urban scholarship that share affinities with feminist approaches to the urban. The chapter then turns to three key lineages of feminist research to study the urban in a comparative fashion, bringing it into dialogue with provocations from decolonial scholarship. Finally, GenUrb's feminist approach to comparison is briefly outlined, showing the ecologies of relationality that make transnational comparative urban research possible.

Section 1. The global South

Transnational feminist urban research indicates that the multiple political, economic, social, and cultural encounters that take place across the globe cannot be read simplistically through the binary prism of **global North** and **global South**: to do so would obscure the dynamic intersections between multiple differences. However, the dichotomous **geographic imaginary** (a taken-for-granted spatial ordering) of the global North and the global South is a key spatial axis of difference around which issues of power, resources, and epistemic authority coalesce. Much has been written about the ways in which global economic and social inequalities cemented through historical structures of colonial dispossession, exploitation, and extraction, which concentrated wealth in former imperial centres of the global

DOI: 10.4324/9781032668727-3

North, continue to shape global economies of knowledge production (Blaut, 1993; Connell, 2007, 2014). In this section we briefly delineate how terminology relating to the global South developed and outline how various usages of the term have been employed before turning to a discussion of the urban global South.

(i) Evolution of the terminology of the global South

Post–World-War-II modernisation theory characterised the world economy as divided between 'traditional' and 'modern' national economies, and countries and regions as 'developed' and 'developing'. The rise of dependency theory in Latin America in this period challenged the determinism of this view. As Marlea Clarke (2018) reports: 'dependency theorists saw [the world] as divided between an industrial "core" and an agricultural "periphery". For them and related "world systems" theorists, the extractive and exploitative relationship between the core and periphery was established by colonialism, and then reinforced by the post–WWII world trade system'. While many Latin American and neo-Marxist scholars continued to use the core/periphery terminology, other activists and scholars turned to the term the Third World, a concept first devised in 1952 by French demographer Alfred Sauvy (Clarke, 2018). In this classification, 'the term "First World" referred to the advanced capitalist nations; the "Second World," to the socialist nations led by the Soviet Union; and the "Third World," to developing nations, many at the time still under the colonial yoke' (Heine, 2023, no page).

After the Cold War ended, the term Third World gradually became less acceptable, partly because the socialist Second World had ceased to exist as such and partly because of a desire to replace the pejorative connotations with a more neutral label. In addition, there was a need for a term that addressed processes of formation as opposed to descriptive categories (Heine, 2023). The concept of the Global South was coined in 1969 by Carl Ogelsby, an activist of the New Left. It began to gain support in the 1970s with the call for a New International Economic Order and especially with the 1980 publication of the Brandt report, recognising 'North' and 'South' as broadly synonymous with 'rich' and 'poor', 'developed' and 'developing' countries (Patrick and Huggins, 2023). Most of the latter are located south of the Brandt line, a north–south divide encircling the globe, running at a latitude of 30° N from the Rio Grande into the Gulf of Mexico, across the Atlantic Ocean and the Mediterranean Sea, continuing north of Africa, the Middle East and most of East Asia, then veering southwards to leave Japan, Australia, and New Zealand north of the line (Brandt, 1980).

Support for the term Global South came in large part from scholars and activists in the South, from those engaged in understanding processes of geopolitics that would recognise their struggles, knowledge, and experiences, further influenced by the discourse of **globalisation** in the 1990s (Mahler, 2017). Increasingly, the term has also become institutionalised. It has become synonymous with the United Nations (UN) Group of 77, now 134 countries strong, a collection of post-colonial and other countries that formed to promote their collective interests and to condemn their objectification (Haug, 2021). Arguing against the Rostovian proposition that Southern countries simply needed to 'catch up' with their Northern counterparts, they countered that underdevelopment was a colonial legacy.

(ii) The employment of global South terminology

Global South terminology has been employed across fields of development and post-colonial studies in three distinct ways: geographically, institutionally, and geopolitically.

First, the term has been employed widely as a descriptive geographical term to refer to low-income and/or socio-economically marginalised nation-states in Africa, Asia, Oceania, Latin America, and the Caribbean. Despite its limited analytical purchase, this definition is commonly used and is often implicit in the understandings outlined in what follows. It has also had significant material implications. For example, efforts to categorise countries according to economic development levels commonly rely on indicators such as income per capita and GDP, used by the World Bank to classify countries as low-, lower-middle-, upper-middle-, and high-income (Hamadeh *et al.*, 2022), determining their eligibility for financial aid.

Secondly, the term Global South has been used in an institutional capacity within global development organisations such as the UN and cross-regional and multilateral alliances to refer to nation-states considered less economically developed. Developing out of the post–Cold War détente, and initially with reference to the Third World, groups such as the Non-Aligned Movement and the Group of 77 formed to resist and influence the distribution of power and resources in global institutional systems. As Stewart Patrick and Alexandra Huggins (2023) point out, this alignment is constantly evolving. For example, when, in February 2022, the UN General Assembly passed a resolution demanding the withdrawal of Russia from Ukraine, the 'institutional' Global South was divided, with more than three-fifth of the countries siding with Ukraine and approximately one-third abstaining.

Thirdly, and most recently, the Global South has been conceptualised geopolitically, as a way of being in the world for the groups of nation-states identified (geographically and institutionally) above, but also spatially, i.e. as no longer tethered to the nation-state, and thus inclusive of sub-national areas within countries of the global North. This employment of the term refers to nation-states in the global South and sub-national regions in the global North that have been historically marginalised by slavery, colonialism, and global capitalism (Mahler, 2018). As Anne Garland Mahler (2017) writes, these shared experiences of subjugation and subordination have led to the formation of subaltern resistance:

> forged when the world's Souths mutually recognize one another and view their conditions as shared . . . . In this sense, the Global South may productively be considered a direct response to the category of postcoloniality in that it captures both a political subjectivity and ideological formulation that arises from lateral solidarities among the world's multiple 'Souths' and that moves beyond the analysis of colonial difference within postcolonial theory.

In a decolonial move there has also been a deterritorial employment of the term Global South that has given analytical purchase to both the global and local (Mahler, 2017). With the nation-state no longer the base unit from which understanding begins – indices of inequality show greater levels of inequality within than between countries (UN75, n.d.) – the term gives recognition to all subjugated and subordinate peoples, acknowledging the presence of the 'South' in the 'North' and vice versa (Roy *et al.*, 2020). One of the first to adopt this approach was the Canadian activist and Indigenous leader George Manuel (Manuel and Posluns, 1974), who expanded the term Third World to include the Fourth World in a reference to the settler colonisation of Indigenous peoples within First World nations (Clarke, 2018).

It is this geopolitical understanding that has led to the practice within GenUrb of referring to the 'global' with a lowercase 'g'. This usage symbolises that the global South is a way of being in the world, inclusive of geographically situated nation-states and other spatial orderings or divisions, but dissociated from any descriptive understanding of the 'global' as geographically determined; rather it is a referent for the historically determined incorporation of spatialities of division into globally uneven processes and structures of capitalist development.

GenUrb example 1.1 discusses the varying attitudes to the term global South and its deployment in the GenUrb project.

GenUrb example 1.1: The standing of the term 'global South'

In GenUrb, there has been no agreement over the term global South: some prefer it not be used, while for others it stands as a geopolitical rubric for decolonised people and places. Its usage is not an issue, however, that has raised any strong feelings or been taken up in extended discussions. Neither has this divide been obvious across particular constituencies, such as those in GenUrb living in the global South versus those living in the diaspora.

Arguments against use of the term 'global South'

For those who preferred not to use the term, feelings ranged from indifference to ideological opposition. Those in question pointed out its unsuitability as an analytical framework, highlighting the constant slippage between its usage as a geopolitical term and its institutional designation, as well as its conceptual incoherence given the economic, political, and cultural diversity the global South encompasses and its high internal differentiation, including as it does approximately two-thirds of the world's population and the economic and political superpowers of China, India, and Indonesia, as well as very small states. There was also caution about the propensity of such a broad term to be associated with sweeping generalisations, such as the assumption that all countries or peoples within the category are *per se* destitute, or to make it a superficial and reductive framing, such as referring to all studies related to Africa, Asia, or Latin America as Global South research (Wa Ngugi, 2012).

For those ideologically opposed to it, the global South is a term that risks reinforcing hierarchies and stereotypes. It still implies a teleological standard of linear progress that classifies nations along a spectrum, according to how closely they approximate a Western mode of development. Finally, the preference by some not to use the term should not be read as preference for a liberal alternative, such as 'global development' (the view that propels the UN Sustainable Development Goals). Those adopting this position accept a geopolitical understanding of the global South and yet prefer not to deploy a term that still tends to obscure as much as enlighten.

Arguments for use of the term 'global South'

For those who chose to engage with the term, it emphasises the construction of the global South as a geopolitical category that steeps political economy in territory and time, its global positionality read through its hierarchical relation to the global North and the resultant historical geographies of slavery, colonialism, and globalisation. As Nour Dados and Raewyn Connell (2012, quoted in Clarke, 2018) contend, the 'term Global South functions as more than a metaphor for underdevelopment. It references an entire history of colonialism, neo-imperialism, and differential economic and social change through which large inequalities in living standard, life expectancy, and access to resources are maintained' (see also Connell, 2007).

While there is a growing awareness of the global nature of the major challenges facing the planet, asymmetric opportunities for development in the global South continue to be structured by the political economy of underdevelopment. Made possible by the extraction of labour and resources from the global South and made visible in racialised divides and migration crises, underdevelopment is indicative of a need for approaches to research that can analytically account for these enduring asymmetries: as Nikita Sud and Diego Sánchez-Ancochea (2022, p. 1144) observe, 'the South cannot simply merge with the North, or into an indistinguishable 'globality'.

Notwithstanding that the spatial formations of the global North or South binary are neither static nor unchangeable or that the boundaries and relationship between the two are subject to remapping, this binary remains a heuristic device that is strategically used by scholars to capture the persistent and historically contingent reality of uneven global development (Sheppard and Nagar, 2004; Peake and Rieker, 2013; Raghuram, Noxolo and Madge, 2014; Simone, 2020).

By Linda Peake

Reflection exercise 1.1 asks you to consider where you stand in relation to usage of the term global South.

Reflection exercise 1.1: Determining your usage of the term global South

Considering the arguments for and against the use of the term global South, determine where you land in your usage of the term. Consider the following questions.

- Who is using this term? For whom is it being used?
- Would your usage of this term differ depending on your geographical location?
- How does your positionality factor into your usage of this term?
- Are there occasions or ways in which you can imagine using this term and others where you would not use it?
- What does the use of capital letters in this term (Global South, global South) achieve?

(iii) The urban global South

As with the term global South, its urban counterpart has passed through various delineations – less developed cities, Third World cities, urban global South – corresponding to the trajectories described earlier. The latest moniker of the urban global South received an even more blasé reception in GenUrb than its global South counterpart, with the same divide apparent as outlined already. However, the urban global South has the advantage of eschewing a focus purely on cities, avoiding the 'methodological cityism' (Angelo and Wachsmuth, 2015; Brenner and Schmid, 2015) of much of the field of urban studies and recognising the 'infinite variety' of urban places in the global South (although agreement over what constitutes the urban is another focus of intellectual debate: see Peake *et al.* [2019] in the theme issue of *Society and Space* on planetary urbanisation). Despite recognition of this variety, the urban global South is still primarily associated with the tropes of the slum and the megacity (Dupont *et al.*, 2015; Chant and McIlwaine, 2016), both of which have received disproportionate academic attention. While it is widely reported that over one billion people reside in slums and that one in eight people live in 33 megacities, it is also the case that 'close to half of the world's urban dwellers reside in much smaller settlements with fewer than 500,000 inhabitants', for which there is far less academic attention (United Nations, 2018).

While there is agreement that the status of informal settlements is defined by their relationship to the state, the diversity that characterises these urban places spills out of the categories that attempt to capture the realities of urban lives. There has been a decolonial move away from a discourse that describes the nature of these settlements in terms of substandard living conditions and their inhabitants in terms of deprivation to a focus on the innovative livelihood practices used by residents to engage in placemaking characterised by the sharing of resources, knowledge, and experiences (Roy, 2011; Jiménez, 2017; Simone and Pieterse, 2017; Simone, 2018, 2022). This move arises from the 'southern urban critique', i.e. the study and theorising of southern cities by post-colonial scholars. Mary Lawhon and Yaffa Truelove (2020, p. 3; see also Lawhon, 2021) elaborate upon this critique revealing three interlocking strands of Southern urban theory based on empiricism, alternative traditions, and deconstruction.

- *Empiricism*. The South is deployed as a geographical location that is relationally produced with the North. It is internally empirically different in social, cultural, political, and economic ways, and this difference matters, raising questions about the ability to generalise across difference.
- *Alternative traditions*. Northern **hegemony** works to displace diverse intellectual traditions from the global South. Decolonising practices have led to new concepts and approaches being woven into or displacing existing understandings (Roy, 2009; Bhan, 2019).
- *Deconstruction*. The South is a metaphor for the need to deconstruct knowledge – power relations about how the urban global South is known. The embedding of colonial relations and rationalities in the present requires the interrogation by global Northern scholars of the 'production both of their own worldview and of the worldview that produces their data' (Lawhon and Truelove, 2020, p. 12). The opening-up to scrutiny of Southern concepts (such as AbdulMaliq Simone's [2004, p. 407] 'people as infrastructure' or Gautam Bhan's (2019, p. 640) 'squat, repair, and consolidate' as modes of Southern urban practice) enables the provincialising of Northern urban theory. In other words, the post-colonial encounter requires the critical interrogation of Northern research through practices of reflexivity (see Chapter 11, Section 4(i) Reflexivity) and unlearning through praxis (see Chapter 9, Section 3(i) Transnational feminist praxis).

While all three strands were understood across shifting coalitions of GenUrb members, they were unanimous in believing that the global South speaks to a way of being in the world, to what Sophie Oldfield and Susan Parnell (2014, p. 3) term a 'sensibility'. There was a strong desire to conduct research across cities in the global South, to explore South–South relations and processes, and to 'extract the Global South from the tyranny of the development discourse' (Rigg, 2007, p. xv). This was with the aim of contributing to the understanding of the South as the epicentre of **urbanisation** and urban futures and of working across the grain of large-grant-funded research programmes being the prerogative of the global North, as part of the larger concerted post-colonial effort to shift the geographies of urban theory towards the global South.

As feminist and post-colonial urban scholars, members of GenUrb advocate for comparative transnational research not with an aim to sublimate or 'ignore' the North but as a necessary corrective to the legacies of modernity and developmentalism in urban studies, which have historically relegated the global South to the margins of urban theorising (Roy, 2009; McFarlane and Robinson, 2012; Sheppard, Leitner and Maringanti, 2013; Lawhon, 2021). Hence, both analytical and political factors played a part in determining the comparative lens of the GenUrb project. The following section turns to the comparative and specifically the ways in which post-colonial scholars have approached **comparison** as a methodological strategy for decolonising urban knowledge production.

Section 2. Comparative urban research

Comparative analysis has long been a methodology utilised in urban studies, although comparativism fell out of favour in the latter part of the 20th century due to a range of perceived shortcomings concerning earlier approaches (McFarlane, 2010; Parnell and Robinson, 2012; Sheppard, Leitner and Maringanti, 2013; Robinson, 2022). These included:

- a limited understanding of scale and 'methodological territorialism', i.e. the idea that all social relations are contained within a city as a discrete object (Ward, 2010, p. 479);
- scientific modes of analysis that controlled for predetermined independent variables;
- reductionist approaches to causation that limited the identification of relevant case studies (for example, privileging economic and financial factors in research on global cities);
- a focus on systemic incorporation or convergence leading to the creation of hierarchies, divisions, and assumptions of incommensurability between different cities;
- the generation of universalising knowledge claims based on prioritising parochial (Western), theory-driven hypotheses.

With the ascendance of post-structuralism, interest in comparison re-emerged during the late 1990s as the field of urban studies turned to debates about globalisation and relational conceptions of space (Nijman, 2007; Jacobs, 2012). The emphasis on relationality accompanying the 21st century's comparative turn broadened intellectual horizons regarding connections between cities and how these could contribute to building urban theory (Çağlar and Glick Schiller, 2018, 2020). The exploration of these relational geographies has been informed by a range of theoretical traditions and readings of the urban, including Marxist political economy (Hart, 2018), post-structuralism, post-colonialism, feminism, new materialism, and post-humanism (Jacobs, 2012). Scholars have come to conceive of the city as embedded in flows and relations at multiple scales (Ward, 2010) and of the urban as mobile, embodied, and co-constituted by non-human agencies and forces (McGuirk, 2015; Simone and Pieterse, 2017). Researchers working towards seemingly different epistemological and theoretical ends – from planetary urbanisation (Brenner and Schmid, 2015; Schmid *et al.*, 2018) to urbanisation 'from below', studied at the scale of everyday lifeworlds and subjectivities, in line with feminist theoretical concerns (Buckley and Strauss, 2016; Ruddick *et al.*, 2019) – have welcomed these moves to challenge the spaces and modalities through which urban theory is generated.

In other quarters, however, reception of these critical interventions, and particularly of post-colonial urban studies, provoked alarm over 'a growing sense of disarticulation, dissipation and fragmentation' in urban theory more broadly (Peck, 2015, p. 162), resulting in calls for a defence of theoretical generalisability (Storper and Scott, 2016). Underlying these critiques is a latent anxiety about the unmaking of urban theory and the burden of reconstructing it after the so-called crises induced by post-colonial and poststructural interventions (Roy, 2016). These critiques seek to restore the hegemonic centre of urban studies from which assessments can be made as to the relevance and value of competing epistemologies and ontologies of the urban. Such critiques precisely reflect ongoing South–North 'geometries of power' (Massey, 1993, p. 59) and speak to the high political stakes involved in collectively re-writing the South as its own 'centre'.

For post-colonial urban scholars comparison is the key methodological strategy in this rewriting. However, despite calls for urban comparative work, there has until recently been no consensus on 'how to design and realize' critical comparative studies (Ren and Luger, 2015, p. 145), or on what a 'good comparison is or should be' (Deville, Guggenheim and Hrdličková, 2016, p. 32).

In the 21st century it is the authoritative oeuvre of Jennifer Robinson (2005, 2006, 2011b, 2016a, 2016b, 2022; Le Galès and Robinson, 2023) that has engaged in such a rewriting, heralding a renaissance in comparative urban research. It is accompanied by calls to deconstruct urban theoretical knowledge production purporting to be global but based on research in a limited range of cities in the global North and to theorise the urban in more experimental, provisional, and revisable ways (Roy, 2009; Jacobs, 2012; Lawhon and Truelove, 2020). Northern theory, Robinson argues, typified by its universalising impetus, confers epistemic privilege to Northern cities by reading the global North and global South through respective hierarchies of modernity and development. Robinson thus calls for the provincialising of Northern theory, employing comparative approaches in which the theorising of the urban can take place from anywhere, from 'ordinary cities' (Robinson, 2006, 2022).

Robinson's (2022) comprehensive schema for comparison is outlined in Table 1.1. She states that 'at the heart of a comparative imagination is the assumption that concepts can be critically interrogated and revised through exploring difference across cases' (2022, p. 196). This is her starting point for (i) recognising the theoretical grounds of thinking with spatialities of the urban and the desire of the researcher to understand a phenomenon and (ii) the 'genetic' and 'generative' comparative strategies that these generate, with their comparative tactics of tracing, composing, and launching comparison. To be sure, this schematic presentation should not obscure Robinson's (2022, pp. 20, 126) insistence that genetic and generative explorations intertwine in practice.

Table 1.1 Schema of Robinson's comparative framework

Comparative strategy	*Ways of thinking with urban spatialities*	*Theoretical starting points*
Genetic	*Differentiation*	*Materialities approach* (of making connections)
Genetic	*Diversity/Variegation*	*Political economy* (of relational comparison)
Generative	*Distinctiveness*	*A concern with variation and individuality*

Source: Adapted from Robinson, 2022

Genetic comparative strategies (p. 11) begin with urban spatialities of differentiation and diversity.

Differentiation results from 'interconnected processes [that] are associated with repetition and differentiation of elements of urban form and social process' (p. 5). It has been taken up theoretically by a materialities approach, which is primarily concerned with making connections. The materialities approach involves 'tracing the interconnected genesis of repeated, related but distinctive, urban outcomes as the basis for comparison' (p. 124). It has commonly been employed in studies of policy mobilities and urban forms (for example, in relation to satellite cities, urban megaprojects, high rise, the bungalow, gentrification, suburbanisation, and the operational landscapes of planetary urbanisation, such as, for example, the spatial extension of large-scale land-use systems devoted to resource extraction, energy, and water and waste management (Brenner and Schmid 2015)).

The primary tactic involved is that of *tracing*, which involves:

(i) following connections (i.e. ubanisation processes) to trace how a specific urban outcome emerges (pp. 138–146); or
(ii) engaging with interconnected or serial cases (conceptual or empirical), which can be studied in three ways (by addition, by subtraction, by attending to each case) (pp. 146–159).

Diversity (or variegation) results from outcomes that are 'specific in different places because of long historical trajectories of development and emergent socio-spatial dynamics' (p. 5). Studies engaging with diversity have been taken up theoretically by political economy approaches, which go beyond making connections to 'work with the (translocal) social processes which link and jointly produce places and phenomena' (p. 14), i.e. with relational comparison. For political economy approaches, genetic comparative strategies involve 'interrogating and understanding "wider processes", such as urban neoliberalisation; and working comparatively with "contexts" ' (p. 162).

Generative comparative strategies begin with *distinctiveness*, with the understanding that 'we can only come to know it [the urban] through its individuality' (p. 5). This strategy involves the inextricable links between 'the curiosity and practices of the researcher to understand a phenomenon or problem' (p. 14) and the researcher's engagement with 'the agency of the urban world' (p. 17) or, in other words, 'the fullness of urban territories to provoke comparative experiments' (p. 18). It follows that for generative strategies the subjectivity of the researcher matters (see Chapter 11, Section 4(ii) Positionality). Two main tactics, *composing* and *launching*, are employed in generative strategies.

Composing is used to generate new concepts and interrogate inherited terms. Composing comparison involves:

(i) *starting with the researcher*
How the researcher conceives of the urban 'as a context, a socio-political conjuncture, a spatial form, as an inexhaustible and distinctive reality, or as a diversity of processes and outcomes – can inspire different kinds of generative urban comparisons' (p. 248).
These comparisons can emerge in relation to what Robinson refers to as:

- 'conjuncture', which 'highlights the significance of contingency, and yet also insists on the importance of structures and wider social processes' for different outcomes;
- or 'specificity', i.e. not in terms of the urban being unique but of 'specificity as an outcome of the way in which the urban is produced' (p. 267), such that 'each urban context will have different spatial dynamics shaping urbanisation' (p. 268);
- or 'diversity', which can involve starting with 'issues and topics which affect or are present in a wide range of cities but are not approached as repeated instances or parts of putative wholes' (p. 272) but rather as enabling '"conversations" across distinctive urban contexts in which each informs and enriches analyses of the others' (p. 17).

Thinking with specificity and diversity moves comparative analysis away from the strictures of interconnection, relations, and conjunctures into a 'looser' and more experimental mode.

(ii) *starting with urban territories as distinctive* ('as "assemblages", events or singularities' (p. 308)
This aspect of 'composing' comparison includes critical engagement with 'any urban territory' (p. 308), moving beyond cities to the 'dispersed, fragmented outcomes of contemporary urbanisation processes' (p. 308), such as suburbs, peripheries, and the operational landscapes of planetary urbanisation.

In addition to this tactic of 'composing' comparison, other methodological innovations include comparison as situated practice (see also Section 4), experimental comparison, bringing cities into conversation, and the exploration of mobile concepts, such as gentrification.

Robinson proposes the tactic of *launching comparison* to allow conceptualisations that arise from this engagement with territories and consideration of the urban as distinctive to be imagined. This requires launching concepts from 'whatever an urban territory entails' (p. 308) to be put to work elsewhere (for example, as has happened with the concept of informality). Starting anywhere, with any singularity or incomparability, can lead to insertion of cases into analysis and wider conversations.

Although Robinson's comparative schema touches on dimensions of feminist urban scholarship – the feminist critique of planetary urbanisation, the importance of feminist conceptualisation, the concern with the positionality of the researcher, and the 'imperative . . . for new subjects and authors of urban theorising' (p. 2) – feminist comparison is relegated to the figurative margins in Robinson's schema, and yet it has a long heritage, to which we turn in the following section.

Section 3. Lineages of feminist comparative urban research

Although aspects of feminist methodologies and approaches to knowledge production have contributed to urban comparative research, comparison as a research methodology and mode of theory-building has had an understated presence on the feminist urban agenda (see Chapter 11, Section 4 Feminist methodologies). To call attention to its long heritage and feminist scholars' ongoing engagement with comparison in urban research, this section identifies three latent 'tracks' of feminist comparative urban research since the 1970s: feminist policy-oriented and development-driven comparison, comparison through transnational epistemologies of the gendered production of urban places, and comparison as feminist urban global theory-building.

(i) Track one: Policy- and development-driven comparison for women's inclusion

The first track comprises studies that compare cases of women's incorporation into the urban and that place particular emphasis on descriptive, evidence-based studies and policy interventions aimed at 'women's right to the city' objectives (i.e. measures to ensure that cities are designed and governed to be accessible for women). These studies are a departure from mainstream urban studies in that they both include gender analyses to make policy prescriptions for urban development and take urban places in the global South as their origin for comparison. The urban spatialities in this track comprise the towns and cities of nation-states, empirically demarcated and bifurcated from their rural hinterlands, usually as free-floating administrative units to which gendered policy prescriptions to increase women's engagement in employment and urban life more broadly can be applied.

These studies adopt a descriptive understanding of comparison, in which the same research design is applied across two or more places to observe similarities and differences therein using classificatory schemes populated by taken-for-granted categories or data variables (see, for example, Cohen, 1986). Having established a problematic that guides the comparison, units of analysis and variables for comparison are selected and variations and commonalities drawn out. Grounded in positivist urban epistemologies, these approaches have not been exempt from reproducing the colonial impulses of comparison in the late 20th century.

This track's provenance lies in feminist development studies from the 1970s, the cornerstone of this lineage being *Women's role in economic development* by Esther Boserup ([1970] 2007; see also Jelin, 1977). This book represents the first attempt by a woman scholar to provide a basis for comparing urban types according to quantitative measures of female-to-male ratios in the cities of 'developing' countries. Albeit a rudimentary format (identifying two types of 'male' and one type of 'semi-male' city!) her categorisation formed the basis of her analysis of the 'role of women in urban development' (Boserup, [1970] 2007, p. 75). It provided the first comprehensive overview of the transformations in women's economic and social roles in the mid-20th-century transition to urbanisation in the global South. Boserup's liberal feminism, however, equated development with Western ideals of progress, employing econometric measurements relating to the sphere of production. Her lack of engagement with issues of social reproduction also led to critiques of her work by feminists working in development studies (Benería and Sen, 1981). In the field of feminist urban studies, however, Boserup's work has remained invisible.

By the 1990s, the household had become an important unit of comparative focus within this framing of the city and, in development policy there arose a preoccupation with its vulnerability and economic viability (Moser and Peake, 1987). The feminist move away from a universalist approach to conceptions of gender as variegated could be seen reflected in feminist urban research that understood subjectivity as embedded in place, as exemplified in Caroline Moser's edited book *Confronting crisis* (Moser, 1996). This study involved extensive data collection, both primary and secondary, in three urban global South communities (Chawama, in Lusaka, Zambia; Cisne Dos, in Guayaquil, Ecuador; and Commonwealth, in Metro Manila, the Philippines) as well as Angyalfold, in Budapest, Hungary. Although the urban was still primarily understood as a discrete container, discursive frames were broadening to consider the urban beyond an 'administrative unit' to one that was incorporated into dense networks of flows of capital, talent, labour, culture, commerce, and governance. Still, this work was geared towards adopting a Eurocentric and humanitarian, asset-based management approach to urban poverty in the 1990s.

Work in this track continues in the 21st century as scholars use comparison to develop evidence-based responses to the challenge of surviving everyday life in cities. These studies range from small national or regional projects to large-scale, global projects. For example, the United Nations' global flagship programme 'Safe cities and safe public spaces for women and girls' (begun in 2010 and slated to involve 50 cities across the globe by 2025) began with a concern to develop a comprehensive understanding of the different types of violence against women and girls in urban public spaces, the prevalence of this violence, and action that can be taken. More commonly, studies compare women's urban experiences across a small set of cities and often focus on a particular aspect of women's urban life, such as access to transportation (Yong *et al.*, 2020). Sylvia Bashevkin's (2006) study, for example, compares women's experiences in London, UK, and Toronto, Canada, exploring the challenges to women's inclusion in city-making and the implications of municipal restructuring experiences for democratic citizenship for women.

(ii) Track two: Transnational epistemologies of the gendered production of urban space

A second track of feminist comparative urban research is characterised by transnational epistemologies, which implicitly challenge the assumption that urban spaces are mere containers of urban life and displace the focus on economic development and poverty common to track one. Feminist scholars have increasingly turned to more complex spatial registers, such as transnational circuits of power, to study the tangled and variegated spatialities of the urban, reconfiguring the feminist comparative imagination away from a universalising liberal feminist voice. While the scholarship making up this track does not explicitly take up the methodology of comparison, it deals with the porosity of borders, the relationality of scale, and the transnational restructuring of social reproduction and, as such, makes a significant contribution to the comparative project (although it is not until the third track that we find an explicit interrogation of comparison itself as a methodology for generating new urban theory).

The analytic focus of studies in the second track did not gain traction before the creation of journals like *Gender, Place and Culture* in the early 1990s. These venues consolidated feminism's place in academic geography and allowed feminist geographers to confront the hegemonic Euro-American construction of knowledge dominating their discipline, by integrating the poststructuralist insights of post-colonial critiques and allowing a focus on migration and the transnational to inform their epistemologies of the urban.

Overall, feminist transnational and migrant epistemologies illuminate how urban placemaking is constituted through gendered, multi-scalar networks of kinship, care, and social reproductive as well as productive labour. Influential scholarship in this field emphasises how gendered ideas of work and care travel and bridge the North–South divide, mediated by national regulations such as migration laws. It also points to studies that concern the gendering of dimensions that are constitutive of the urban, such as cosmopolitanism, citizenship, and activism (Willis and Yeoh, 2000; Ehrenreich and Hochschild, 2003; Pratt and Yeoh, 2003; Glick Schiller and Çağlar, 2009; Glick Schiller, 2012). These travelling gendered social relations play out at the everyday level of the household, where, for example, experiences of migration and displacement cut across class to complicate discourses about the ease of access to and provision of care for elders (De Silva, 2018); or where, in the same city, varied commitments to religion across cultures can produce competing views about what transnational family arrangements ought to look like (Trovao, Ramalho and David, 2015). Transnational and migrant epistemologies thus critically reveal the simultaneous stretching out and collapsing of spatiotemporal and cultural boundaries in urban placemaking by pointing out how the micropolitics of life-making are inextricably linked with broader geopolitical arrangements.

(iii) Track three: Comparison as feminist global urban theory building

The third track comprises feminist urban knowledge production that makes comparison qua methodology explicit, moving beyond the urban as container to urban as process. Through this reframing, comparison is cast as a tool for the development of new anti-colonial and decolonial urban theory. As such, feminist urban comparison makes the multi-scalar, transnational, and multi-sited characteristics of urban placemaking explicit, such that new urban imaginaries emerge from porous understandings of the urban. One example is Faranak Miraftab's (2016) ethnographic study of Beardstown, Illinois, which reveals how, in order to revitalise itself, this small deindustrialised town, dependent on meat processing, has come to rely on transnational labour from Mexico and West Africa, revealing how global networks of social reproduction (transnational care work and support) serve to reproduce labour and contribute to the production of urban places. Comparative research by Sage Ponder (2021) provides another example. Her feminist and anti-colonial scholarship on municipal bond markets in the 21 largest majority-Black cities in the United States reveals them as 'sites for global capital extraction more . . . than concentration' (Ponder, 2021, p. 2113). Drawing on Katherine McKittrick's (2013, p. 2) understanding of 'plantation futures' as linking past and present forms of anti-Black violence, Ponder suggests that 'plantation spatialities are also what financially and ecologically connect majority-Black cities to the rest of the global city hierarchy' (p. 2113). They put these cities into analytical relation with other cities in the global urban landscape, documenting the role of racial capitalism in the production of urban space.

Such unbounded spatial epistemologies of the urban as wrought through transnationalism, racial capitalism, globalisation, and neoliberalism speak to the way that a relational comparative approach brings into focus not only topographic depictions of space – factories, houses, banks – but also their entanglement with relational topological spatialities – bank transfers, informal practices of lending, caring across distance – allowing us to grasp the enfolding of proximity and distance in ways that are more or less durable in time and constitutive of urban (and other) spaces (Harker, 2020).

In the hands of feminist scholars actively seeking to disempower the dominant signifier or 'comparator' of the Western city or its models of social and economic organisation, comparison can be used to make implicit or emergent comparative epistemologies across 'ordinary cities' or cities studied from the 'bottom up' explicit (Gough, 2012, quoted in Jacobs 2012, p. 911; see also Hansen and Dalsgaard, 2008). For example, for Kate Gough, the value of working collaboratively in an ethnographic study across three cities – Lusaka, Zambia, Recife, Brazil, and Hanoi, Vietnam – on young people's differing experiences of urban life was 'not simply to conclude something about urban-induced difference or youth-based commonality [but to] challeng[e] the complacent security of place-centrism, which in turn forced methodological innovation' (Jacobs, 2012, p. 911) (see also Koleth *et al.*, 2023).

Before turning to the ways in which GenUrb considers comparison, Reflection exercise 1.2 asks you to consider our usage of the terms post-colonial, decolonial, and anti-colonial in this chapter. We provide some context for this and ask you to think about the implications of the differences and similarities of these terms for producing comparative feminist urban knowledge.

Reflection exercise 1.2: The post-colonial, anti-colonial, and decolonial

Each of the terms 'post-colonial', 'anti-colonial', and 'decolonial' calls up its own historical, geo-political, and discursive particularities, although these contexts often overlap. In one way, shape, or form, all three share an interest in critiquing and resisting Western notions of modernity, from which (as this chapter has elaborated) the notion of the global South emerges. Yet each term expresses a different geographic starting point, calling up the many parallel histories of global South practices of resistance to Euro-American hegemonies.

Post-colonialism emerges especially out of the histories of colonialism and imperialism in India, Southeast Asia, and the Middle East region (Said, 1978; Spivak, 1999); decolonialism, from Latin America and African struggles for independence, travelling to other settler-colonial nations such as Canada and Australia (Quijano, 2000; Parry, 2004; Tuck and Yang, 2012). Anti-colonialism has a less fixed regional focus. Some of the most influential anti-colonial scholars, such as Frantz Fanon, Aimé Césaire, and Sylvia Wynter, wrote, resisted, and thought anti-colonialism through the making and unmaking of Blackness as an invention of colonial and early capitalist modernity (Fanon, 1986, [1963] 2004; Césaire, [1972] 2000; Wynter, 2003).

These varied points of departure, each with its own historical texture and 'structure of feeling' (Williams, 1961, p. 64), contribute to differences in the way these terms have evolved discursively (Bhabha, 2012; Bhambra, 2014). Post-colonialism has often been regarded as a movement in literary and cultural studies, rooted in discourse, to engage with the ways that even after national independence from colonial powers, remnants of the so-called 'modernising' project remain in culture or among leadership. Like post-colonialism, anti-colonialism has been framed as both a historical period and political project, especially of Black political histories (such as the Negritude movement of the 1930s (Rabaka, 2015)), that has coalesced into a political sensibility that is geographically mobile across regional space. It brings the insurgent and militant impulse of resistance to colonialism into the realm of the discursive. Decolonialism, which also began as a tangible political project, has in its journey to the North seemingly been reduced to the epistemological practice of rethinking modes of thought and living imposed by the West, or a praxis of undoing (Mignolo and Walsh, 2018). As Eve Tuck and K. Wayne Yang (2012) have pointed out, going into debates in the sphere of global North academia over the varied meanings and usage of these terms can have the effect of metaphorising terms to such a degree that their historicity is dissolved and emptied of their political force.

As this chapter has suggested, however, academic scholarship is also a political practice when we actively produce knowledge and concepts that displace hegemonic epistemologies. The approach to comparison that we are elaborating in this chapter is meant to give a sense of the way that a hegemonic and colonial epistemology can be decentred, deconstructed, and reconstructed as a methodology through which researchers can shape urban knowledge (see also Chapter 11, sections 3 and 4, Feminist epistemologies and Feminist methodologies).

- Given the view of feminist urban comparison as a **methodology** being elaborated in this third track, how, in your opinion, would each of the terms post-colonial, decolonial, and anti-colonial affect the terms of this methodology (its framing principles, rationale for use of methods, and the theories of knowledge it draws on)?

In the concluding section, we turn to the ways in which the GenUrb project addresses comparison.

Section 4. GenUrb and the comparative

The GenUrb approach to feminist urban comparison puts life-making at its centre, displacing the centrality of production in urban studies to emphasise the intertwined relations of social reproduction and production, and mobilises the impulse to praxis underpinning the history of transnational feminisms to further the aim of decolonising knowledge. Even as feminism continues to be a contested discursive field among us, our engagement with the urban has developed in dialogue with critical traditions of post-colonial theory to critique the heteropatriarchal and white nature of the knowledge production systems that buttress it while privileging the everyday situated knowledges of women and other marginalised groups.

Decolonial feminist critiques demand that we also contend in a more insurrectionary mode with the ways that comparative approaches to the urban might serve to 'perpetuate and interrupt the symbolic and epistemic logics of racial capitalism' and dispossession that underpin current crises (Roy *et al.*, 2020, p. 925). When processes of decolonising and feminising knowledge production underlie research, then as 'the purpose of the research is about more than gathering data to generate findings – it is political and highly relational and not devoid of the self' (Moewaka Barnes and McCreanor, 2022, p. 227). In other words: who the researcher is matters. Within the comparative context of GenUrb this 'matter' is an orientation made possible through transnational feminist praxis, counter topographies, and partnerships of feminist scholars and activists based across the global North and global South (see Chapter 4, Section 4 The challenges of feminist scholar-activist research in transnational contexts and Chapter 2, Section 3 Circulations of spatial and temporal difference). As an alignment across difference, our understanding of comparison as a decolonising methodological practice is relational, intersectional, materialist, grounded in everyday life, and embedded in an ethics of care (see Chapter 7, Section 2 Towards a feminist ethics). Ontologically, we centre the everyday in the historical geographies of life-making and their shaping of the urban at the same time as we interrogate cracks, slippages, contradictions, and surprises that reveal possibilities for change and different ways to imagine the urban. Ethically, we are committed to a praxis that can engage as much with everyday micropolitics as with the epistemological directive to engage in the co-production of knowledge that is attentive to relational accountabilities.

As such, we find ourselves in agreement with Aram Yengoyan (2006, p. 4): there is a 'vitality' to the practice of comparison which 'requires us to reflect on our different ways of knowing things'. It is this notion of comparison as a way of seeing, as an 'epistemological register', that motivates practice within GenUrb. The modes of comparison in which we have been engaging so far include relational and materialities approaches as well as comparison as situated practice.

Elaborating on Christopher Harker's work (2020) on place ecologies of debt we engage in relational comparative analysis to analyse both topographic and topological dimensions of urban placemaking that create place ecologies of economic precarity and violence. The GenUrb project employs geo-ethnographic research to examine the everyday experiences and practices of women living in marginalised neighbourhoods across cities in the global South. This, in turn, enables us to think through the intersections of post-colonial and gendered difference and to decentre Eurocentric conceptualisations of urbanism. GenUrb's relational methodology recognises, as Kevin Ward (2010, p. 281) puts it:

> both the territorial and the relational histories and geographies that are behind their production and (re)production . . . . Stressing interconnected trajectories – how different cities are implicated in each other's past, present and future – moves us away from searching for similarities and differences between two mutually exclusive contexts and instead towards relational comparisons that uses different cities to pose questions of one another.

We also deploy a materialities approach, using case studies to make connections across the cities concerning the ways in which the Sustainable Development Goals (SDGs) have been implemented, specifically SDG 5 (Achieve gender equality and empower all women and girls) and SDG 11 (Make cities and human settlements inclusive, safe, resilient, and sustainable). Following in the footsteps of the Mistra Urban Futures project (Simon, Palmer and Riise, 2020) with its concern with the SDGs and with producing comparative and transdisciplinary research as well as building on feminist critiques of the Millennium Development Goals (MDGs) and SDGs (Kabeer, 2005; Sen, 2018; Mölders, 2019), the GenUrb research on SDG 5 and SDG 11 is a longitudinal comparative study allowing for an assessment of the extent to which the synergy between these goals in the GenUrb cities is manifested in their respective local contexts.

Thirdly, we employ comparison as situated practice in which the emphasis moves from epistemological concerns to those of how comparison is done and the methodological, practical, and ethical challenges it presents (Deville, Guggenheim and Hrdličková, 2016). While there is a long tradition of feminist scholars producing self-reflexive and autoethnographic accounts of their research, ranging through footnotes to separate book-length publications, understanding comparison as a situated practice goes further than such reflexive accounting to address the logics of comparison. It involves taking the practice of comparison itself as an object of analysis. For Joe Deville, Michael Guggenheim, and Zuzana Hrdličková (2016), it is the infrastructures of comparison that are of interest. It is the achievements of combinations of research teams, funders, being in the 'field', disciplinary methods, practices and conventions, feelings, and technology, played out within an array of national, cultural, and historical power relations – the 'how, what, and where' of comparison – that allow us to see what can be done in its name.

Section 5. Summary

In this chapter, we have addressed the long history of hierarchical terminology that has been used to categorise and divide peoples and places between the global North and global South. The ways in which the urban global South has been imagined have been explored to show how post-colonial urban scholars, and to a lesser extent feminist urban scholars, have been and are challenging, however imperfectly, the universalising and Eurocentric impulses in urban research. The subsequent need to decolonise urban research and the attendant move to a comparative methodology has been outlined. We concluded by unpacking three strands of feminist comparative research and outlining GenUrb's approach to the comparative.

2 Decolonising feminist knowledge production

Elsa Koleth and Linda Peake

Learning objectives

After completing this chapter, readers should understand

- how multiple axes of difference affected the GenUrb project;
- and how members of GenUrb negotiated difference and decolonisation in the everyday practices of knowledge production.

Further, readers should be able to

- think through issues of difference and decolonisation in their own research projects;
- and be aware of the multiple ways in which knowledge production is taking place in their own projects.

This chapter situates the GenUrb project within the context of decolonising **knowledge production** in feminist urban research, starting with a brief discussion of what GenUrb understands by 'the decolonisation of knowledge production'. It then turns to an interrogation of GenUrb through the micropolitics of transnational feminist praxis deconstructing presumptions of feminist affinity and collectivity by attending to intersecting lines of spatial and temporal **difference** that disrupt imagined geographies of the global North and South. The reflections it shares are drawn from activities in GenUrb between 2017 and 2019, prior to the global onset of the COVID-19 pandemic, from the perspective of a specific geographical location, namely, the administrative headquarters of the project at the City Institute, York University in Toronto, Canada. The chapter traces multiple threads of difference and circulations of the transient workers (junior scholars and students) in GenUrb to examine how peer-mentoring, training, and co-learning can be reflexive praxes. These feminist research praxes can actively engage difference, forge affinity, and foreground the generative possibilities of knowledge production with early-career and student researchers. As such, this chapter focuses on academic spaces in a feminist transnational project that rarely receive the level of ethnographic attention given to sites of fieldwork in decolonising feminist scholarship.

Section 1. Decolonising knowledge production in feminist urban research

Feminist researchers take seriously the politics of knowledge production in their research fields. That is, they are attuned to and aim to subvert the relations of power that hegemonic citational, discursive, and material practices (for example, hiring practices) reproduce within academic disciplines. In their foregrounding of the politics of knowledge

DOI: 10.4324/9781032668727-4

production, feminist urban scholars owe a debt to post-colonial theorists, such as Gayatri Spivak (1988) and Edward Said (1978), who question the epistemic authority of European modes of reasoning for producing common analytical frames. Following in this tradition, contemporary post-colonial scholars such as Jennifer Robinson (2006), Ananya Roy (2016), AbdulMaliq Simone (2017; Simone and Pieterse, 2017), and Tariq Jazeel (2019), have been instrumental in articulating how urban scholars are complicit in legitimising European modes of reasoning through their research models and processes. Their critiques call for urban researchers to grapple with the ways in which conventional methodologies and epistemologies can reproduce colonial hierarchies of power, extract the knowledge of marginalised communities, and reify infrastructures of state violence.

Decolonial thought has emerged not only from Indigenous philosophies but also from African, Caribbean, and Latin American philosophies and contexts (de Leeuw and Hunt, 2018). Emma Velez and Nancy Tuana (2020, p. 366) point out that while decolonial thought engages with anti-colonial and post-colonial theory (see Chapter 1, Reflection exercise 1.2 The post-colonial, anti-colonial, and decolonial), it also seeks to differentiate itself from them and to 'emphasize the importance of the still lingering structures of **colonialism** in power, ontology, epistemology, and its entanglement with the imposed categorial logics of race and gender', or what Sarah de Leeuw and Sarah Hunt (2018, p. 4) refer to as 'the lived and living voices and experiences of colonized subjects, especially in Indigenous and settler-colonist occupied spaces'.

In relation to Indigenous research, Linda Tuhiwai Smith ([1999] 2012, p. 1) argues that research has historically been 'one of the dirtiest words in the indigenous world's vocabulary'. And as Angela Moewaka Barnes and Tim McCreanor (2022, p. 211) point out: 'research in colonial settings is always at risk of deliberately or inadvertently broadening, intensifying and entrenching the status quo of unjust social relations'. While feminist researchers have been vocal in calls to decolonise the knowledge produced through research as an ethical imperative, feminist research is not immune from serving to embed, multiply, or re-draw existing axes of power. The boundary between 'colonial' research and decolonising attempts can be fuzzy and exploitative, and ethnocentric attitudes can persist alongside intellectual arrogance (Howitt and Stevens, 2016).

A feminist ethical commitment to decolonising knowledge production relates to the belief that researchers should be critically aware of the ways in which historical geographies of colonialism and whiteness have fundamentally shaped the production of knowledge about the world (Daigle and Ramírez, 2019; Smyth, 2023). Doing decolonial feminist work in this context requires a clear-sighted appraisal of the materialities of historical and structural conditions and transparency about the harms associated with them, as well as enactments of accountability as part of feminist praxis. For the Indigenous scholar Linda Tuhiwai Smith ([1999] 2012, p. 8), decolonial knowledge production involves 'a recovery of ourselves, an analysis of colonialism, and a struggle for self-determination' ([1999] 2012, p. 43). To take Smith's words seriously, decolonising knowledge production involves interrogating, actively dismantling, and seeking to move beyond colonial modes of knowledge production (see also Chapter 12, Section 1 (ii) Disrupting colonial practices in fieldwork). In practice, for feminist and other critical scholars this means:

- connecting knowledge production with practical projects for reparations, redressing inequality, overcoming disadvantage, and enabling self-determination and rights;

- developing new knowledge by drawing attention to new and emerging forms of **imperialism** (for example, systems of domination tied to the expansion of global capitalism) and new possibilities for solidarities to resist the oppressive impacts that result;
- drawing critical attention to the operation of coloniality in knowledge production, including forms of epistemic violence that result from racist and ethnocentric forms of thinking and representations of colonised peoples and places;
- finding new routes to creating knowledge in ways that are inclusive of the knowledge and desires of people who have been historically excluded from knowledge-production processes, expanding the parameters of what counts as 'knowledge' and who can produce 'theory', including the transfer of research skills and tools;
- honouring local research protocols and creating less exploitative and more respectful relations between researchers and those who are the subjects of research, as well as between white and racialised researchers, including those in hierarchical research teams;
- critically evaluating citation practices to reflect: on whose knowledge is referenced and circulated; on the sites where the sources cited are written from; on who and which regional knowledges are being erased or marginalised in citation practices; and on how citation practices challenge or reinforce hierarchies in knowledge production around the world.

One small example of such decolonising practices in GenUrb was with the Cochabamba City Research Team that comprised Indigenous Quechua women who had long been affiliated with CETM. Their participation in the project was negotiated through CETM, who determined the terms of their engagement. This approach through an intermediary facilitated consent being given and trust being established.

While the points listed relate to ways in which academics can address the **decolonisation** of research, in the GenUrb example 2.1, we turn to those on the receiving end of feminist academics' attempts to engage in decolonisation. Specifically, we turn to the grassroots women's organisation Red Thread, one of GenUrb's partners, based in Georgetown, Guyana. Their coordinator, Karen de Souza, thinks aloud about Red Thread's long-term engagement with academic feminist researchers and the extent to which those with whom they have worked have been successful in decolonising their research practices.

GenUrb example 2.1: Decolonising feminist research: the perspective of Red Thread

In Red Thread we have a decades-long history of engaging in research with both Guyanese and non-Guyanese feminist scholars, those who live in the country and those who live elsewhere (see Peake and de Souza, 2010; also see the Red Thread website). We have also worked with several graduate students, primarily from North America, who have conducted their PhD or MA research either with or on Red Thread. Overall, our experiences have been positive as the women we have worked with have been conscious of the need to treat us as equals and to recognise our authority. In other words, they did their homework on decolonising their research and how to work with others outside the academy. This is not to say that hundreds of glitches – some small, others large, some absurd, others serious – have not arisen, and time and emotional energy has had to be taken out of our schedules to address them.

A few of the issues that Red Thread has had to address with researchers from the UK, Canada, and the USA include the following. One researcher, who came from a city in the USA with a large visible and organised lesbian population, assumed that because Red Thread was a women's organisation, all of us were lesbians. Another argued that she could not be expected to share a desk, when this was a normal practice at that time for nearly everyone in Red Thread. Another researcher assumed that her research could be used by United Nations bodies without first securing permission from Red Thread. And more than one researcher wanted to employ Red Thread members to work on their research projects but assumed that they could be paid less than university students from the global North working alongside them (although women in Red Thread have built up their own proficiency and research skills, having been trained in research by Linda Peake and other researchers over a number of years). Other examples cannot be given because they risk identifying individual researchers, but they involved extremely serious situations that put lives at risk and could have led to imprisonment.

While at least one of these events might pose an occasion for a wry smile, some were exasperating and dangerous. In short, time and energy have been taken up in agreeing to host researchers on Red Thread premises, leading us to ask: what do poor women in the global South gain from these transnational feminist engagements? More pertinently, what do they stand to lose?

One of the most difficult aspects of working with feminist academic researchers has been the challenging of classist attitudes and behaviours. In this case I include some Guyanese researchers, invariably middle class, who have been unable to refrain from deploying stereotypical views of the abilities of grassroots women. Also of concern has been the preoccupation of some feminist researchers from the global North with self-reflexive concerns with their whiteness, which at the same time appeared as appeals for reassurance. For the women in Red Thread the emotional and political **labour** involved in its interrogation was a diversion from the 'real work' of daily processes of survival.

Perhaps the most important question I have for feminist researchers who engage in the global South is: how much of their privilege are they willing to put on the line? Unless those with privilege are prepared to give up some of their power, unless they are prepared to make a political commitment to put social change into practice, then whose interests is their research serving? Given that many of them work in institutions that reward acquiescence and not challenging the status quo and in which connecting action and research is rarely encouraged as a cultural practice of academic production, the prognosis for research as an exercise in transnational feminist praxis is not good. But for those researchers who have stayed with us, who have used their research to educate beyond the academy, to equip Red Thread members with transferable skills, and to transform the Northern academy, well, those folks we are prepared to keep on working with.

By Karen de Souza, Red Thread

In the following section we turn to an interrogation of the GenUrb project based at York University in Toronto, turning the spotlight on ourselves and the challenges to, and possibilities of, engaging in decolonising practices in a transnational feminist project. We focus specifically on one of the points raised earlier, about finding new routes to creating knowledge in ways that are inclusive of people who have been historically excluded from knowledge-production processes. We push beyond the proclivity to locate alterity and politics in a distant, elsewhere, 'out there' in the field. Instead, we turn the gaze of feminist reflexivity inwards to those more opaque sites of academic encounter and labour that build the scaffolding of transnational feminist research projects. We highlight that negotiations undertaken in these academic spaces are as fraught, politically charged, and consequential for decolonising knowledge production, and thus as liable to scrutiny, as the more visible aspects of such projects.

Section 2. Decoding affinity and difference in a feminist project

The history of feminist scholarship and activism make it abundantly clear that there are many different 'feminisms' around the world and that the ongoing dynamism and relevance of feminism is sustained in part by contentions across various forms of difference, including **gender**, **race**, class, sexuality, ability, and geographic location, that push feminist thought and action in new directions (Mohanty, 2003a; Nnaemeka, 2004; Chowdhury, 2009; Dhamoon, 2015; Browne *et al.*, 2017; Asher and Ramamurthy, 2020). Indeed, contestation, disagreement, and dialogue across difference are central to cultivating the reflexive and critically engaged communities necessary for transnational feminist praxis (Nagar and Raju, 2003) – not least in projects that stretch across the global North and global South (Dempsey, Parker and Krone, 2011). Nevertheless, feminist collectives are, at some level, bound together by presumptions about shared affinities that provide the grounds for consensus about the aims and methodologies of collective projects. How then does difference disrupt the scaffolding of presumed affinities in transnational feminist projects such as GenUrb?

While transnational projects may invariably continue to be underwritten by material and epistemic disparities between the global North and global South, it cannot be assumed that matrices of power are solely structured by the material and symbolic dominance of the global North over the global South. The landscapes of power in GenUrb are far more complex than the binary formation implied in imagined geographies of global North and global South. Nor can it be assumed that affinities will be naturalised along regional or other geographic lines or through performative avowals of presumed feminist solidarity across those geographies. Indeed, decoding this transnational landscape requires a more nuanced examination of encounters within and across those geographies and proliferating assemblages of difference (Puar, 2012). To name difference(s) here is not to 'fix or ontologise difference along such lines as race, ethnicity, caste, class, gender, [and] sexuality' but rather to proceed with an awareness that 'the multiple border crossings that define human lives, histories, knowledges, and bodies, defy neat typologies of difference' (Nagar and Shirazi, 2019, p. 237). In GenUrb, differences manifest across multiple, intersecting lines of race, class, sexuality, geography (in-situ versus diaspora), generation, personality, academic hierarchies, locations within and beyond the academy, and related differences in interests, epistemic assumptions, political priorities and approaches to feminist praxis.

The specificities of the histories and geographies within which such differences are embedded fundamentally shape the ways in which feminists are positioned in relation to the agenda of a feminist project: to their colleagues in the project, to institutional structures, and to various post-colonial and settler-colonial spaces (Mani, 1990; see also Chapter 1 Feminist comparative urban research). The intersecting temporal and spatial lines of differentiation in a transnational feminist project, in turn, shape the structures of power and privilege and the corollaries of disempowerment and marginalisation that characterise scholarly spaces, academic labour, and knowledge-creation processes (Mahtani, 2014; Henry *et al.*, 2017).

Reflection exercise 2.1 encourages you to explore these lines of difference and affinity in relation to your own research.

Reflection exercise 2.1: Lines of affinity and difference

While you may not be undertaking a transnational feminist research project, you should still consider issues of affinity and difference in your research, especially in relation to your own positionality.

- In relation to those who will be participating in your project: Do you share identities? Membership in social groups? Experiences of place?
- How may these affinities or differences affect your relationship with participants?
- Will you be perceived by participants as an insider or an outsider or both?
- How may these affinities and differences affect the interpersonal dimensions and material aspects of the research process, such as data analysis, knowledge mobilisation, and finances?

In the rest of this chapter we turn to three specific instances of feminist enactments of labour for decolonising knowledge production, namely peer-mentoring, training, and co-learning across difference, and how these circulate across spatial and temporal difference. In the following section, we address peer-mentoring, and specifically that of early career scholars who are women of colour.

Section 3. Circulations of spatial and temporal difference

In a transnational context, it might be assumed that feminist scholars from the global North share affinities and privileges relative to their counterparts in the global South. However, not all bodies are similarly positioned in the circulations and mobilities within transnational feminist projects, and some bodies are reinscribed as they move (see Chapter 4, Section 4 The challenges of feminist scholar-activist research in transnational contexts). In GenUrb this becomes clear, for example, in the subjectivation of racialised scholars who are based in the global North but claim diasporic identifications that continue to bind them through genealogical and intimate ties to the global South. As Black geographies scholarship highlights, diasporas as spatial formations are constituted through the production of difference (Hawthorne, 2019). For racialised diasporic scholars moving between spaces constituted through histories of racial domination and coloniality, possibilities for affinity may be undercut by the embodied designation of being *out of place* across multiple geographies. Indeed, to be in this liminal position can induce a sense of alienation from presumed communities of affinity, whether in the global North or global South (Chua and Mathur, 2018).

The interaction of identities such as racial, ethnic, religious, and class background, gender and sexuality, and concomitant positionalities in relation to structures of white supremacy and other formations of racial patriarchy ensure that feminist collectives occupy highly differentiated spaces. Scholars who seem privileged in a global context may be divested of some or all of that privilege as they move in and between the global North and South (Mullings, 1999; Miraftab, 2004). Similarly, differences of race, caste, class, ethnicity, religion, gender, and sexuality (among other factors) also structure feminist spaces within the global South, such that scholars who seemingly may not possess the material institutional privilege of their counterparts in the global North may nevertheless hold considerable privilege in their countries of origin, which can, in turn, inform their transnational engagements with feminists from other geographies.

In addition to historical contingencies that structure raced and gendered differentiation, another axis of temporal differentiation within feminist spaces relates to generational differences and class formations. Where a feminist project brings together different generations of feminist scholars and people at various stages in their scholarly careers, from senior professors who are well-established figures in their respective disciplines to early-career pre-tenured faculty and scholars on limited-term contracts, and to students, class works in concert with other lines of difference to structure hierarchies of power. Class operates as a temporal form of differentiation, first, in that people's access to secure, livable incomes is shaped by the conditions prevailing in the specific historical period in which they enter the labour market and the time they have had to accrue financial resources, and second, in that the viability of intergenerational reproduction of feminist scholars in the academy depends on the availability of opportunities for junior scholars to secure a sustainable living (see also Anwar and Viqar, 2017).

Racialised junior scholars in the global North navigate the twin conditionalities of indefinite precarity that structure contemporary academic labour and pressure to become an entrepreneurial neoliberal subject to survive in the academy. This necessitates the cultivation of canniness to parse out the ways in which their bodies and their labour can be readily appropriated and exploited to serve the neoliberal university and its various governmental and disciplinary grammars for domesticating alterity and extracting value from difference (Ahmed, 2012; Duncan, 2016).

Drawing on Black feminist Barbara Christian's cautionary critique of the fetishisation of Black feminism in the university functioning alongside and facilitating the persistence of systemic violence against Black women, Grace Hong draws attention to 'the effects of racialized and gendered premature death on academic generationality' (Hong, 2015, p. 129). Christian's work, she argues, highlights the need for the analytic of 'difference' to be directed 'toward a meta-analysis' of academic spaces to ensure that the enlistment of 'difference' serves to enable more Black and Indigenous women and women of colour to enter and remain in the academy, 'rather than as a ruse to effect their necropolitical excision' (Hong, 2015, pp. 127, 131). When class difference encounters embodied racial and gender difference in feminist academic communities, the stakes are not merely about the rupturing of affinities within these communities but even about who survives the gauntlet of extraction and expulsion – who is seen as worthy of intergenerational preservation in the normative reproduction of feminism in the academy.

To counteract these prevailing forces, feminist scholars have adopted a variety of practices to open up space for racialised feminist scholars. Practices of intergenerational mentoring, pedagogy, and training, for example, operate as key modalities of community building (Mullings and Mukherjee, 2018; Johnston-Anumonwo, 2019). In GenUrb example 2.2 we provide a specific example of how difference and decolonisation circulated within GenUrb for three junior women-of-colour scholars in the Toronto-based GenUrb team.

GenUrb example 2.2: Feminist enactments (i) Peer-mentoring

The multiple and shifting axes of difference that shape the landscape of the transnational feminist collective were brought into sharp relief for three early-career women-of-colour scholars in the Toronto-based GenUrb team in the early stages of the project, namely, myself, Elsa Koleth, a postdoctoral fellow and two research assistants (a PhD student and undergraduate student – both prefer not to be named). Upon joining GenUrb in late 2017, I worked on the creation of the GenUrb research training modules on feminist urban research (these modules were later developed into this book). In the early stages of this process, we worked together to prepare a set of guiding principles for the development of the modules, including commitments to transformative and decolonial knowledge production, accessibility, and linguistic multiplicity. The commitment to decolonial knowledge production became a source of ongoing debate between the three of us as we grappled with what it meant for us as three junior women-of-colour scholars to create a decolonial pedagogical resource from within a university in a settler-colonial context of the global North. As our discussions evolved, it became clear to us that to engage substantively with the idea of decoloniality, we had first to parse out the axes of difference through which we were situated in GenUrb and to grasp the significance of historical difference for decolonial feminist praxis.

We traced our respective genealogies to account for the different historical and geographical routes by which we had come to the settler-colonial context of Toronto and to account for the ways we related to the project of decolonisation. The three of us were each born in formerly colonised places in the global South, and we each had different journeys, across disparate geographies, from the global South to the global North. Across the racial, cultural, linguistic, and class differences that we embodied, a common ground of affinity between us was that our diasporic identities disrupted the binary geographical imaginary of global North and global South. There was also a generational affinity between us based on our experiences as first-generation migrants who had grown up in the last two decades of the 20th century and our positionalities as racialised women employed on a temporary basis in the political economic machine of the corporate university. Each of us, however, had different relationships to the varying temporalities and iterations of colonialism and, by extension, to the varying temporalities and spatialities of decolonisation.

The respective histories and geographies traversed in our individual biographies fundamentally shaped the way we encountered the disciplinary canons we were working through in constructing the research training modules. They shaped our understanding of how the epistemic mandates of white supremacy that undergird coloniality continued to be reproduced in those disciplinary canons and the limited ability to challenge the structure of coloniality while leaving those canons intact. We recognised the limitations of the academic setting for enacting a meaningful decolonial praxis in a settler-colonial context built on the ongoing dispossession of Indigenous peoples (Tuck and Yang 2012; de Leeuw and Hunt 2018). By parsing through the meanings of our embodied historical differences, we arrived at a more complex understanding of the multiple spatialities and temporalities of coloniality and, as a corollary, the necessity of contemplating multiple itineraries of decolonisation.

Working through the implications of multiple intersecting differences was a necessary foundation for forging affinities. In the process of deliberating on the relationship between decolonisation and difference, we cultivated a generous and generative space of peer-mentoring and community between the three of us through what María Lugones (2010, p. 755) terms 'an ethics of coalition-in-the-making'. This space facilitated co-writing, and it also subsequently became a crucial site for supporting each other as we navigated the disciplinary gatekeeping structures represented by review processes in academic publishing. It enabled us to contend with the epistemic violence perpetrated through these structures.

Importantly, the space we created was one in which we could be candid in speaking the rebellious thoughts, doubts, and questions that decolonial praxis demanded; for example, drawing on critical Indigenous and Black radical thought to contemplate the imperative to 'refuse' the university and the colonial recognition it conferred (Simpson, 2007, 2017; Kelley, 2016; Grande, 2018). We could also be candid about the costs involved in attempting to do critical decolonial work within the university, particularly for precariously employed racialised and gendered bodies in the academy who operate under the ever-present threat of rejection, censure, or expulsion (Rodríguez, 2018). The space of affinity that we created across our differences was an organic response to the structural and institutional context in which we found ourselves. It captured the dynamism, intelligence, humour, creativity, and friendship that could be fostered in a collegiate space free from punitive oversight. The cultivation of this space and the praxis of peer-mentoring within it was, then, crucial in enabling us to contribute to feminist knowledge creation (Alarcón *et al.*, 2020; Fernández, Hisatake and Nguyen, 2020).

By Elsa Koleth

Reflection exercise 2.2 asks you to think about the issues discussed earlier, and how they speak specifically to temporal differences between generations of feminist scholars, and how mentorship can create spaces of affinity.

Reflection exercise 2.2: Generational positionality and mentoring

Consider the ways that generational positionality and mentorship influence academic life.

- Are resources and opportunities in your discipline/institution allocated based on seniority?
- What mentorship networks are available in your field? At your institution? How can you connect with these networks?
- How can you work to help others in more junior or similar positions in your field?
- What challenges do you face when working across generational differences?

The following section contends with hierarchies of class and generational difference (as they operate in concert with racial and gender difference) in the feminist academic space of GenUrb, bringing into view the conditions through which labour has been enlisted in our transnational feminist research project.

Section 4. Circulations of labour in feminist knowledge production

Contending with hierarchies of class and generational difference in feminist academic spaces (as they operate in concert with racial and gender difference) brings into view the conditions through which labour is enlisted in transnational feminist research projects. Multi-sited transnational research projects necessarily require the input and collaborative labour of a wide range of people. In GenUrb, this includes scholars with faculty positions in academic institutions of the global North and South, postdoctoral scholars and administrative staff in those institutions, members of civil society and policy organisations, grassroots feminist activists, students and others employed as research assistants, and local informants, not to mention those who provide the logistical infrastructures to move researchers to and around research sites and to support GenUrb events. In the context of the Toronto-based GenUrb team, in addition to a full-time project manager, several students have been employed as research assistants in every semester since its commencement (see Chapter 0, GenUrb example 0.3 Members of GenUrb's York University Team). The engagement of students constitutes an important part of the education and training objectives of GenUrb, which are both required by the project's funding body and in keeping with commitments to feminist pedagogy and community building. Undergraduate and graduate students are employed on a sessional basis with opportunities for contracts to be extended on occasion, providing an important source of income for students and academic opportunities to expand their skills and knowledge.

Student research assistants can support transnational feminist research by conducting research, co-writing academic papers, analysing data, translating, transcribing, providing web support, organising events, supporting networking, and assisting with knowledge mobilisation. Because students are often casual sessional workers, their administrative or organisational labour can often seem ancillary to the loftier goals of feminist knowledge production. Except for instances where students are listed as authors in published knowledge outputs, other forms of labour that students engage in can become invisible at the higher echelons of theorising and peer-reviewed publication. In this way, training and education activities can be rendered as collateral addendums to the more sacralised goal of generating theory rather than being a visible element of the project's intellectual work.

Feminist literature is rich in discussions of participatory and collaborative methodologies to incorporate non-academic collaborators, informants, or research participants into knowledge-creation processes through, for example, co-writing, co-creating, or data-sharing practices (Mountz *et al.*, 2003; Nagar, Ali and Sangtin Women's Collective, 2003; Peake and de Souza, 2010; Nagar, 2013, 2014; Banerjea, 2015; Chakravarty, 2015). Less is said about the role in the knowledge-creation process of people employed on feminist projects on a transient basis, such as research assistants or translators (but see Wittmer and Qureshi, 2023). This elision contributes to the reification of academic knowledges as the preserve of 'experts'. However, feminist praxis should arguably include a commitment to making visible as intellectual contribution the labour of the various people who contribute to different stages of transnational feminist research. Such a praxis rejects the modality of disposable labour that underpins the contemporary university, to invest in feminist research as an intergenerational form of community building. In GenUrb example 2.3, we explore the experiences of four student research assistants based at the City Institute at York University, showing how methodological praxes which value, foster, and make visible the intellectual contributions of the various bodies that move through a transnational project are arguably key to the intergenerational reproduction of feminist communities.

GenUrb example 2.3: Feminist enactments (ii) Training

Between late 2018 and early 2019 four student research assistants at the Toronto-based GenUrb team (two undergraduates and two master's students) worked to translate interview transcripts from Mandarin and Arabic to English. We do not refer to these students by name as this is their preference. During their work they raised questions about their struggles in translating ambiguous or complex terms into English, which were not merely technical issues but rather went to the heart of epistemic frameworks for understanding gender and the urban. It was clear that to answer many of their questions they needed to be in conversation with members of the respective CRTs who conducted the original fieldwork and had local knowledge of those field sites. However, the possibilities for such exchanges were mediated by availability for engagement across different time zones and limited by the time people had at their disposal for additional meetings beyond their existing work commitments.

In the meantime, resonances in the issues raised by the students suggested that it might be productive for them to engage in collective reflection and dialogue about the challenges of translation in a transnational project. I convened an informal workshop with the four research assistants during which they were invited to share with each other their questions, the challenges they encountered, and the strategies that they used to deal with them. In addition to their multi-lingual skills, the students leveraged their experiential knowledge of living between cultural and linguistic worlds, their socio-cultural and political understandings of the regional and national contexts of fieldwork sites, and their critiques of the epistemic and political implications of parsing through ambiguity and untranslatability to illuminate empirical complexities within the lexical limitations of the English language. Translation was for them a full-body experience, informed by a lifetime of embodied knowledge.

Three of our four research assistants were international students with experience of living and moving between the global South and the global North. However, none of them had experience of living in the respective GenUrb fieldwork sites, and they reflected on the limitations of their understandings of the specific historical, spatial, cultural, and linguistic contexts of these places that at times made it difficult for them to resolve the meanings of unfamiliar or ambiguous local vernaculars or dialects. For example, two of the three students who were translating transcripts from Ramallah were Egyptian; they noted that the Arabic they knew was different from the Arabic spoken in Ramallah. They also reflected that their lack of knowledge of the topography of the research site impinged on their ability to translate references in interviews to the ways in which participants moved through urban space or to how they organised their day. At the same time, these students also brought their own embodied knowledge of mediating between the Middle East and the anglophone North American context in which they now lived to critically reflect on the implications of the choices they made in the translation process. For example, one noted that she was keenly aware of the way in which the language she used to translate the gendered experiences of women from the Middle East, particularly regarding issues such as family roles and autonomy, could be read or interpreted through a colonialist lens by a Western audience if the complexities of meaning in the original text were not fully conveyed in English.

In this workshop the intelligence and enthusiasm of our research assistants was palpable, and their insights were rich in theoretical import. However, there was no clear avenue for their critical insights to be explicitly incorporated into the knowledge outputs that would eventually be published by more senior members of the project. To record their insights, the workshop was filmed (with their permission) for the ethnographic records of the project. As a supplement to the workshop, they were asked to document words from transcripts that they struggled to translate. In the process, they shed light on the complexity of some of the key framing concepts in the project regarding gender, power, and urban spatialities and temporalities and brought to light the complexities of meaning that could get lost in translation. For example, a Mandarin-speaking student highlighted that the use of gendered pronouns in Mandarin – (它) it, (他) he, (她) she – did not translate directly into English. These terms are all pronounced as 'ta' in Mandarin, hence the correct usage of the term must be deduced from the context and could be misread. Three Arabic-speaking students further conveyed the complexities of translating terms in Arabic that could contain multiple meanings or refer simultaneously to multiple spatial scales but which would effectively be flattened when translated into English. For example, البلد (El-balad) is translated as 'country' but can be used to refer to a town, village, city, city council, or town council. It is used variably, and its meaning can only be deduced from the context. Its various usages are indicative of the ways in which spatialities are understood and that common usage of terms may not align with academic definitions. The problematic nature of employing translators outside the CRTs led to a decision to transfer all translation work back to CRT members (see Chapter 15, Section 3 (iii) Fieldwork: working with translators on interview guides and transcripts).

As an initial strategy to make their labour visible, a brief excerpt from this workshop recording was played as part of a presentation delivered by members of the GenUrb team at an international conference on comparative urbanisms held at the University of Georgia, Atlanta in March 2019. Playing the video recording before an international gathering of urban scholars, animated as it was by the hand gestures, exclamations, humour, and thoughtfulness that the students displayed in sharing their insights, was a meaningful feminist act; it honoured and dignified their labour and the value of their embodied and intellectual contribution to the collective work of the transnational project. Rather than merely presenting a synthesised, theoretical report of 'findings' we were, thus, militating against the modality of 'expert' knowledge making, which has historically colonised and appropriated the knowledges of multiple lifeworlds and which continues to obscure the labour and intellectual vitality of the many bodies involved in making transnational research possible.

Following the workshop, research assistants were invited to continue the discussions they had at York University by engaging in virtual conversations with members of the Shanghai and Ramallah CRTs and by co-authoring a collaborative article about translation. The invitation to ongoing collaboration did not come to fruition in this instance due to time constraints and the fact that some of the research assistants moved on at the end of their sessional contracts and graduated from their degrees. However, the students were able to share their reflections as invited speakers in a roundtable session on the social, cultural, and embodied aspects of translation held at the GenUrb conference of September 2019.

Feminist praxis that actively creates space for training and pedagogy as knowledge-creation processes through facilitating discursive engagements between student research assistants, recording their deliberations, making their intellectual contributions visible, and, ideally, putting them in conversation with the broader transnational network of more established and senior scholars in the project democratises the knowledge-creation process, particularly in connecting such labour to the higher echelons of theory making. In so doing, pedagogy and training become incorporated into the core of feminist knowledge making rather than rendering the former a mere subsidiary to the latter.

By Elsa Koleth

The final section turns to co-learning across difference and the experiences of the GenUrb's Early Career Network (ECN). Its discussions allowed researchers to reflect on their affective and bodily experiences of conducting research, their ethical dilemmas and doubts about inequalities and fissures within the structures of transnational feminist projects, and the political implications of these fault lines for feminist knowledge production.

Section 5. Co-learning across difference

Encounters between feminists across transnational lines are events in which the proliferating lines of difference highlighted earlier are rendered visible and negotiated. In reflection of the complexities characterising a network of this size and nature, transnational encounters in GenUrb have proved to be by turns rewarding and fertile and also fraught with misunderstandings, silences, or confrontations over divergent viewpoints, with negotiations across difference occurring in both formal and informal settings through exchanges that are at once intellectual, embodied, and affective. GenUrb example 2.4 draws on an example of a transnational meeting in GenUrb bringing together participants from different CRTs, to highlight the generative possibilities for learning through encounters across difference.

GenUrb example 2.4: Feminist enactments (iii) Co-learning across difference

In addition to the examples discussed earlier, GenUrb's ECN is another forum that has facilitated feminist practices of sharing and co-learning. The ECN, which draws on a mix of pre-tenured faculty, postdoctoral fellows, research assistants, and grassroots women, was convened in the second year of the project as a voluntary space for exchange between more junior members of the project. Convening this network across disparate geographies involved negotiations with diverse cultures in different places regarding the organisation of relations between senior and junior scholars – for example, in relation to hierarchies of seniority and representation – and, in turn, differing expectations around practices such as training, mentoring, or networking. One of the ECN's areas of discussion in the first three years of GenUrb concerned experiences of trauma in fieldwork. This issue emerged during the project's annual general meeting in Mumbai in September 2018, as each CRT reported to the broader network about the progress of their fieldwork. During these reports, it became apparent that some teams had explicitly addressed the logistical and ethical challenges that researchers and assistants faced in conducting fieldwork in contexts of urban poverty. While the issues raised were not discussed further at the general meeting, the members of the ECN were invited to explore their experiences further in dialogue with each other. As the convener of the ECN, I extended this invitation to members of the network based on my reflections on insights shared by junior GenUrb members and facilitated discussions about the contributions of ECN members from the CRTs. Initial discussions occurred through a virtual meeting and were followed by a longer conversation in which members of the GenUrb network met in person at a writing retreat in April 2019.

This meeting involved my participation as well as that of Grace Adeniyi-Ogunyankin, Monica Orisadaare, and Sylvia Bawa (Ibadan), Penn Tsz Ting Ip (Shanghai), Joy Marcus (Georgetown), and Mai Al-Battat (Ramallah).

The ECN proved to be a productive space for these scholars and activists to give voice to challenges they faced in conducting fieldwork. It was clear that researching the life histories and lifeworlds of marginalised women demanded the mobilisation of significant emotional, psychological, and, in some cases, financial resources from researchers in the field. It was also clear that conducting fieldwork in such contexts often took a mental and, in some cases, a physical toll on researchers during and after fieldwork. However, there were few organised opportunities to discuss strategies for navigating such challenges prior to fieldwork and few opportunities to debrief with other CRTs about the affective, intellectual, and bodily experience of conducting fieldwork after returning from the field. In the ECN conversations, researchers discussed the affective circulations of emotions such as sadness, discomfort, abjection, grief, frustration, anger, and depression in fieldwork. They reflected on how these affects resulted from encouraging the women they interviewed to recall the various traumas of their own lives and from the vicarious upset of hearing these stories as researchers. They also reflected on the distress they experienced in wrangling with the implications of their research, in the context of their relative privilege, their intervention in the lives of the women involved in the research, and their limited capacity to contribute to alleviating the conditions of poverty they encountered in the field.

However, the GenUrb CRTs comprise a combination of diasporic scholars based in the global North, those working and living between sites in the global North and the global South, those based in the cities being researched, and CRT members who were not themselves conducting GenUrb fieldwork. The issues raised by the ECN thus had different valences for each of the participants in these discussions, even as they resonated with all participants. For example, Joy Marcus, a member of the Guyanese grassroots women's organisation Red Thread, highlighted how the everyday priorities and goals of the women activists in Red Thread, which is embedded in long-term relationships with the community in which they work and conduct research, are necessarily different from those of scholars based at universities in the global North. In enabling researchers to find resonances between their experiences across the disparate research sites and extend support and empathy to each other, the ECN allowed participants to exercise radical vulnerability as what Richa Nagar and Roozbeh Shirazi (2019, p. 240) call a 'mode of togetherness' while also sharing their critical reflections on feminist methodologies without fear of judgement.

The content of these discussions spoke to the uneven landscapes of power and accountability that mediate encounters between academic researchers across the global North and global South and grassroots communities (Nagar, 2014). The ECN created a space for researchers to engage in critical reflexivity as part of transnational feminist praxis. In recognition of the broader resonance of the issues discussed within the ECN for feminist praxis and methodology, members of the ECN (plus Nasya S. Razavi, who joined GenUrb's Cochabamba team in mid-2019) held a roundtable session on 'Trauma in the Field' at GenUrb's international conference in September 2019, which served as a pedagogical and discursive forum to engage other feminist scholars and extend these conversations beyond the GenUrb network.

By Elsa Koleth

The labour discussed in this chapter, undertaken within the academic context of York University, is only a small sample of the varied forms of labour and collaboration involved in transnational feminist urban research. This decolonising feminist work is needed, however, to attend to historical and other related forms of difference that structure relationalities in a transnational feminist project, first, because it is necessary to contend openly with those lines of difference to do the work of forging affinities across difference in a meaningful way, and second, because the terms of engagement across difference fundamentally shape the kinds of decolonial and feminist knowledge produced through transnational feminist research projects.

There is much more to be said about the relationalities that comprise the networks of a transnational feminist project, not least regarding the mechanics of forging solidarity and exchange beyond the academy, between different research sites, and between different actors involved in the different stages and aspects of the project from planning to fieldwork, to processing and analysis of findings, and to mobilisation of knowledge with various audiences (see Chapter 4, Section 3 The challenges of feminist scholar-activist research in transnational contexts). Each of these conjunctures of exchange and labour are significant and warrant close ethnographic attention to develop a fuller picture of the conditions of possibility for decolonising the contemporary urban through a lens of transnational feminist praxis.

Reflection exercise 2.3 asks you to think across identities of being in the university, being an activist or ally, and being a friend. How do they come together in ways that might spill out beyond the academy?

Reflection exercise 2.3: Collaboration and community in and beyond the academy

Academic researchers are conventionally portrayed as independent, but feminist researchers have demonstrated how formal and informal networks shape knowledge production. Even if you are researching alone, it can be beneficial to participate in an intellectual community beyond formal institutions.

- What peer networks are available to you for knowledge exchange?
- How can you pool resources (intellectual or practical, social or material) to work collaboratively and aid each other's work?
- How can you engage perspectives across academic disciplines to enrich your collaborative networks?
- How can you engage with existing scholarly networks that cross national boundaries?
- How important is activism and allyship to you in your intellectual networks?
- How important to you is friendship in your intellectual networks?

Section 6. Summary

This chapter has explored practices of knowledge production in the feminist urban research project of GenUrb. It started with an overview of how we understand practices of decolonising knowledge production and then applied one practice in particular to GenUrb, namely that of finding new routes to creating knowledge in ways that are inclusive of people who have been historically excluded from knowledge-production processes, in this case, early-career scholars and students who are women of colour in the GenUrb project based at York University in Toronto. It has discussed three examples of feminist enactments of labour in GenUrb that demonstrate the necessity and potential of engaging with multiple intersecting lines of difference in the uneven landscapes of power that form transnational feminist research. These enactments revealed attempts to deconstruct and remake affinities across those differences by creating spaces that invited reflection and radical vulnerability as part of feminist praxis. They highlight the need to ground decolonial feminist knowledge creation in a robust engagement with the fissures produced by difference, particularly for more junior scholars and students involved in less visible aspects of research. This chapter has discussed these issues largely from the perspective of activities undertaken in an academic context of the global North, in which the combination of persistent inequalities of power and privilege, precarious labour conditions, and the marginalisation of the intellectual contributions of transient workers undercut presumed affinities of feminist collectives such as GenUrb.

3 Feminist engagements with translation

Wiley Sharp

Learning objectives
After completing this chapter, readers should understand

- the distinction between translation and interpreting;
- the hegemony of the English language in the academy;
- how histories of colonisation affected practises of translation;
- and why decolonial-feminist scholars practise translation as transnational praxis.

Further, readers should be able to

- evaluate the role of translation or interpreting in their project;
- and consider how their citational practises may reproduce epistemic injustice.

This chapter introduces the politics of translation and interpreting, building upon the insights of the GenUrb project to theorise a decolonising feminist praxis of translation. It unpacks the issue of anglophone hegemony in urban studies and its reproduction of epistemic injustice, underscoring the epistemological and political necessity of research that critically engages with translation. It turns then to the colonial roots of anglophone hegemony, reviewing decolonial critiques of translation. It concludes with an account of decolonising feminist approaches to translation as key to dismantling colonial hierarchies and addressing the material legacies of colonisation, paying particular attention to the intersectional critiques of decolonial feminist scholars. In doing so, it foregrounds three key themes central to decolonising feminist translation studies: anglophone hegemony, the role of difference in colonial power, and translation as decolonial feminist praxis.

Section 1. 'Translation', 'interpreting', and making meaning across difference

> *Translation could be described as a means of transport: a mobile vehicle carrying meaning from place to place.*
>
> (Haddour, 2019, p. 202)

What is translation? A provisional definition could be that translation is the practise of meaning-making across different languages. But this is not quite precise enough. Language professionals (e.g. translators and interpreters) use the term **translation** to refer to meaning-making across *textual* languages, while they use the term **interpreting** when referring to meaning-making across *verbal* languages. The feminist geographer and

DOI: 10.4324/9781032668727-5

professional translator and interpreter Sara Koopman (forthcoming, p. 2) writes that 'translation is a skill more like chess, with slow strategic deliberation, and interpretation [is] more like volleyball, where the focus is on keeping the ball in play'. Yet language is amorphous and relational: nonverbal, subtextual, and cultural meaning can dramatically shift the meaning of text or speech (Hatim and Munday, 2004; Valero-Garcés, 2015). Translation, then, does not always fall neatly into distinctions between textual and verbal modes. As we discuss further in Chapter 15 Feminist practises of translation and interpreting, technological developments have served to further blur this distinction. The extent to which the distinction is upheld varies across fields. Unlike translation scholars and language professionals, other scholarly fields (especially in the humanities) employ the term *translation* to refer at once to translation, interpreting, and the abstract practise of meaning-making across linguistic difference.

Rather than resolve the differences between these approaches, we hope that the friction between them serves as an occasion for you to reflect on how you do or do not engage with translation, interpretation, and difference in your work. As we will see, the language around translation and interpreting is ripe with political and ethical tensions – down to the very words we use to refer to these practises of communication. In this chapter (and in Chapter 15) we adopt a mixed terminology. Following the lead of language professionals, we use the term 'translation and interpreting' to refer to the practical activities of negotiating linguistic differences in spoken languages and written texts. At the same time, we abstract the term 'translation' to refer to myriad practises of meaning-making across difference.

Since the late 20th century, feminists have engaged with translation across difference to build both critical analysis of gendered oppression and political movements challenging it. The academic field of 'feminist translation studies' was started by Quebecois literary scholars in the late 1970s in response to Canada's particular politics of colonial bilingualism (Eshelman, 2007). This early scholarship critiqued modern paradigms of translation, which erase gendered difference by performing the 'god trick' of masculinist objectivity (Santaemilia, 2017, p. 15), and, in doing so, brought feminist standpoint theory to bear upon the practise of translation. In the words of the feminist scholar Barbara Godard (1989, p. 50), a feminist practise of translation 'would involve the replacement of the modest, self-effacing translator. Taking her place would be an active participant in the creation of meaning, who advances a conditional analysis. Hers is a continuing provisionality, aware of process, giving self-reflexive attention to practices'. In this way, early feminist scholars challenged the notion of a stable, objective translation, emphasising the role of its practises of meaning-making across difference in reproducing epistemic and political injustice.

In the decades since, feminists from across the world have brought their own perspectives to translation as a transnational project, provincialising early interventions from the global North and extending debates about language, power, and positionality (Irshad and Yasmin, 2022). A plethora of such feminist engagements with translation have developed in India and Latin America. The work of the decolonial feminist scholar Richa Nagar is exemplary in this regard: in collaboration with the *saathis* (friend, comrade, or co-traveller) of the people's movement *Sangtin Kisan Mazdoor Sangathan* (henceforth SKMS) and the experimental performance collective Parakh Theatre, Nagar has articulated a notion of 'hungry translation' as 'a forever evolving entanglement' with others across difference – linguistic, geographic, classed, or otherwise – in pursuit of a journey for political transformation that can never be charted in advance (2019, pp. 41–42). We will return to Nagar's work in section 4; for the time being, suffice it to say that translation is a messy, difficult practice that is nevertheless essential for a radical, transnational politics.

Thus, it is no surprise that translation was essential for the GenUrb project: researchers conducted 127 in-depth interviews and 100 life-history interviews with women participants, as well as 143 interviews with policy shapers. These conversations were held in more than two dozen languages and dialects across six cities in the global South (see GenUrb example 3.1). During the project, researchers had to grapple with the entanglements of language, difference, and power. In this chapter, we ask: how does language and translation reproduce hierarchies of power across difference, gendered, and otherwise? And how can scholars employ translation to challenge domination and address social change? To answer these questions, we turn to the work of decolonial and feminist scholars, who understand translation as a historically and geographically contingent process – one laden with unequal power relations.

GenUrb example 3.1: Languages and dialects spoken by members and participants in the GenUrb project

Cochabamba: Spanish and Quechua

Spanish and Quechua Sureño (or Southern Quechua, the most widely spoken of dialects in the Quechua language) are the dominant official languages spoken in Cochabamba. About 40 percent of the population speaks Quechua and most Quechua speakers are bilingual, also speaking Spanish. There is a smaller group of migrants (largely from the north, for example, from El Alto-La Paz) who have settled in the region and who speak Aymara. Quechua terms are peppered into everyday conversations: for example, many food items are referred to by their Indigenous names.

Spanish was the *lingua franca* of the Cochabamba City Research Team (CRT). Participants were Quechua women, but they were mostly bilingual (Spanish and Quechua), and they chose to hold their interviews in Spanish. A few people at our partner organisation, Centro de Estudios y Trabajo de la Mujer, speak Quechua, but we only conducted one interview in Quechua. When this participant presented her story to a focus group in Quechua, almost all the research team and other participants were able to understand her. No one in the project spoke Aymara.

Delhi: English, Hindi, Malayalam, and Tamil. Dialects: Awadhi, Bangla, and Bhojpuri

The official language of Delhi is Hindi, although other regional languages, such as Punjabi, Tamil, Bangla, and Malayalam are also spoken in the city, sometimes clustering in different localities and indicative of different streams of migration.

By strict definition Hindi is a language, and Awadhi, Bhojpuri, Haryanvi, and Pahadi are regional dialects. However, they have common roots, and there are political implications of being recognised as a language, resulting in the fluidity of the ways in which languages are reported in official channels like the Census of India.

The Delhi CRT members speak to each other in English, Bangla, and Hindi. Members of the CRT also speak Tamil, Malayalam, Bhojpuri, and Awadhi. Interviews with participants were conducted primarily in Hindi.

Georgetown: English and Creolese

The official language in Guyana is English, a result of British colonisation, which lasted until 1966. The vernacular language is Creolese. It is similar to other English dialects in the English-speaking Caribbean, containing words and phrases from the languages of the original inhabitants of Guyana – Arawak (Lokono) and Carib (Karina) – as well as from the languages of those brought to Guyana under slavery – Akan, Kikongo, and Yoruba – and indentured labour programmes – Bhojpuri – and from colonisers' languages – Dutch. There are many sub-dialects of Guyanese Creole, and in practise, there is a speech continuum between standard English and Creolese based on class, geographical location, urban–rural location, and the racialised ethnic group of the speakers. There are also distinct accents, such as in Georgetown, the capital city.

English and Creolese are spoken by all members of the Georgetown CRT. Most of the participants in Sophia, the neighbourhood in Greater Georgetown where research was conducted, gave their interviews in Creolese.

Ibadan: English, Yoruba, Etsako, Hausa, Ibie, Igede, Igbo. Dialects: Ika

Ibadan is a Yoruba city. The CRT members speak a variety of languages, including English, with all but one speaking Yoruba. CRT meetings with participants were conducted in Yoruba to ensure that they were accessible to virtually everyone.

Some of the research participants speak the language of the region from which they migrated – Etsako, Hausa, Ibie, Igede, Igbo and Ika (Ika is a dialect of Igbo). While these languages are spoken in Ibadan, they are not common, and some, such as Hausa, are restricted to particular ethnic quarters. All the participants also understand and speak Yoruba, and two-thirds of participants speak English. The ability to speak English is heavily age dependent. All the younger women speak English, whereas only a minority (n = four) of the older women do so. Consequently, our interviews were conducted in both Yoruba and English.

Ramallah: Colloquial Arabic (central mountains dialect), Fus'ha modern standard Arabic, English

Among the members of the Ramallah CRT we spoke dialect Arabic, Fus'ha Arabic, and English.

In everyday life Palestinians speak colloquial dialects of Arabic that significantly differ from Fus'ha, modern standard Arabic, which is only used in writing, by news channels, and in official speeches. Palestinian dialects combine words from Aramaic and other languages of powers that colonised the region (Italian, Turkish, English, Hebrew, and some French in areas bordering Lebanon). The dialects differ from one area to another, and in Ramallah the dialect is mainly that of the central mountains' region. English is taught in schools as the second language, and therefore many speak it, and it is widely used in some sectors, such as academia, and in those dependent on foreign funding, such as non-governmental organisations.

In the interviews that we conducted in Ein Qiniya and with policy shapers in Ramallah we used the local dialect to reduce the feeling of formality and to allow interviewees to best express themselves. In some of the interviews with policy shapers the responses included English terms and expressions.

Shanghai: English, Mandarin. Dialects from Anhui, Changzhou, Chongqing, Jiangsu, Jiangxi, Shanghai, and Shandong

In China, the official language is Mandarin, with ten major Chinese dialects (fangyan).

Within the CRT, the lead researcher is fluent in several southeast Asian languages, including Mandarin, Cantonese, and Korean, as well as English and intermediate Dutch. One research assistant is a fluent Shanghainese speaker, and the lead researcher learned Shanghainese to better communicate with the research participants. All the CRT members speak Mandarin and English. In addition, one speaks a Shandong dialect, and another speaks Taiwanese Hokkien (Taiyu) dialect.

All research participants spoke Mandarin (n = 22), and approximately a half also spoke Shanghainese (n = twelve) to varying levels of ability, with this divide heavily influenced by age. Older Shanghai women tend to speak Shanghainese at home and often cannot speak fluent Mandarin. Consequently, in 2018, when we conducted the first round of in-depth interviews, we recruited a Shanghainese-speaking research assistant, who was a graduate student at Shanghai Jiao Tong University. At the same time, because the PRC promotes Mandarin in schools, Shanghai's younger generation are less able to speak and understand Shanghainese.

The participants' local dialects (tuyu) are officially recognised dialects that correspond to their hometown origin. In addition to Shanghainese, these include the dialects from Anhui, Changzhou, Chongqing, Jiangsu, Jiangxi and Shandong. In all these cases the dialect in question was spoken by only one woman, with the exception of the Jiangsu dialect, spoken by three women.

By Nasya S. Razavi, Anindita Datta, Karen de Souza, Grace Adeniyi-Ogunyankin, Natasha Aruri, and Penn Tsz Ting Ip

In the following section we turn to the stranglehold of anglophone **hegemony** on academic knowledge production and its domination of feminist urban studies.

Section 2. Anglophone hegemony

Practises of translation in the social sciences are strongly influenced by the political economy of research and publication. For example, in the Anglo-American academy, the dynamics of academic translation are shaped by anglophone hegemony: the exercise of power not through explicit violence, but through linguistic norms and the naturalisation of hierarchy. For example, the most prestigious academic journals are dominated by anglophone scholarship, and anglophone gatekeepers in the publishing industry reproduce English-language hegemony (Kong and Qian, 2019; Müller, 2021). While key English texts are frequently translated into French, German, Spanish, Portuguese, and Chinese (among other languages), research written in these languages is not often translated into English. Furthermore, texts are rarely translated between such languages – a practise that reinscribes English as the *lingua franca* of academic research (Germes and de Araújo, 2016). In 2021, the urban geographer Martin Müller noted that 'the UK, the US and Canada still produce more than 60 percent of the total articles in [geography] journals' and that no other country reaches 'even a 5 percent share in authorship' (Müller, 2021, p. 1450). It is not only journals: the authors and editors of academic books predominantly reside in anglophone countries. In this section, we turn to strategies for unsettling the hegemony of English within the academy. As Müller (2021, p. 1459) remarks, 'what is at stake is who, what and how we can know in the world – no more and no less'.

Both feminist and decolonial theorists have criticised the anglophone hegemony of the academy. Beginning from a critique of enlightenment-era universalism, feminist scholars argue that knowledge is situated in the historical and geographic context whence it came (see Haraway, 1988). Because of its long and bloody relationship with projects of colonial domination, the Western academy as we know it today produces necessarily partial knowledge. The professionalisation of knowledge production is integral to the reproduction of these exclusionary epistemologies. As Nagar argues, 'there is no field of intellectual and creative engagement that is untouched by the messiness and contradictions of professionalization and commoditization' (in Nagar *et al.*, 2016, p. 505). Professionalisation devalues the knowledge of the global majority without degrees, certifications, or institutional backing, positioning them as objects yielding empirical data, not subjects with knowledge to share.

A fuller understanding of the world requires two transformations: the elevation of historically marginalised knowledges and the provincialisation of Euro-American knowledges (Noxolo, 2017; de Leeuw and Hunt, 2018; Daigle and Ramírez, 2019; Jazeel, 2019). As translation scholars Olga Castro and Emek Ergun (2017, p. 3) write: 'Without confronting the epistemic privileges of the Global North, intervening in the geopolitics of FTS [Feminist Translation Studies] by simply making room for knowledge and theory produced in the Global South would be an incomplete as well as risky venture'. This is especially true for the disciplines of anthropology and geography, which have been entangled with colonisation from their genesis (Faria and Mollett, 2020; see Chapter 12, Section 1 Fieldwork and Chapter 13, Section 1 Ethnography).

It is well established that anglophone scholarship from Europe, North America, and Oceania dominates the field of feminist urban studies (see Chapter 2 Feminist comparative urban research). In 2006, a team of researchers found that over 90 percent of publications in the feminist geography journal *Gender, Place and Culture* (*GPC*) were published by authors at Anglo-American universities, with US- and UK-based scholarship accounting for 64 percent of the total. The only countries from the global South then represented in the volumes of *GPC* were Hong Kong, India, Singapore, Thailand, Sri Lanka, and Uganda (García Ramón, Simonsen and Vaiou, 2006). Despite this geographic imbalance, journals such as *GPC* claim global scope, contributors, and readership at the same time as they mask their own provinciality. Introducing the 25th-anniversary issue of *GPC*, the decolonial feminist scholars Marianne Blidon and Sofia Zaragocin (2019) note that these inequalities persist: despite the vibrance of communities of urban feminist scholars in the global South, their dialogues do not always translate to the pages of Anglocentric feminist journals (see Koopman [forthcoming] for a review of the translation policies for various geography journals). Anglophone domination excludes those who are not fluent in English, those who may not be familiar with anglophone academic conventions and debates, and those who do not cite the anglophone canon.

The geolinguistic disparities of knowledge production are (re)produced through an individualistic practise of citation that lends credibility to the author by invoking the geographic canon – which is overwhelmingly white, male, heterosexual, and cisgender (see Peake and Sheppard, 2014; Oswin, 2020; Roy, 2020; Kinkaid, 2023; See also Chapter 2, Section 1 Decolonising knowledge production in feminist urban research). Authors and texts are vested with authority by virtue of how widely and deeply they are cited, while the degree of engagement with an author or text is affected by its perceived degree of authority. Of course, it is not only geographical research at stake: the field of urban studies is also shaped by an exclusionary politics of citation. To break this vicious cycle that marginalises women, queer and trans scholars, and people of colour in urban studies, we must make intentional, political choices about whose voices, ideas, and research we wish to carry forward. As the feminist geographers Carrie Mott and Daniel Cockayne (2017, p. 956) argue, 'To ignore the politics of citation risks the continued hegemony of white heteromasculine knowledge production incongruous with the nuance and richness of other understandings of and perspectives on geographical phenomena'. In contrast, a critical citational practise enacts a politics of refusal: by declining to (re)produce the hegemonic geographical canon, we can support alternative genealogies of research and, in turn, elevate knowledge that is marginalised by the colonial anglophone academy.

Logics of exclusion continue to gatekeep access to the international academic community writ large. The cost of anglophone domination of the academy more generally is the impoverishment of our knowledge of the world. In this book, we have striven to challenge this anglocentric epistemology by fostering dialogue between anglophone urban studies and feminist urban scholars from around the globe. As Nagar (2019) reminds us, this form of translation is a difficult and imperfect process. Reflection exercise 3.1 invites you to evaluate the extent to which we have been successful in moving our citationary practises beyond anglocentrism.

Reflection exercise 3.1: Citational politics

Turn to the references of this book and consider the following questions.

- How successful have we been in moving beyond the anglophone canon in this book?
- What practical or theoretical challenges affect scholarly dialogues across linguistic difference in your area of expertise?

The remaining sections of this chapter explore the thorny problems of translation through two different academic *corpora*: decolonial theory and decolonial feminist studies. Language, culture, and territory are intimately interconnected, and Section 3 reviews decolonial theorists' critiques of translation and their theory of translation as the practise of moving across difference, whether linguistic, cultural, or geographic, exploring the role of these practises in histories of colonisation.

Section 3. Decolonising translation

Since antiquity, imperial powers have employed translation to make the different familiar: assimilating foreign knowledge and culture, promoting religion to colonial subjects, and *interpellating* – naming and producing –landscapes and resources as objects of colonial domination and extraction. Translation has historically functioned, not exclusively, but significantly as a tool of colonisation: the annihilation of Indigenous worlds – including their polities, languages, cultures, ecologies, and ontologies – for the profit of the imperial nation. This logic of eradication or elimination also drives **settler colonialism**: the ongoing occupation of Indigenous lands and the assault on Indigenous peoples by settler groups – a process actively contested by Indigenous polities (see Tuck and Yang, 2012; de Leeuw and Hunt, 2018; Daigle and Ramírez, 2019). As translation studies scholar Douglas Robinson (1997, p. 10) argues, the control of language through translation

> has always been an indispensable channel of imperial conquest and occupation. Not only must the imperial conquerors find some effective way of communicating with their new subjects; they must develop new ways of subjecting them, converting them into docile or 'cooperative' subjects. One of the earliest areas of concern in the history of translation as empire was the selection and training of interpreters to mediate between the colonizer and the colonized.

Indeed, as cultural theorist Tejaswini Niranjana (1992, p. 63) remarks, '*Traducir* in Spanish means to translate as well as to convert'.

The paradigm of translation employed by European colonial powers from the 16th century assumes both an objective relationship between language and meaning and a racialised hierarchy between cultures and languages. Colonial translation works to repress the difference inherent to meaning making across languages, projecting difference onto subaltern Indigenous languages to position them as inferior (Cheyfitz, 1991). As a result, colonial practises of translation render the tongue of the motherland pure and superior and Indigenous languages, contaminated and imperfect (Robinson, 1997). This colonial ideology drives the dual practises of training interpreters for the practical necessities of colonisation – governing the colony, understanding colonial territory, and extracting wealth from the landscape – while also working to suppress and sometimes even eliminate Indigenous languages (Niranjana, 1992; Rafael, 1993). In other words, colonial translation seeks to harness difference that is useful for the colonial project but contain or eradicate difference that threatens it.

The dynamics of colonial translation still have strong effects upon dominant practises of translation in the present (Rafael, 2015). For instance, popular forms of machine translation, such as Google Translate, frame their services as an objective action. Despite presenting multiple translations for a given word, machine translation glosses over the complexities and cultural differences between the source and target languages and enables users to receive answers to their queries without experiencing the difficulty and sometimes discomfort that follows from translating across languages and cultures.

Yet, translation can also be a powerful tool for decolonisation. A wide array of scholars including those situated in the 'post-colonial' nations of the global South (Fanon, 2004 [1963]; Lugones, 2007, 2010) and Indigenous scholars situated in settler-colonial nations in the global North have proposed a heterogenous range of strategies for actualising decolonisation (Smith, 2012 [1999]; Tuck and Yang, 2012; de Leeuw and Hunt, 2018). Towards the end of the 20th century, decolonial scholars began to challenge the ideas that underpin the colonial paradigm of translation and advance alternatives that work towards decolonisation (Cheyfitz, 1991; Rafael, 1993; Robinson, 1997; Chapman, 2023). In her landmark intervention, Niranjana (1992) deconstructs the idea of 'pure' language and the idea of an objective relationship between language and meaning. While colonial translation denies difference between the 'original' and the translation, Niranjana argues that we should view difference as the catalyst that makes new meanings, analyses, and worlds possible. In this view, 'translation, from being a "containing" force, is transformed into a disruptive, disseminating one' (Niranjana, 1992, p. 186). From this perspective, we can see how translation can be used to undermine colonial containment. For example, the historian Vincente Rafael (1993) demonstrates that Tagalog people in the colonial Philippines employed a practise of translation drawn from a Tagalog ethics of relational reciprocity to push back upon colonial ideology under the auspices of the 'dominant' language. Catholic Tagalog converts would interpellate 'untranslatable' theological and ritual language through an Indigenous lexicon to 'elude the interiorization of the interrogative language of the Law carried by the insistent voice of the dominant other' (Rafael, 1993, p. 135). These tactics of anti-colonial translation demonstrate how the possibility of difference inherent to translation can be leveraged in ordinary and extraordinary ways to challenge colonial power.

In short, we can say that decolonial theories of translation work to address the question: *In which tongue can the subaltern speak?* As the translation scholar Paul Bandia (2008) argues, decolonial translators ought to challenge dominant practises of translation by employing a 'polylingual' approach to translation: one that retains some of the 'foreign' lexicon of the source language in order to bring out rather than suppress difference. In this way, decolonial scholars aim to develop alternative paradigms of translation that can transform the languages, cultures, and territories of both the colonial periphery and the metropolitan centre.

For decolonial theorists, however, the problem of how to practise translation in the shadow of colonial empire remains unresolved. Some scholars have critiqued the decolonial approach on the grounds that it abstracts translation away from the actual practises of translation and interpretation, distinguishing between a more theoretical 'translation-as-metaphor' and a more grounded 'translation-as-practise' (Murphy, 2015; Shamma, 2018). Perhaps, then, what is necessary is an analytical frame that can grapple with translation at different scales, such as the scales of the individual sign and signifier, of the text and its authors, and of broader cultures.

Use Reflection exercise 3.2 to investigate translation, difference, and power in the texts you are reading and citing in your own research. In the next section, we turn to the work of decolonial feminist scholars, whose novel praxes of translation grapple with the classed, gendered, and racialised legacies of colonisation.

Reflection exercise 3.2: Translation, difference, and power

Consider the key theoretical or empirical texts supporting your project.

- Are you engaging with any translated texts?
- Where were those texts written?
- What languages are they written in?
- What is the positionality of the authors?
- How do these texts shape the questions you are able to ask in your research?

In the final section we turn to the work of two bodies of work on decolonial feminist translation, from India and from Latin America, exploring how translation can work as a feminist political praxis.

Section 4. Decolonial feminist approaches

Decolonial feminist scholars have developed paradigms of translation that challenge the legacies of colonial domination by muddling through the messy process of meaning-making across myriad axes of difference: not just linguistic difference but social, political, and geographic distances as well. These practises are part and parcel of broader feminist commitments to decolonisation and transformative politics (Lugones, 2010; Nagar *et al.*, 2017; Zaragocin, 2023). Without translation, it is impossible to build a political movement broad enough to challenge the classed, raced, and gendered oppressions of what feminist theorist bell hooks calls 'imperialist-white-supremacist capitalist patriarchy' (hooks, 2004, p. 29). Here, we explore two major bodies of work on decolonial feminist translation that have emerged in the last two decades: the work of Richa Nagar – in conjunction with SKMS and Parakh Theatre – which explores the relationship between translation and transformative politics in India, and the work of feminist scholars such as Sara Koopman, Sofia Zaragocin, Verónica Gago, and Liz Mason-Deese, which considers the relationship between translation and transnational politics in Latin America.

Throughout their two decades of work together, Nagar and her collaborators have developed a praxis of translating knowledge and solidarity across difference that subverts the colonial notion of objective translation between dominant and subaltern tongues. They call this *hungry translation* (Nagar, Meier and Spathopoulou, 2023, p. 5):

> A non-stop striving for ethical retelling – an ever-evolving relation between self and other – where each one consistently works to listen, feel, trust, and retell ethically, despite the challenges . . . and despite an understanding that each retelling with be incomplete and imperfect.

We can understand this sort of hunger as the will for knowledge and power common to those marginalised by epistemic and political injustice. For Nagar, hungry translations are animated by radical vulnerability, or the collective grappling with 'the violent histories and geographies that we inherit and embody, despite our desires to disown them' (Nagar and Shirazi, 2019, p. 239). Building relationships of radical vulnerability entails translating knowledge, politics, and desires across differences. Nagar shows how these relationships can come together to form what they call situated solidarities, '"*a blended but fractured we*" [that works] across multiple axes of power and difference' (Nagar and Shirazi, 2019, p. 237; emphasis mine).

Because of the ambiguity of language and the differences between interlocutors, hungry translations are messy and imperfect. Indeed, not everything can be translated. The literary theorist Damien Tissot (2017) argues that feminist practises of translation should work to preserve the 'foreign' specificities of the source text. This sense of *foreignness*, rather than an analytical reduction, points towards the radically different ontologies, epistemologies, and ethics that are always-already encoded in different languages. For Tissot (2017, p. 39), 'the translator should find a way of making textually accessible to the target reader what is precisely different from their reality'. In this way, the process of translation can create a situated objectivity by engaging interlocutors in the contradictions between what can – and cannot – be known about each other's worlds.

The outcome of these hungry translations can never be known in advance: the transformative power of situated solidarities is grounded in this very indeterminacy. As Nagar explains (2019, p. 22):

> It is through this possibility of dwelling, breathing, touching, and creating between and across worlds that hungry translations can become productive of alternative worlds, and where academic intellectuals can participate not merely as detached analysts or do-gooders of the world but as active political beings who labour to co-create just worlds by sharing epistemic agency with those we accompany.

By challenging the hierarchies of colonial translation and building relationships of radical vulnerability with others, then, translation of knowledge and power does not just change how we see the world but transforms the world itself.

This praxis of hungry translation resonates with conceptions of translation developed by decolonial feminists working in and from Latin America. As Zaragocin *et al.* (2022, pp. 8–9) write, 'in Latin America, decolonial feminist geographies are defined as centring knowledge production from the diverse experiences of the "others" of hegemonic feminism – queer and trans, Indigenous, Afrodesendants [sic], campesinas, migrants, working-class peoples'. Like Nagar's (2018, p. 20) notion of 'situated solidarity', these decolonial feminist analytics are predicated upon the translation of knowledge across linguistic, classed, raced, and gendered difference.

Koopman was an early voice raising these political formations in the anglophone academy. Trained as a professional translator and interpreter, she works to translate a politics of solidarity between the global North and South through her research on Colombian peace movements (Koopman, 2008, 2011, 2020, forthcoming). As Koopman (2008, p. 832) explains, this praxis of translation manifests from her experiences of building relationships across difference:

> I am on stage. My voice follows that of survivors, as I interpret their testimonies into English. I am in the meeting room. I have been a core activist for eight years on the Translation and Interpretation Working Group, and for several years have represented our group on the vigil coordinating committee. I am in the movement. This is the primary gathering of the North American movement in solidarity with Latin America, which I have been active in for 20 years. I theorise this work not to distance myself from this activism, but as a form of engaging in it.

For Koopman, political work with Latin American migrants in the global North was an opening to a situated solidarity with peace movements in the global South. This work is complimented by the recent work of Latin American feminist scholars. For example, Sofia Zaragocin's (2019a, 2019b, 2023) work with the Critical Geography Collective of Ecuador engages Latin American theory in dialogue with the anglophone geography canon. Zaragocin (2021, p. 236) employs the rather queer concept of *transloca* to explain her practise of translating across the particularities of language, space, and culture: a 'framework of translating theory [that] speaks to what happens when one or more place-based systems of thought encounter one another, affecting not only theoretical frameworks but also those involved in creating those dialogues'. By putting feminist standpoint theory in dialogue with post-colonial notions of language, Zaragocin encourages us to reject cultural and geographic essentialisms about which knowledges can help advance a decolonial politics and to use *transloca* to create new dialogues, alliances, and theory across difference (see Chapter 11, Section 2 Feminist ontologies).

Somewhat differently, Verónica Gago and Liz Mason-Deese (Gago and Mason-Deese, 2019; Gago, 2020) employ a process of creative co-translation in their work on Latin American feminist movements and women's strikes: during the same period when they were participating in political mobilisations, they worked side-by-side to practise translation through situated objectivity. As Mason-Deese explains in the 'Translator's foreward', 'the manuscript was translated as it was written, chapter by chapter, over the course of nearly a year, accompanied by a constant dialogue between the author and translator as we both participated in feminist mobilizations, assemblies, seminars, exchanges, and debates' (Mason-Deese, 2020, p. ix). Through this radically vulnerable practise of translating text, Gago and Mason-Deese hope to translate the insights from these movements across the globe, building situated solidarities across space and time. Again, their political imperative is inseparable from the 'practical' activities of translation (Mason-Deese, 2020, p. vii):

> If there is one thing we have seen in the process of constructing a feminist internationalism, it is that internationalism . . . [is] only possible through practices of translation . . . . Translation from one language to other, but also from one context, and its associated histories and geographies, to another, translations across asynchronous temporalities and uneven spatialities.

In this way, decolonial-feminist praxes of translation foreground the difficulties of communicating across difference at the same time as they underscore how these practises can lay the groundwork for political transformation.

Decolonial feminists challenge colonial modes of translation grounded in the 'god trick' of masculinist objectivity by building messy, imperfect relationships across difference. Indeed, the feminist scholar Cornelia Möser provocatively argues that 'the task of the feminist translator has to be treason' (2017, p. 91). In her view, 'The famous saying *traduttore traditore* that was used to highlight the perils of translation should therefore be resignified and reclaimed'. The purpose of this reclaiming of the idea of translator as traitor is to emphasise the radical capacity for transnational solidarity to disrupt the hegemony of the nation-state. From a decolonial feminist perspective, *traduttore traditore* is an ethical duty to subvert colonial paradigms of translation and embrace the messy practise of negotiating difference. To return to the question of hungry translation,

Siddharth Bharath, Sara Musaifer, and Richa Nagar (2019, p. 246) explain the nature of their collective decolonial feminist praxis:

> Grappling with the poetics and politics of radical difference implies a careful engagement with what we deem to be legible or intelligible, and it includes decisions about which stories are told and untold, by whom, how, and under what circumstances. It leaves open the possibilities of disagreements within the collective, and in relation to the stories and storytellers we find ourselves in conversation with.

This ambivalence – the discomfort and excitement of not understanding, of being misunderstood, and the imperative to try to communicate anyway – is precisely the source of translation's radical, transformative power. Researchers need to be aware of the thorny problems of translation and interpreting because they will be 'translating' their work even if their participants and their audiences seem to be speaking the same language. They also need to take care not to employ the concept of translation as mere metaphor for working across difference – instead, we must grapple with the entangled ways in which language, epistemology, and power shape our world. Careful engagement with translation is well worth the trouble: by engaging across language, across cultural difference, researchers have the capacity to transform their understanding of the world and evoke new forms of social and political resistance. For decolonial feminists, then, translation is not merely a research practise but a praxis for invoking other worlds.

Section 5. Summary

In sum, translation and interpretation are power-laden processes of meaning-making across difference – linguistic, geographic, classed, or otherwise – that demand careful consideration. Feminist practises of translation were not initially connected to decolonial politics, but since the late 20th century scholars have connected feminist insights about the role of language in the production of classed, gendered, and racialised hierarchies of difference. Decolonial feminist scholars argue that translation is key to dismantling colonial hierarchies and addressing the material legacies of colonisation. Colonial practises of translation are predicated upon racial hierarchies that reify colonial languages as the 'pure' source of meaning and subordinate Indigenous languages as inferior modes of communication with only instrumental uses. The hegemony of the English language in the academy reproduces these epistemic injustices, excluding other languages – and those who speak them – from the production of knowledge. Transnational feminist movements demonstrate that new political formations can emerge in and through translation. This praxis of translation enables decolonial feminists to build solidarity across social, geographic, and linguistic contexts as they grapple with the challenges of working across difference.

4 Feminist scholar-activism

Mantha Katsikana

Learning objectives

After completing this chapter, readers should understand

- modes of feminist scholar-activism within and beyond the academy;
- feminist scholar-activist practices and their challenges;
- the impact on feminist scholar-activism of neoliberalisation of the academy;
- and the challenges of feminist scholar-activism in transnational feminist research.

Further, readers should be able to

- assess whether a scholar-activist approach is suitable for their own research;
- challenge exploitative and extractive tendencies within academic research;
- and anticipate how to negotiate cultural, national, or other kinds of difference in their own research.

This chapter discusses **feminist scholar-activism** and its diverse actions, spaces, and scales within and beyond the neoliberal academy. It highlights how feminist scholar-activists engage in research that involves collaboration with grassroots movements, creating communities creating spaces for difference and decolonisation within the academy. The various methods feminist scholar-activists deploy in their work are explored, as well as the ways they forge ethical relations with communities and adopt a feminist ethics of care attuned to power dynamics in relations. Specifically, the chapter addresses the challenges of engaging in feminist scholar-activism in the neoliberal academy. It also illuminates the challenges of doing feminist alliance work, including that of co-authorship, in antiracist and decolonial struggles, as well as in transnational feminist and academic communities, and the challenges of navigating differences across geographical, socio-political, and cultural contexts.

Section 1. The academy and feminist scholar-activism

Academic institutions have always been places where power and interlocking systems of oppression are both upheld and challenged. The academy as an institution varies, and its positioning within national and international structures, politics, and economies, such as neoliberalism and globalisation, manifest differently across different places (Brajkovic, 2018; Hoang, 2019; Ramohai, 2019; Li and Li, 2022). While globalisation and internet communications technologies have precipitated further collaboration between

DOI: 10.4324/9781032668727-6

universities and individual scholar-activists, each academic community has its own institutional framework, as well as a position and visibility in a broader hierarchy of knowledge production which favours the Anglo-American global North.

While it is impossible to define a single history of the academy, patterns in the evolution of academic institutions across the world reveal the complex role of the university as a space of maintaining and upholding systems of power and oppression in relation to race, class, gender, disability, and other relations of difference. Through its institutional status within broader systems of domination, the academy has been associated with exclusionary practices, with their roots in elitism and in compliance with monarchy, imperialism, and colonialism. Western academia, specifically, has been a place directly profiting from the oppression of marginalised peoples in both the global North and the global South, associated with a colonialist gaze and **settler colonialism** (Lee and Ahtone, 2020), slavery (Wilder, 2013), and Eurocentrism across disciplines. More recently, the academy has perpetuated globalisation and the negative consequences of economic development (including education-related migration flows linked to the international brain drain) (Münch, 2014).

The academy is also a space of critical thought, knowledge production, and radicalisation, as human rights movements across the globe have increased and universities have become more accessible, signalling a shift in the possibilities of changing the institution of the academy from the inside out. Often, both locally and internationally, universities have been hubs of organising for political movements, including as sites for the formation of political consciousness and mobilisation against oppressive regimes (Vrana, 2017; Ordorika, 2022). In its various forms, the academy has become a place of knowledge production where marginalised communities, women, Indigenous people, and queer and differently abled people demand space and engage in practices of decolonisation at the intersections of **activism**, teaching, and transformative **praxis**, making knowledge production and its liberatory possibilities accessible to the world. Yet its discriminatory past and the pressures of neoliberal capitalism still pose challenges and intensify existing injustices (Della Porta, Cini and Guzmán, 2020; Kidman, 2020; Kushnir, 2020; Nuttall, 2021).

This chapter focuses on how feminist scholar-activists navigate their situatedness within the academy, committing to political ideals such as a feminist ethics of care or decolonisation in an institution that is historically embedded in imperial and colonial projects of **dispossession**. **Feminist activism** refers to the heterogeneous practices that work towards liberation from gender-based oppression, often taking the shape of collective action by women in **grassroots** social movements. This activism is grounded in praxis, which is not reducible to a theory/practice binary but is forged through ethical relations with communities and by the ways in which theory is grounded in everyday life, and reciprocally, by the way everyday life informs and produces knowledge and pedagogy (see Chapter 13, Section 3 Feminist geo-ethnography). Feminist praxis does not happen in a void; it is only from the viewpoint of particular places that feminists can address how ethical relations take shape (Haraway, 1988) (see Chapter 7, Section 3 Feminist ethics and research). Feminist praxis is rooted in an ethics of care, the building and sustaining of relationships of trust, reciprocity, collaboration, and friendship among scholars and the people they work with, as well as between scholars themselves, across race, class, sexuality, ability, and political and geographical borders (Cahuas, 2023). Feminist scholar-activism is thus produced through reciprocity, usually following horizontal structures of collaboration with communities, grassroots organisations, and social movements, as well as with non-governmental organisations (NGOs) and international organisations.

In the following section we explore the many ways in which feminist scholars have introduced activism into the academy.

Section 2. Being a feminist scholar-activist

The form that feminist scholar-activism takes varies; it can have a multitude of goals, impacts, and final products. Feminist scholar-activists work collectively in their various capacities – as a scholar (in an educational institution or independently) practising activism in and/or beyond academia, adopting an activist ethos in their research, pedagogy, and service. Individual feminists' paths to activism and grassroots engagement differ. Some scholars, for example, are involved with activism before entering academia and go to university to advance their skills for activism. Those from marginalised communities may enter academia to help their own communities and make issues heard (Zerai, 2002). Others only become involved with activism once they enter the academy.

Being a feminist scholar-activist is not an identity that is or can be openly embraced by everyone. Not adopting the label of activist often relates to politics and the cost of being visible in that role to the academy, the state, and other institutions. It can also serve as a refusal of the appropriation of activist research by the neoliberal university. For activist scholars from the global South who work on the ground with marginalised communities, the label of **feminism** is sometimes seen 'as a foreign export mobilised by elite women who are unconcerned with "bread and butter issues" ' (Okech and Musindarwezo, 2019, p. 78). And for those wanting to work within the label of feminism, the cost is often too high to pay, with governments – for example, in India, Iran, and China – stamping down on freedom of expression for those who identify as feminists (see Chapter 13, GenUrb example 13.3 Security issues) through the repression of feminism, including the criminalisation of dissent and activism (on China see Stewart *et al.*, 2023).

While political visibility is often a vital part of Northern feminist praxis, under authoritarian and patriarchal contexts, visibility may have different consequences: navigating the surveillance of authorities and regimes does not always allow Southern feminist scholar-activists to render their work visible in the same manner as those in the global North. Similar issues can, however, also apply in the global North, where antiracist organising and protesting can be a complicated issue for individuals who hold fewer privileges and are thus hypervisible and disproportionately targeted by racists. Akwugo Emejulu and Jo Littler (2019) argue that such hypervisibility can lead to women-of-colour feminist scholar-activists stopping frontline work or engaging in any kind of activism as a strategy of survival to protect themselves and their families. Feminist praxis can thus vary according to the lived realities of feminist scholar-activists and the communities they work with (Martínez and Agüero, 2021).

Feminist scholar-activists may pursue social justice within the university through service, pedagogy, and teaching, as well as through labour organising and union politics. These objectives can be achieved by claiming and carving out spaces for change through **knowledge production**, communicating knowledge and research outcomes, building bridges between the university and communities, and creating access to the university for those who are marginalised (see Chapter 19, Section 2 Feminist approaches to knowledge mobilisation). These practices can take the form of utilising the privileges of universities (for example, access to technology or funding for the dissemination of knowledge) to enhance issue visibility, building coalitions and creating local, national, and international scholar-activist networks.

These multiple routes to realising a feminist scholar-activist life often depend on an individual's goals for scholarship and reformist or revolutionary activity (Zerai, 2002). Many believe that their academic research is their contribution to revolutionary social change and that critical scholarship has historically proved crucial to organising against oppressive regimes and dictatorships. For others, a key part of being a feminist scholar-activist is activism through the radical potential of teaching: 'developing different relationships with students that are not distant, hierarchical, and tied to power imbalances; [but] working alongside student activists . . . to help students mobilize their own anger into activism' (Quaye, Shaw and Hill, 2017, p. 385). For others for whom pedagogical engagement in the classroom is not enough, activism necessitates working beyond the academy. This can take the form of working towards short-term goals, engaging directly with marginalised communities, including connecting communities and grassroots groups with the academy to provide access to funds and technology or to aid future collaborations. For others, it involves focusing on long-term goals that may involve transnational organising or working with international organisations practising consultation and advocacy to effect changes in policy.

Notwithstanding the amount of time, energy, and labour that goes into activism, much of this work is routinely undervalued, if acknowledged at all. Yet this work takes its toll on the scholars who perform it on both a professional and a personal level. Such is often the case with practising feminist practices of care within the university or of the labour of carving out space through one's embodied difference while experiencing microaggressions. This invisible radical labour is often performed by women, queers, and differently abled people in the academy, by many scholars from the global South working and studying in institutions in the global North, and by feminist scholar-activists in male-dominated academic environments. In the academic context of the global North, it has been recognised that racialised faculty members, particularly women, assume significantly higher loads of emotional labour in mentoring racialised students, often at a cost to their own health and often with this labour left unaccounted for in formal remunerative structures of reward and promotion (James and Chapman-Nyaho, 2017; Dhamoon, 2020). Such labour may only come to the attention of the academy in a negative fashion, for example, when it threatens to destabilise the very power structures that the university guards, as a space of privilege.

Reflection exercise 4.1 presents some necessary steps to help you either start thinking about your own engagement in activism within and beyond the academy or to engage in further self-reflection within existing activist practices.

Reflection exercise 4.1: Situating the self in feminist activist praxis

This exercise helps you situate yourself within (and in relation to) configurations of power and systems of domination within academic institutions as a first and fundamental step towards mobilising and developing a feminist activist praxis, within and beyond the academy. (It may help to first read Chapter 11, Section 4 Feminist methodologies, which outlines understandings of reflexivity and positionality.)

Situating the self draws on the feminist principle of reflexivity, part of an ongoing feminist praxis that prioritises empathy, accountability, and critical consciousness while thinking about the complex ways you may be experiencing privilege and/or oppression due to your position in various systems of domination and oppression. Reflexivity should be exercised frequently alongside awareness of the significance of intersectionality in every step of your journey in feminist activism, as they are central to the building of solidarity and alliances. The process of situating the self should not be reduced to merely situating yourself in identity categories but should include how you have come to be the person that you are through your experiences, community membership, engagement with social issues, and the ways you contribute to or are being affected by systems of domination and oppression.

Thinking critically about your own positionality is often the first step to discovering the political frameworks that align with your sense of self, ideals, dreams, and lived experiences. It can also be a step towards finding community through praxis, direct action, and organising. Do not expect this process to be something that you only do once: this sense of yourself might change throughout your life based on your experiences.

Additionally, in the context of knowledge production and research, this exercise will help you develop a critical position in relation to your choice of a research framework, agenda, and ethics that align with your activist values.

Here are some questions to help you situate the self.

- What are the identities I identify/align with and present myself as?
- How did I come to these identities? (Through your family, community, culture, education, religion, etc.).
- Are these identities related to specific communities in the places I live my life? What is my relationship/involvement with these communities?
- How are these communities embedded in local, regional, national, and international contexts?
- Are identities attributed to me that I do not identify with? If so, why?
- Within the identities that I attribute to myself, do I feel that my experiences are fully expressed? If not, how do my experiences differ? What aspects are missing?
- How do my identities intersect and affect each other?
- In the context of my identities, do I hold privilege in relation to other groups/communities/identities? If so, how is this privilege manifested?
- Are there any identities/groups/communities I am not part of but consider myself allied to? If yes, which are these?
- In what ways am I being an ally? How do I use my privilege in favour of the groups to which I am an ally?

Now think about the institutions you belong to and interact with. List them and, based on your responses in the previous questions, consider the following questions.

- How are my identities and communities represented within these institutions?
- Do these institutions' public statements and policies align with my own lived experiences?
- How is my privilege/marginalisation manifested in my position/treatment within those institutions?

This reflection exercise will help you situate yourself in a broader context of power dynamics and identify your common ground with social justice causes. Such awareness can assist you in identifying organisations, communities, and causes that align with your research objectives. This exercise can also help if you are not participating in scholar-activism, as it includes critical reflection on your own positionality within academic institutions.

The following section builds upon these points, focusing on how the neoliberalisation of academia manifests and hinders the work of feminist scholar-activists, who, in turn, mobilise on the offensive.

Section 3. Defending space for feminist activism in the neoliberal university

Neoliberalism is both 'a theory of political economic practices . . . within an institutional framework characterized by strong private property rights, free markets, and free trade' (Harvey, 2005, p. 2) and an 'order of reason' remaking social life (Brown, 2015, p. 9). Globalised yet differentiated across local contexts, neoliberalism is an 'economizing' practice whereby 'both persons and states are construed on the model of the contemporary firm' and 'are expected to comport themselves in ways that maximize their capital value in the present and enhance their future value, and both persons and states do so through practices of entrepreneurialism, self-investment, and/or attracting investors' (Brown, 2015, p. 22). Failing neoliberal subjects, whether person or state, face fiscal crises, downgraded credit, and a loss of legitimacy at the least or bankruptcy and dissolution at the extreme, thus risking impoverishment.

Since neoliberal ideology permeates all aspects of life, academia has become a neoliberal project too. Public institutions of higher education have been facing budget cuts as well as increasing privatisation (Nixon, 2017). The rise of neoliberal ideology has also transformed how academic institutions operate, prioritising the university as a profit-making competitive institution-enterprise for income generation through tuition, research, and external funding. As Claire Polster and Janice Newson (2015) have observed, in Canada, neoliberalism has made universities more commercial in orientation, business-like in knowledge practices, and corporate in self-presentation, resulting in the reproduction of existing inequalities and unequal power dynamics. Henry Giroux (2014, pp. 11–12) stresses that 'under such circumstances, education becomes obsessed with accountability schemes, redefining students as consumers, deskilling faculty, governing through the lens of a business culture, and dumbing down the curriculum by substituting training for a critically informed education'. For example, the financialisation of the student experience, as the increasing cost of attendance drives the student debt crisis, disproportionally affects low-income and first-generation students as well as students of colour, deepening racialised accumulation of debt (Mayers, 2021). Academics are increasingly expected to bring research money to their departments (Christou, 2016), and research and its outputs have become measurable products. In this context, faculty and student protests, strikes, and demands have emerged as resistance against the neoliberalisation of higher education (Maira and Sze, 2012).

The intensifying neoliberalisation of academic institutions hinders the work of feminist scholar-activists not only in terms of how knowledge production and its impacts are regulated, monitored, and measured but also through the rising precarity of labour and a research culture that prioritises competitiveness above collaboration. Thus, research agendas risk being depoliticised, appropriated in the name of measurable outcomes and of servicing the needs of job markets. To ask what it means to be a feminist scholar-activist in the neoliberal academy is to ask what it is like to produce knowledge, provide service, build communities, and commit to social justice within a place of institutional power linked to various aspects of privilege while simultaneously being evaluated, monitored, and often only precariously employed.

And yet these very conditions within the neoliberal academy have also provided the ground for an increasing commitment to activist research. Thus, being a feminist scholar-activist in this context requires creating and defending spaces from which to undertake activist scholarship within often inhospitable environments, pushing back against multiple barriers to generate possibilities for institutional change (Hale, 2008). This feminist scholar-activism takes many forms and uses a variety of tools to create spaces that defy neoliberal dogma and the patriarchal, colonial roots of the academy. Its key aspects are:

- creating spaces for feminist scholar-activist research in the neoliberal academy for the creation of community, solidarity, and alliances. This is particularly important where gender, feminist, or women's studies are marginalised (Flores and Garcia, 2009; Pereira, 2010; Gouws 2012; Kašić, 2016; Decker and Baderoon, 2018);
- developing an ethics of care across intersectional lines of gender, race, class, sexuality, and disability, often including the creation of community within the university to cope with the costs and the demands of neoliberal academia, ensuring wellness and providing guidance and mentorship while also informing critical understandings of how privilege and oppression work in complex ways within the academy (Mullings, Peake and Parizeau, 2016; Decker and Baderoon, 2018; Singh and Mathews, 2019);
- on-campus activism, often incorporating gender issues into wider social movements in the context of campus life and community and addressing issues such as sexual harassment, safety, and violence at universities (Ampofo and Arnfred, 2010; Hosein, 2011);
- engaging in teaching as feminist pedagogical praxis and the classroom as an activist space (Ampofo and Arnfred, 2010; Pereira, 2010; Hosein, 2011; Underhill-Sem, 2017);
- working on the ground with marginalised communities and grassroots group for the co-production and dissemination of knowledge and the development of transnational feminist praxis (Benson and Nagar, 2006; Peake and de Souza, 2010; Nagar, 2013);
- advocacy and consultation with and for NGOs and grassroots groups (Jauk, 2017; Kagal and Latchford, 2020);
- building and maintaining co-operative international knowledge networks with other feminist scholar-activists not only to build coalitions but to make issues visible at a global scale (Olutayo and Yalley, 2019; Tambe and Thayer, 2021);
- (in certain cases, for strategically placed feminist scholar-activists) working within international institutions and organisations, such as the United Nations, the International Criminal Court, and the World Bank, to contribute to the incorporation of critical feminist perspectives in human rights and international development agendas (Kelly, 2005; Casimiro and Andrade, 2010; Quataert, 2014).

Feminist scholar-activism in the neoliberal academy does not come without challenges. The requirement of being measured against neoliberal metrics of academic influence and success can often pose obstacles for activist research, especially for untenured feminist faculty whose work involves participatory projects and for whom it is not possible to measure the amount of time as well as collaborative effort put into maintaining relationships with communities (Underhill-Sem 2017; Reynolds, Block and Bradley, 2018). In the context of 'publish or perish', the privileging of individual efforts and merit, and traditional means of recognition (i.e. awards from professional associations, grant procurement, tenure, and promotion), activist research is often discouraged because of its supposed inability to generate new income streams and the long periods of time it may take to reach fruition (Underhill-Sem, 2017; Davis *et al.*, 2019). Most recently, for those feminist scholar-activists who do pursue social justice, the neoliberal commodification of knowledge production can add the extra complication of how to deal with the appropriation of their research and labour as indicators of equity, diversity, and inclusion (EDI) agendas (see Chapter 8, Section 2 (ii) 'Equity, diversity and inclusion'). Aziz Choudry (2020, p. 31) remarks that the declaration of 'scholar-activist' identities 'is not necessarily linked with participation . . . in social and political struggles' but can still function as a ' "currency" . . . [for] universities promoting themselves as community engaged'.

This problem is often accompanied by extractive attitudes towards marginalised communities that are over-researched, especially by global North scholars (even feminists!) who may reduce political engagement to methodological issues of reflexivity and positionality (Vélez, 2019). Hence, feminist scholar-activists' position in the academy can be met with suspicion from communities they approach as a result of previous experiences with exploitative and extractive behaviour by researchers and also because of the inaccessibility of the knowledge that is produced (Smyth, 2023).

This situation creates further tensions for scholar-activists who need to be accountable to the communities they work with. Healing relationships between academia and marginalised communities is crucial within the activist praxis of feminist scholars, yet the burden of doing so can fall disproportionately on the shoulders of those who are themselves members of these communities. In the case of activist researchers from minority linguistic communities, gaining academic recognition by complying with demands to publish in English and securing access to often scarce funding impose further hurdles. The failure to scale these hurdles risks devaluing their work and furthering the invisibility of the issues they research, resulting in both these scholars and their work being sidelined.

Given the costs and constraints of working in the neoliberal academy, Yvonne Underhill-Sem (2017, p. 337) stresses the importance of 'not giving up' as a 'radical practice' that scholar-activists must engage in within its institutions. Building on her ideas, GenUrb example 4.1 outlines how GenUrb has attempted to develop space for feminist activist practice in the context of a transnational academic project. Beyond the academy we have aligned the project with activists, acted as an ally, and prioritised participant safety and well-being. Within the academy, our efforts have focused on the following areas: pedagogy, supporting emerging scholars (particularly those working with GenUrb), assisting with the enrolment and progression of marginalised students, addressing decolonisation within the project, and exploring creative, imaginative, and just outputs and outcomes that advance a feminist research practice.

GenUrb example 4.1: Types of engagement adopted by GenUrb inside and beyond the academy

Types of engagement	*GenUrb practices*
Beyond the academy	
Continual involvement with decolonial and radical practitioners, activists, and scholars	As feminist scholar-activists we work with others, including women's activists outside the university, most specifically the grassroots women's groups Red Thread in Georgetown and Centro de Estudios y Trabajo de la Mujer in Cochabamba, who are both partners in our research and who work respectively with women living in poverty and Quechua women. Connections with Red Thread, through Linda Peake, go back over three decades.
Being an ally	We have provided intellectual support to early-career feminist scholars in the global South seeking to publish their work; engaged students who have been historically excluded from knowledge-production processes, through processes of peer-mentoring, training, and co-learning across difference (see Chapter 2 Decolonising feminist knowledge production); provided financial resources to support feminist scholars in the global South to attend international meetings; and engaged our non-academic partners such as Red Thread, CETM, and the Centre for Human Development in Ibadan in research, providing them with research skills and financial resources to help them attain a more secure future.
A commitment to participant safety and well-being	We prioritise more than anything else the safety of members and participants engaging in research. Where necessary, we have held meetings in secret and concealed the identity of GenUrb members, and we have always been prepared to remove GenUrb members from the field. We have also helped participants financially to remove themselves, temporarily and permanently, from tricky situations.
Inside the academy	
Retaining radical pedagogy	We have taught undergraduate and graduate courses at York University, Toronto, based on GenUrb material, in which we have introduced material on decolonising pedagogy.
Supporting emerging scholars	Over the life of the project, we have employed seven undergraduates, 21 graduate students, and four post-doctoral fellows and research associates. We also created an Early Career Network for co-mentorship of PhD students, postdoctoral fellows, newly tenured faculty members, and activists (see Chapter 2, Section 5 Co-learning across difference). We have organised writing retreats bringing together a range of GenUrb members to think and write together.
Engaging with university administrative processes that ensure the enrolment and progression of marginalised students	We have employed undergraduate students who are enrolled in the Research at York (RAY) programme for financially challenged students at York University.

Challenging inequitable processes	We have turned the spotlight on GenUrb to address our efforts at decolonising practices (see Chapter 2, Section 2 Decoding affinity and difference in a feminist project).
Exploring creative, imaginative, and just options	Our outputs include academic publications, publications for the general public, and resources for local populations such as web-based information on COVID-19 and radio programmes and exhibitions.

By Leeann Bennett and Linda Peake

Reflection exercise 4.2 describes how you can engage in feminist scholar-activism by locating and reaching out to communities, both within and beyond the academy, exploring ways to connect the two.

Reflection exercise 4.2: Connecting with communities, creating solidarities

The aim of this exercise is to enable you to identify the type of connections you can build between the different locations you occupy and collective causes. It will help you locate and reach out to communities and grassroots initiatives that align with your positionality, allyship, and values within and beyond the university.

Step 1. Make a list of groups, community projects, collectives or other initiatives in your neighbourhood, university, and city that, based on your positionality, you see as allies. Besides issues of oppression, representation, and identity, this allyship could also be based on local issues of access, infrastructure, environmental degradation, or other local needs. Similarly, in the context of the academy, there might be campus initiatives such as study groups, reading groups, and associations that pique your interest. People in your classes might already be part of such initiatives – don't hesitate to talk to people you feel comfortable with about their engagement.

Step 2. Research the websites, social media, and other outlets of these initiatives/projects to obtain a more complete idea of their demands, goals, agendas, and practices. Do these align with your own values about activism? Could you see yourself taking part in any of these?

Step 3. If possible, attend in person a public event or action organised by any of these groups. Introduce yourself to activists and community members and ask for any information you might need to assess if you are interested in joining. Issues such as time commitment, required availability, and skills could be part of the conversation.

Step 4. After discussions with activists and witnessing their work in person, you will be able to assess whether you can join or support them. Levels of involvement may vary, especially when you need to become familiar with the inter-personal dynamics in activist communities as well as find your place within them.

Step 5. Think about all the initiatives and projects you are participating in. How do their goals and practices relate to your studies, research, and academic trajectory? Could you introduce connections between projects or initiatives from the academy to those outside it?

After completing these five steps, you should be able to identify communities and organisations that align with the type of activist work you want to engage in.

In the following section we turn to specific issues that transnational research can pose for feminist scholar-activists.

Section 4. The challenges of feminist scholar-activist research in transnational contexts

In addition to the challenges of operating in the neoliberal university, specific difficulties can occur when engaging in feminist scholar-activist knowledge production across transnational differences. In this section, we discuss a range of issues that the power dynamics of transnational partnerships can give rise to as well as focusing on one particular issue, that of co-authorship.

(i) Feminist scholar-activist research and transnational power dynamics

It is rare if not impossible for transnational research allegiances to be partnerships of equals. Structural biases and institutional agendas resulting in inequalities of access, finance, language, and mobility abound in global North–South and Indigenous and settler-colonial contexts, as well as in others. Many feminist scholars have bitter stories of transnational encounters that started out well but soon succumbed, the best of intentions proving an inadequate foundation for such endeavours. For example, decisions questioned and overturned without discussion, lack of transparency in budgets, and work produced by members lower down 'the hierarchy' or by those who have been paid (as consultants) not being acknowledged because they are not seen as knowledge producers, have all been recognised as problematic issues (see Chapter 2, Section 1 Decolonising knowledge production in feminist urban studies research). Tensions also emerge from the assumption that certain Northern positionalities are inherently 'feminist' with no need for reflection on notions of privilege or the significance of different socio-political, geographical, and historical contexts. Amanda Gouws (2012, p. 532) outlines other uneven conditions feminists from the global South face while working with those from the global North, such as having to engage in feminist theory (mostly generated in the North) 'in order to show that they are "well read"' and that their research meets the standards set in the North. Moreover, the privilege granted to Northern analytical frameworks, despite their inability to travel well across different places and situations, can serve to reduce the value of global South feminists' analytical frameworks.

Additionally, when working in transnational teams, the labour and effort of those who lack the 'Anglo privilege' of being native English speakers can be dismissed (Lunny, 2019, p. 67): the hegemony of the English language as the default for publishing often remains unproblematised. Thus, a significant part of the work and research conducted by those in transnational partnerships and that is not translated into English remains inaccessible and ultimately invisible to international audiences (see Chapter 3, Section 2 Anglophone hegemony). This argument does not apply in the global South alone. Biljana Kašić (2016, p. 133) situates the issue of knowledge production in southern Europe within what she calls 'the globalisation of feminist excellence' in academia, which marginalises scholars in peripheral or semi-peripheral countries. Local knowledges, not only in the context of the global North/global South but also in the less-visible post-socialist reality of the East/West, are subject to an academic hierarchy whereby local theory, methods, perspectives, and canons are rendered inferior. As Kašić (2016, p. 133) asks: 'what constitutes local scholarship in order [for it] to be taken into account within a global setting or feminist transnational networks?'

While attempts by feminist and decolonial scholars to raise these issues are increasing, such structural biases and institutional agendas still serve to benefit the career goals of Northern-based scholars, the volume of whose research outputs dominates academic production. However, while the positionality of the Northern-based researcher is tied to certain privileges in the context of the neoliberal academy, it is not a homogenous positionality; it also includes global South and Indigenous feminist scholar-activists working and studying in Northern universities, with increasing numbers having only a fragile and temporary toehold in the academy (see Chapter 2, Section 3 Circulations of spatial and temporal difference). This diversity does not serve to render Northern positionalities unproblematic. For example, while many feminist scholar-activists identify themselves as producing decolonial work and/or participate as allies in demonstrations and protests around Indigenous peoples' struggles, Sarah de Leeuw and Sarah Hunt (2018, p. 6) stress that white settler and non-Indigenous scholar-activists 'often continue to engage *concepts of indigeneity* rather than Indigenous peoples themselves', producing scholarship that is disengaged from the actual Indigenous communities whose issues are being discussed.

Scholar-activist research by Northern feminists may be motivated by intentions to help communities, but the typically small-scale participatory research projects that predominate most commonly cannot solve structural problems or issues relating to poverty and everyday living conditions, while they do take up participants' time and involve emotional and physical labour that can create feelings of exploitation. Healing communities from these extractive relationships, sustained by 'parachute researchers', as Thelma Vélez (2019, p. 216) (among others) describes them, is work that disproportionately affects local feminist scholars who may belong to or are close to these communities (Mwangi, 2019). Even in the context of research that is aligned 'in solidarity' with communities and their agendas, Jamie Gagliano (2021, p. 3) stresses that solidarity 'should not be treated as a methodological end goal, but yet another complex set of social relations which profoundly shapes research without necessarily undoing its ethical dilemmas'.

The different geopolitical and institutional contexts across the globe, as well as the conditions of life in local and national economies, can significantly limit the ability of feminist scholar-activists not only to engage in transnational research alliances but also to secure funding and a livelihood. For many such feminist scholar-activists unable to secure a place in the academy, working in consultancies for NGOs and other organisations may be their only alternative (to the extent that such work is available) (Pereira, 2010). Maintaining a research profile and engaging in activism is difficult, not least because of the time required to raise funds for research and the competing claims on time that arise from activism.

The location of many feminist urban researchers in the global South in the development industry is a primary issue of concern for those who are also activists. Practical, political, and ethical constraints often challenge feminist scholar-activists in countries of the global South where UN consultancies or working for internationally funded NGOs can be the only safe options for making a living in an economy characterised by underdevelopment (Ampofo, 2010, p. 29). The NGO-isation of development and the professionalisation of activist movements into NGOs often capitalise on the co-opting of grassroots efforts (Bernal and Grewal, 2014; Trotz, 2020). Feminist research can also be marginalised or depoliticised in the name of gender mainstreaming (True, 2003), while the need for funding often requires abiding by agendas and completing projects to produce a final outcome that may have little to do with the everyday lives of the participants and communities that helped produce the knowledge in question (Peake and de Souza, 2010).

Not all feminist activist-scholars eschew working in the field of development. Many have engaged with advocacy through working with NGOs and multilateral organisations such as the United Nations (UN) (Quataert, 2014; Jauk, 2017). Feminist scholar-activists have participated in UN conferences, advancing policy agendas and rights in relation to the lives and rights of women and girls at urban, national, regional, and global levels. However, these activities have also contributed to what Chandra Talpade Mohanty (1988, p. 61, 2003b) and Awino Okech and Dinah Musindarwezo (2019, p. 257) call the production of the 'Third World woman' or 'Third World women', respectively. Moreover, the homogenisations embedded in UN discourses, the NGO-isation of development, the

Audio clip 4.1: Podcast on differing relationships to activism, featuring Natasha Aruri, Karen de Souza, and Nasya S. Razavi

Audio clips of interviews with GenUrb members Natasha Aruri, lead of the Ramallah City Research Team (CRT), Nasya S. Razavi, team lead of the Cochabamba CRT, and Karen de Souza, co-lead of the Georgetown CRT, are available on the book's website www.routledge.com/9781032668680.

Natasha Aruri discusses the links between activism and research work that affects the on-the-ground lived experiences of communities in Ramallah. Natasha also discusses intergenerational activist praxis and practices of participatory research along with the challenges of conducting such work in Palestine.

Karen de Souza discusses political activism in Georgetown through the collaborative work of the grassroots organisation Red Thread with GenUrb as well as through personal trajectories, together with the meanings of activist praxis and community.

Nasya S. Razavi discusses feminist scholar-activism and international development and the challenges of collaborating across different institutional structures and geopolitical contexts. Nasya also discusses her GenUrb-related work with Quechua women in Cochabamba, in collaboration with Centro de Estudios y Trabajo de la Mujer.

By Mantha Katsikana

proximity of Western feminists to policymakers in the global North, and their easier access to funding have all created tensions within transnational relationships of feminist solidarity (Okech and Musindarwezo, 2019).

These issues extend to deep differences of practice that feminist scholars are assumed to have 'dealt with' already and to which they are presumed to be immune. In transnational collaborations, reflexivity can be a tool for the renegotiation of differences and pre-established hegemonic relations. However, as we outline elsewhere (see Chapter 7, Section 3 Feminist ethics and research), reflexivity is often insufficient to translate into accountability and the transformation of dynamics rooted in systems of oppression.

GenUrb audio clip 4.1 features interviews with some of the GenUrb feminist scholar-activists in which they discuss how they navigate power dynamics and engage activist praxis in Cochabamba, Ramallah, and Georgetown.

(ii) Co-authorship as feminist alliance work

For research to be useful to communities, it needs to be accessible. Western academia, however, prioritises specific forms of knowledge production and mobilisation. For example, academic publishing becomes the end goal of research projects, while the collective labour of knowledge production through collecting data in the field, of building and maintaining relationships, though essential, is seen as separate from and secondary to that end product. Academic outputs, moreover, are usually written in languages or formats that can have little value for – or even be incomprehensible to – the communities the researchers worked with. The peer-reviewed publications required of scholars are seen by community partners as 'largely inaccessible documents that serve dominant interests' (Gustafson, Parsons and Gillingham, 2019, p. 8) and that take time away from more inclusive engagement. While alternative and creative deliverables such as theatre performances and videos can be produced with and for local communities, they are rarely undertaken, with few academics having the skills and time required to engage in them (but see Nagar and Selmeczi, 2018; Pratt and Johnston, 2017). Furthermore, even when promised in grant applications, alternative plain-language reports, reports co-authored with participants in local languages and dialects, or return visits to the field to discuss the research results and their implications can fail to materialise. They are often relegated to the last activity of the research process, and when time and other resources run out, they can be sacrificed to the demand to move on in the cycle of yet more grant applications and publications.

At the same time, co-authorship and especially collaborative storytelling are powerful forms of feminist alliance work that bring together academic and non-academic knowledge that can serve to disrupt persistent hierarchies of knowledge production (Nagar, 2013, 2014). Co-authorship is a tool for mobilising intellectual spaces in which stories from multiple locations can speak with one another so that they can evolve into more nuanced and effective critical interventions. Richa Nagar (2013) emphasises the need to make interventions through which the privileges and resources afforded by the academy can be mobilised to advance knowledge from sites that are systematically excluded from, illegitimised, or rendered invisible in the dominant class-based system of intellectual labour. Co-authorship is, therefore, based on producing knowledge that is not characterised by academic expertise but is accessible to audiences outside the university, for example, through writing in newspapers, pamphlets, poems, and plays as well as in languages that might be seen as 'vernacular' (Nagar, 2013, p. 10).

While co-authorship is a valuable tool that reflects the importance of feminist collaboration beyond the metrics of the neoliberal university, the process of producing co-authored material involves a set of challenges related to uneven power dynamics and geopolitics. Co-authorship requires the work of building non-transactional, ongoing relationships of care, trust, and reciprocity. Through this practice of feminist alliance-building the engagement with difference (such as language, life experiences, and the ways individuals and communities feel heard in such relationships) and resultant disagreements, mistakes, and dissonance mark the collective 'we' and the knowledge produced within that relationship. Inevitably, as seen in the work of scholars engaging in co-authorship, different priorities may create tension around the collective 'we'. These priorities sometimes clash in the various phases of co-authorship, from initial discussions to choosing the form of the final product (Nagar, 2013, 2014). While such work helps the researcher and the researched to produce the new realities they are fighting for, it entails knowledge production in forms that go beyond the publishing economies of neoliberal academia.

In GenUrb example 4.2, we discuss how co-authorship has been addressed in GenUrb, specifically in relation to academic publications.

GenUrb example 4.2: Issues of co-authorship on academic publications

GenUrb has grappled with many discussions involving many of its members about co-authorship. How do we determine co-authorship in a project with so many people, each with their own stake in the publication process? How do we balance the needs of postdoctoral fellows to acquire publications versus the rights of those who collected the data to have their names on publications? How do we agree on whose names go first? And do we provide authorial credit to those who have had no engagement in a publication (including analytical material, the collection and analysis of data, and writing) but are senior to those who have contributed?

These are just some of the questions we have had to ask in GenUrb to which there are no obvious answers or at least an array of answers. It is only through discussion that a resolution is reached. These discussions are continual, as each publication in GenUrb has a distinct set of criteria determining authorship. Here we lay out the questions we had to answer in relation to this book.

The purpose underlying the book was twofold: first, we wanted to provide an open access book that could relate the experiences of the GenUrb project throughout the research process. Secondly, we wanted to provide an opportunity for publication for early-career scholars in GenUrb, based in the global North at York University in Toronto, Canada. It was a dual purpose that risked obliterating the work of CRT members who had conducted the research but were too busy to write chapters. Hence, we came up with the idea of having the GenUrb material in the book highlighted in its own boxes, each one listing those who had contributed to the research discussed.

In keeping with not wanting to privilege the contributions of those based in the global North, our wish was not to have the names of individuals listed as the book editors but rather to list 'The GenUrb Project' as the editor. However, our publisher made it clear that such an entity could not legally be named as editor and that we had to provide the names of individuals. We decided to provide these names alphabetically.

It was also important to us that the CRT members who contributed to the book were also listed on the front cover. But which members? Some had played pivotal roles working on the project for several years, while others had been minor players involved with only specific tasks or for short periods of time. Listing everyone would drown out what virtually all concerned regarded as an important distinction. We eventually agreed that we would list only the core members of each CRT who had participated in the research, with each team themselves defining who these members were. We then had to decide whether these contributors would be listed alphabetically or grouped according to their CRTs. We had support for both options but opted for listing by CRTs, as this was the strong preference of some CRT members. As is our practice throughout the book, the CRT members are listed alphabetically. This does not mean to say that all the core contributors to the research are mentioned – some CRTs had to withhold the names of some of their members for political reasons – or that everyone listed played a central role. Some CRTs chose to list members who had played only an advisory role.

We also negotiated with the publisher that the 'GenUrb examples' in the table of contents, which list members of the CRTs as authors, should not be moved to the less important list of tables and figures. It was imperative as a decolonial feminist practice that the names of CRT contributors appeared on the same pages as the names of the chapter authors. Fortunately, it was a request with which the publisher immediately agreed, creating what may be a new convention in naming who is listed in a table of contents.

Although there were differences among GenUrb members around naming practices in this book, these have not been insurmountable: indeed, they have been amicably resolved. It was with the publishers that we came up against a brick wall, namely the legal proscription on being unable to recognise the entity of 'GenUrb' as editor. Thus, we took one step forward, one step back in the dance that is the decolonising of feminist research through co-authorship practices.

By Linda Peake

Section 4. Summary

This chapter has discussed modes of feminist scholar-activism within and beyond the academy. Feminist scholar-activist research is forged through ethical relations with communities and by the ways that theory is grounded in everyday life, in local and transnational contexts, with a view to building a feminist ethics of care as a foundation of decolonial work in the academy. It has shown that in the neoliberal academy, organising to keep spaces open for feminist scholar-activism has led to extra challenges but has also resulted in increased commitment. Engagement in transnational research partnerships raises challenges with regards to collaborative knowledge production, including: power dynamics between researchers (and their varying positionalities) and the communities researched; navigating international organisations and development agendas and co-authorship. Like many others, these are issues for which there are no easy answers.

Part II

The context of 21st-century feminist urban research and policy

5 Feminist urban research in the time of COVID-19

Mel Mikhail

Learning objectives
After completing this chapter, readers should understand

- that the COVID-19 pandemic, like all pandemics, is a gendered, racialised, and classed crisis best understood through an intersectional lens that can account for the complex geographies of social reproduction shaped by patriarchy and racial capitalism;
- that conducting feminist research during crisis demands key ethical, methodological, and epistemological adaptations;
- that a feminist approach to policy development and implementation can identify and address limitations to pandemic policy;
- and that data disaggregation is a necessary component of feminist pandemic policy.

Further, readers should be able to

- apply feminist research insights from the pandemic to their own research;
- and address the practical limitations imposed by the pandemic or other interruptions on their own research programme.

This chapter concerns the relationship between pandemics, urban society, and the conduct of feminist research. It focuses on the everyday realities of COVID-19 but considers the applicability of the discussion to other pandemics and ecological crises. The chapter begins with a general overview of the urban nature of COVID-19 and related pandemics, followed by a discussion of the geographies of social reproduction that have emerged from the pandemic, underpinned by the interlocking dynamics of patriarchy and global racial capitalism. It then turns to consider how research processes are affected by and can be creatively adapted to meet the challenges of pandemic geographies. The chapter ends with an overview of the role of feminist research in global COVID-19 policy development, with emphasis on the politics of data disaggregation. It explains why intersectional, antiracist feminist urban research is crucial for policy development during pandemics, not least because of the failures of urban policy to respond to intersectional marginalisation in urban space.

Section 1. Pandemics and urban society

In December 2019, the first known cases of the coronavirus (COVID-19) were identified in Wuhan, China, and in the months that followed the virus spread worldwide

DOI: 10.4324/9781032668727-8

(World Health Organization, 2020). Though stabilised at the time of writing due to increased crowd immunity and global vaccine uptake, the global pandemic has been marked by traumas of mass illness and death. In 2020–21, to curb rising rates of infection, international organisations and national governments responded in myriad ways including public education, lockdowns and closures, travel restrictions, and workplace controls and with constantly shifting policy implementation. Despite these and other efforts, cases of COVID-19 infections and deaths attributed to the virus rose sharply, though unevenly, around the world. Policy implementation was irregular, differentially affecting populations from one country to the next. Researchers in the GenUrb project witnessed this firsthand. The pandemic affected each city in the study differently; timelines differed, as did choices regarding appropriate measures to curb the spread of the virus. For example, Shanghai began to limit entry and exit into the municipality as early as January 2020 because of its proximity to Wuhan; Cochabamba, Delhi, Georgetown, and Ibadan would not see the full-fledged impacts of COVID-19 until March 2020, responding with a wide range of measures, from lockdown(s) and stay-at-home orders to curfews and social distancing (Razavi *et al.*, 2023). In some cases, the COVID-19 crisis was seen as secondary to political or economic crises, overshadowing the much less visceral sense of the virus. But as we will show, the economic and the ecological cannot be disentangled.

Spatial analysis of the way pandemics become global (Sugrue, 2020) can help to illuminate their social impacts. Globalisation, since the 1980s, has 'foster[ed] the conditions that produce outbreaks and that turn those outbreaks into pandemics' (Altschuler and Wald, 2020). Urbanisation is key among these conditions for outbreak. In particular, the interfacing of the rural and the urban acquires special significance, along with rural–urban migration and the development of peri-urban and suburban spaces (Ali and Keil, 2006). Urban ecologists and epidemiologists confirm that cities and their suburbs are 'ground zero' for the spread of infectious diseases (Santiago-Alarcon and MacGregor-Fors, 2020). It is clear that 'the global trend towards urbanisation has contributed to the rise in the total number of disease outbreaks per decade since the 1980s' (University of Lincoln, 2021).

Given these spatial realities, ecological crisis – defined here as the persistent and widespread negative outcome of actions or behaviours that throw relations between species and organisms into disarray (leading, for instance, to pandemics, wildfires, and famine) – unevenly and unjustly affects the distribution of social relations across place to create new and exacerbate already existing social and economic inequalities. The knowledge produced by feminist researchers during this pandemic, emphasising an ethics of care and attention to the relational character of the virus, has been indispensable for fostering a collective understanding and sense of responsibility to the social disparities of global urban life that predate this crisis and will continue to persist long after it (Shotwell, 2020) (see Chapter 7, Section 2 Towards a feminist ethics).

As the next section explores, a racialised and gendered intersectional analysis of different social groups illuminates angles of social experience that remain opaque from non-feminist and mainstream perspectives on the pandemic. In particular, the pandemic has further sedimented socio-spatialities of social reproduction underpinned by the interlocking dynamics of race, class, and gender.

Section 2. The gendered and racialised social impacts of COVID-19 on everyday urban life

This section employs an intersectional feminist urban research lens to examine the way that pandemic geographies of **social reproduction** are crosscut by gender, race, and class. Kimberlé Crenshaw's now canonical work (1989), in which she develops the framework of intersectionality, is specific to a tradition of Black feminist epistemology forged in the struggles for labour and political rights in the USA, which can be dated back to Sojourner Truth's 1851 speech 'Ain't I a Woman?' at the Women's Convention in Akron, Ohio. While the framing of intersectionality has since taken on a life of its own in academic and cultural contexts, this provenance has sometimes been lost on those very feminist scholars seeking to describe how several systems of discrimination could bear on an individual's experience in the world. It is thus important to be aware of the way intersectionality is often shorn of its Black epistemological roots in its application to feminist research.

We extend this classical framing of intersectionality to the spatialities of COVID-19 to make sense of the unevenness of the pandemic's 'power-geometries' (Massey, 2005, p. 82) (see Chapter 13, Section 3 (i) It is place-based). A focus on the intersectionality of social reproduction in particular shows how the global system, rooted in systemic patriarchy and racial capitalism, over the course of many epidemics – such as Dengue, Ebola, SARS, and Zika – has proven the disregard that global health institutions have for women and especially for the lives of racialised girls and women in the global South. Intersectional feminist analyses not only contribute to understanding of pandemics but can also shed important light on the neglect of these women and girls in policy development (see Section 4 COVID-19, feminist policy, and data disaggregation).

(i) Gender and the pandemic geographies of social reproduction

We adopt here an expansive definition of social reproduction that captures its complexity and understands it as the practices of life-making that exist beyond and within capitalist circuits of production, serving to make the latter possible. We understand social reproduction to be:

> an historicized and spatialized construct . . . [which] includes the embodied labour (paid and unpaid) in conjunction with the resources, such as those of land, 'nature', time, technology, and increasingly capital, that enable human and non-human life to occur, the emotional and material needs of everyday life to be met, as well as hopes and dreams for the future, and the material social practices that constitute the organization of daily life and life over generations to take place. It is about the process of the **production** of value – both use and exchange value – moulded through the spatialities and temporalities of the everyday and determined through differentiation and struggle.
>
> (Tanyildiz *et al.*, 2021, p. 9)

Feminist economists' and geographers' considerations of various dimensions of social reproduction – care work, paid and unpaid labour, education, food security, and family/

kinship – have informed their analyses of COVID-19 (Heintz, Staab and Turquet, 2021; Neely and Lopez, 2022a, 2022b; Flor *et al*., 2022), revealing a geographically uneven crisis of social reproduction emerging during the pandemic and demonstrating how global capitalist production makes misogyny an economic necessity (Rao, 2021). Capitalist patriarchal ideology surmises that within a social order based around the family unit, women ought to be coercively and/or repressively relegated to the realm of care; otherwise, labour will not be reproduced. Feminist research supports the claim that, in terms of broad social groups, women have experienced a greater degree of overall negative impacts than men, both mentally and physically, due to the loss of paid work, loss of childcare for women who parent, and the stresses of increased family care and domestic duties sprung on women in families because of the pandemic (Al-Ali, 2020; Hammonds, 2020; Wenham, Smith and Morgan, 2020; Adeniyi-Ogunyankin and Peake, 2021a; Kabeer, Razavi and van der Meulen Rodgers, 2021; Flor *et al*., 2022). Placing the 'COVID' burden of extra unpaid caring work on women's shoulders has had a knock-on effect on other social areas such as education (Flor *et al*., 2022). Flor *et al*. conclude from available evidence that numerous girls in the global South have had to abandon their studies to take on the added domestic duties produced by the pandemic (2393). And in a study in South Korea (Ham, 2021) on the multivariate reasons for women's overrepresentation among workers who either took a leave of absence or became unemployed during the pandemic, key factors included job status with a 'higher percentage of women being unstable workers' (145), or an 'overrepresentation in the care and hospitality industries' as opposed to men (146). These feminist urban studies reveal both the global reach of capitalist patriarchal relations and the role they play in shaping gendered geographies of social reproduction during the pandemic.

GenUrb example 5.1 offers a snapshot of the lives of women affected by the crisis of social reproduction.

GenUrb example 5.1: Research on everyday life during the pandemic

In mid-2020, members of the GenUrb network undertook comparative research on the gendered impacts of the COVID-19 pandemic in five cities: Cochabamba, Delhi, Georgetown, Ibadan, and Shanghai. We analysed women's accounts of their everyday lives during the pandemic to understand how processes of production and social reproduction were being reconstituted. We found that the COVID-19 crisis has magnified how unpaid and low-paid social reproduction work, primarily undertaken by women, crucially supports urban economic and social life. For women already facing structural inequities and conditions of precarity, hardships multiplied during this time. The measures to restrict movement increased the time and effort needed to undertake essential activities. Loss of employment and income-earning opportunities were accompanied by increased care work. Furthermore, for some of the women participants' mental health became strained as they navigated the constraints of the 'pandemic city'. In the context of political conflict in cities such as Cochabamba and Delhi, the COVID-19 crisis was seized upon to enact violence, thereby deepening landscapes of urban insecurity. In response to destabilised social reproduction processes, however, women across the cities organised circulations of care through both traditional and new networks of reciprocity and solidarity (see Razavi *et al.*, 2023).

By Nasya S. Razavi

The pandemic's adverse effects have also been felt by women, such as many groups of queer and trans women, whose daily lives are not necessarily conditioned by the care needs of the normative familial unit (see, for example, Banerjea, Boyce and Dasgupta, 2021; Haworth, Cassal and de Paula Muniz, 2023). The vulnerabilities experienced by women thus take different forms in a situation where 'gender-based systemic and institutionalized discrimination interlocks with a set of social categories (e.g. racialization, migrant status, poverty, disenfranchisement, HIV seropositive status)' to produce differential impacts on mobility and access to spaces of work, healthcare, and community (Garcia-Rabines and Bencic, 2021, p. 663). While women as a social group have borne the brunt of the crisis of social reproduction, our views of the way geographies of social reproduction emerge between groups of women are obscured if other axes of social difference are left unexamined. Race, to which we now turn, is one such major axis of social difference along which pandemic geographies of social reproduction coalesce (Andrews, 2021).

(ii) The carceral logics of pandemic geographies

Carcerality is a social, spatial, and political condition characterised by the ideology of capture. This is the notion that our actions, behaviours, and social dynamics are ordered at least in part by the logics of captivity, an inheritance of colonial regimes of territorialisation, including the rationalities and technologies of settler colonialism and transatlantic slavery. Though carcerality as an analytical framework originated in the United States to clarify the historic struggles of Black Americans in the face of an imperialist and white supremacist state, the framing is weighted with a valence apt to explain global pandemic geographies. On a global scale, the formation of carceral spaces is one main way that vulnerability to premature death is produced for racialised people. The Black feminist geographer Ruthie Wilson Gilmore has famously stated that 'racism . . . is the state-sanctioned or extralegal production and exploitation of group-differentiated vulnerability to premature death' (2007, p. 28). Urban scholars adopting a critical race and Black studies approach (Oeur, 2021; Whitacre *et al.*, 2021; Neely and Lopez, 2022b) have shown how, as Alexander Liebman, Kevon Rhine, and Rob Wallace (2020, p. 334) argue, the 'intertwinement of Fascistic power and premature viral death due to race [has] laid bare that COVID-19 is inseparable from the carceral structures of racial capitalism'.

COVID-19's inseparability from carceral structures reflects a well-documented history of infectious diseases being spatially organised along racial lines, often connected to labour migration trends, wherein xenophobic attitudes scapegoat migrant labour as 'a vector for infection' (Boris, 2022, p. 67; see also Chen, 2020). The confluence of the fear of the virus and fears of the foreign 'Other' entering the labour force (and white homes) as care workers results in discriminatory practices and the passing of bills and laws that further relegate the racialised Other to dwelling conditions and work environments with high exposure to viruses.

Furthermore, these spatialities of infectious disease map on to the historical production by the state of racialised and gendered urban spaces. For example, in the late-19th-century USA, the state intensely regulated the migration of Chinese men to prevent them from establishing families. This resulted in groups of single Chinese men practising 'queer domesticities, arrangements that included all male living groups, but also female-headed households networked together' (Boris, 2022, p. 68). These arrangements contributed to the establishment of urban Chinatowns, simultaneously targeted for exclusion and sanitisation, while serving as a site of Chinese placemaking. The trace of these histories of racialised and gendered urban spatial production live on today through anti-Asian sentiment regarding the COVID-19 virus as well as policies restricting urban mobility that have undermined the vivacity of these centres of small business and culture.

In the following section, we look at some of the overlapping methodological concerns and adaptation of methods that the COVID-19 pandemic has occasioned for feminist urban research.

Section 3. Practicing feminist urban research in a time of ecological crisis

The COVID-19 crisis ushered urban researchers into a new reality, one in which the conditions informing our research contexts can change dramatically at the drop of a hat. The constantly changing material realities of the pandemic, such as lockdowns, travel restrictions, and closures, made it increasingly difficult, and even in some cases impossible, to gather data according to conventional research schedules and programmes. Seemingly overnight, our hyper-connected, rapidly urbanising world can amplify localised events into global crises, and political divisions can intensify and create difficulties for researchers working with participants in the field. Your methodology may suddenly be unfeasible, participants may no longer be able or willing to stay involved, and your ability to respond to the needs of stakeholders or partners may be affected. Considerations around mental health issues have also increased during the pandemic and can affect researchers and participants alike (World Health Organization, 2022). These are just some of the ways the pandemic has left an indelible mark on our research practices, demonstrating that ecological, economic, and political unpredictability is a horizon shared globally (see also Nyambura *et al.*, 2021).

This section addresses these concerns in terms of potential shifts that may be required in both methodology and methods.

(i) Shifting methodological perspectives in research during a time of crisis

As conditions shift dramatically due to a pandemic or other widescale crisis, some approaches to field research and data gathering may become inaccessible. Ethnographic methods such as participant observation or face-to-face interviews become very difficult to conduct (Yeager, 2020) – and yet geo-ethnographies are central to a great deal of feminist urban knowledge production (see Chapter 13, Section 3 Feminist geo-ethnography). In GenUrb, by prioritising geo-ethnographic forms of knowledge production, we implicitly valorise the mechanisms, including participation and participant interviews, by which such approaches justify beliefs about the world. In this way, specific methodologies bolster and reproduce epistemologies (ways of knowing and justifying what we know about the world) (see Chapter 11, Section 1 Introduction to the philosophical fields underlying research).

When these approaches become inaccessible, we are left to meditate on how we produce knowledge as social scientists. A space of self-reflexivity can be opened – how do we come to the work we do? – inviting feminist researchers to consider what Zoe Todd (2020) calls 'the stakes of their position': their role in research and knowledge production, the reasons they choose the methods they do, and whose stories are being told, how, and why. So often as academic researchers, the colonial origins of the disciplines we come to and the methodologies they employ (particularly ethnography) remain buried, only coming to the surface through critique after we have reached a certain degree of literacy and fluidity with disciplinary norms or felt their limitations for ourselves. The sudden suspension of these norms, as in a pandemic, can leave room to reflect on the stakes of this work – what it offers despite its originary violences, if anything, what reparative work critique can do, and other similar questions.

Pandemics can lead to more than a reconsideration of methodologies. The pandemic contributed to galvanising a spark of political struggle and unrest, as with the global marches for Black life erupting in summer 2020 in the United States and Canada and struggles in the same year against political and economic reform in Indonesia and Nigeria (Westerman, Benk and Green, 2020). Similar protests started in 2021 in Colombia, and in South Africa the combination of struggles against political corruption and economic conditions and the pandemic produced a 'perfect storm' (Mbete and Goodman, 2021). Building consideration of such events into research contingency plans and how to deal with the uncertainties that arise is important.

Reflection exercise 5.1 asks you to think outside the box and to question the epistemological assumptions we take for granted as feminist researchers.

Reflection exercise 5.1: Collecting data when not in the field

It is often assumed that 'nothing replaces the professional and scholarly value of "being there" to collect data and experience a place firsthand' (Krause *et al.*, 2021, p. 265). A global crisis may, however, make 'being there' impossible in many instances. Reflect on the theoretical implications of this observation and respond to the following questions (you may want to jump ahead to Part IV Feminist methodologies and methods, and skim it for ideas).

- Why is first-person observation valued so highly? Is this justified?
- How, if at all, does the pandemic, in conjunction with the growing realities of climate crisis, war, or other ecological crises, affect your perspectives and practical approach to fieldwork?

(ii) Adapting research methods in a time of crisis

Though researchers are no strangers to changing plans, the pandemic created unusually difficult barriers to conducting empirical research (Radecki and Schonfeld, 2020; Ramos, 2021). As universities shut down in-person classes and pivoted to online teaching, learning, and administrative work, many researchers experienced disruptions.

For feminist researchers, adapting methods based on their circumstances requires an understanding of the conditions for research that their field makes (im)possible. The issue of the digital divide (pitting those who have internet access against those who do not, often due to inequitable socioeconomic conditions), for example, varies within and between places. A lack of access to the internet can create obstacles for researchers who are trying to move to largely 'desk-bound' research alternatives but do not want the validity or quality of the data to be compromised. Recommendations to address the digital divide can include budgeting for cheap mobile phones and/or mobile hotspots to be delivered to participants along with easy-to-follow instructions or longer-term strategies of lobbying governments to subsidise the costs of internet access (Lourenco and Tasimi, 2020).

In response to the limitations that COVID-19 has placed on field research, others have opted to maintain participatory methods through the involvement of 'citizen scientists' (Crimmins, Posthumus and Prudic, 2021). If face-to-face interviews and participant observation become inaccessible, consider other actors related to your participants – such as academics, public officials, or members of social movements – with whom you can communicate via digital means to gain access to data. By providing detailed research guides, creating easy-to-use data-input applications, and carefully coordinating data review processes, researchers can equip community members 'on the ground' to participate more fully in the field research process. Building such participatory research practices is increasingly important, not only to manoeuvre around constraints from pandemics or other major events but also to create an 'opportunity for collaborative agenda setting and knowledge production; designing more nimble research; and rectifying . . . structural . . . inequalities' (Krause *et al.*, 2021, p. 3).

The immersion of fieldwork can be non-negotiable for some. In this case, consider advocating for 'slow science', where the focus is as much on relationship-building as it is on data collection. As a result, you may choose to put your research on hold until you can return to the field. If your research must be put on hold due to restrictions, your skills and knowledge as a researcher may be useful to the community in different ways. Reach out to grassroots networks in your place of research, for example, and offer your energies. Depending on your social location, your ability to do research may become constrained by the amplification of other life demands, such as care work. As crises put different life priorities into focus for research participants, aspects of your completed research may begin to feel irrelevant. Reevaluating your research goals, timelines (for funding, completion, and otherwise), and approaches may be necessary.

Alternatively, a global crisis like COVID-19 may not affect your research at all, but it may affect the conditions in which you do your work. Pandemics can have political impacts that deepen ideological divides and heighten social tension. Research participants' political viewpoints may not align with those of the researchers. It is therefore important to consider the steps one might take to maintain a balance of closeness and distance from participants whose own political relationship to the pandemic means health and safety measures may not be observed. Alternatively, increased isolation and anxiety can make work feel futile and more difficult to accomplish. Staying plugged in to social issues and intellectual community, and, in particular, participating with groups doing solidarity work can help to foster a sense of camaraderie and connection.

While doing feminist research in a time of urban ecological crisis can seriously upend the logistics that make sure standards of care in research are met, ensuring the ethical integrity of data practices can go a long way to ensuring the foundations of research remain stable even as extenuating circumstances arise. GenUrb example 5.2 outlines some of the decision-making processes behind the many adaptations to research scope and foci that the pandemic brought about in the GenUrb project.

GenUrb example 5.2: Adapting research methods during the COVID-19 pandemic

For feminist urban researchers working across national borders and with women in marginalised circumstances, restrictions on mobility and curfews posed a major obstacle to research that a widespread shift to technological methods or 'desk-bound alternatives', as described by Krause *et al.* (2021, p. 265), could not completely remedy. In early 2020, when COVID-19 was declared a global pandemic, members of the GenUrb network had to decide how to address our ongoing research as it became clear that our existing practices had to be adapted to the new and varying limitations and the new realities faced by the women participating in the project. The undeniable impact of the pandemic on everyday life and its role in increasing precarity among research participants on many different fronts (e.g. jobs, housing, relationships, mental health, and so on) necessitated the study of its impacts on women's lives and the building of the new data into our analyses.

As researchers, we had to address how the participants were able to secure their own and their families' social reproduction in a context in which many of them had a reduced flow of income coming into their households. The women who had been participating in GenUrb research were confronted with a new intensity of urban challenges of limited mobility, fewer income-earning opportunities, spikes to costs of living, and increased care work. How could we ask them to continue to engage in the research when their daily lives had become even more precarious? A short conversation led us to agree to supplement the monies we paid women for their time to engage in the research. Each City Research Team (CRT) was given extra money, to be applied in contextually specific ways, including in the form of direct cash transfers and the distribution of foodstuffs and other items such as masks, hand sanitiser, and data cards for cellphone use. A couple of CRTs also supported local grassroots mutual aid efforts taking place in participants' neighbourhoods.

Given the changing conditions in the cities in which GenUrb worked during the pandemic, we had to revisit our field methods while maintaining a feminist ethics of care, prioritising participants' material well-being and not placing additional burdens of emotional labour upon them, enabling them to continue their participation, if they were able and so desired. Given travel bans and restrictions on movement, some CRTs experimented with digitally mediated methods, with varying degrees of success (see Razavi *et al.*, 2023). The final resolution of GenUrb research methods consisted of a mix of virtual and in-person engagement with research participants.

- The Cochabamba CRT collected 'digital diaries' of voice memos, texts, photos, and videos via WhatsApp, followed up by phone conversations.
- Similarly, the Ibadan CRT used WhatsApp to prompt questions and conversations and conduct in-depth phone interviews.
- The Shanghai CRT organised weekly online diary-writing workshops using WeChat and made home visits to participants who did not have internet access.
- In Delhi, the CRT did not use digital platforms. Because the participants did not have access to phones or the internet, they conducted in-person, in-depth interviews outside

women's homes with field assistants having been trained virtually in conducting in-depth interviews.

- The Georgetown CRT also opted for in-depth, in-person interviews. Although participants had mobile phones and regularly used WhatsApp, they found that digital methods proved inadequate to capture the emotional dimensions of women's everyday lives.

While digital media allowed us to continue to connect with some participants more easily amidst the pandemic restrictions, it also limited our ability to fully engage all participants, given the uneven access to technologies among the GenUrb participants. This already uneven access to technology is deeply entrenched along gender lines, where intensified domestic work and care duties put added strains on women's time, making it more difficult to find time to engage in research activities.

With the push to virtual methods, we also encountered findings that did not appear to fit in with observations from other research. For example, some of the GenUrb researchers engaging with digital methods reported that despite the widely reported rises in gender-based violence globally during the pandemic (UN Women, 2020), participants reaffirmed their sense of family harmony. These findings required us not only to consider the nature of the relationships we had built with participants (that they may not always be as trustworthy as we assumed) and the ways that crises can intensify participant vulnerabilities (their sense of safety and security may understandably vary as their daily realities shift, and they may become less willing to be forthcoming about their experiences) but also to recognise that it is not always possible to verify whether participant accounts correspond to their realities. In other words, engaging only digitally, at a distance, our analyses must be sensitive to various possibilities of securing credible responses.

By Linda Peake

Reflection Exercise 5.2 provides an opportunity to begin thinking about how to approach the logistics of a feminist research project that shifts rapidly due to an unforeseen pandemic while also taking social differentials at play in the project into careful consideration.

Reflection exercise 5.2: Doing research in a time of global crisis

Imagine you are a team lead on a three-year research project in Dhaka, Bangladesh, about women's perceptions and experiences of changing daily patterns related to water infrastructure expansion in the Bangladeshi capital. In the first year of your project, you and your team made efforts to build rapport with interested research participants who include cis, queer, and trans women, expecting to conduct in-depth interviews in the second and third years. In the second year of the project, a pandemic starts, and you are forced to leave the field and return home. Your team members are Bangladeshi, and their mobilities are restricted. As a result, you decide to shift to digital means of data collection.

Think about the impacts of these events on your research methods, scope, and foci, including the practicalities of collecting information from the women participants and their uneven access to technology.

- How might you navigate these unexpected changes in the research field and process?
- How might you secure answers to your research questions now that the setting and mode of access to participants have changed?
- How will you account for differentials in risk to members of the research team and participants?
- Could virtual tools help you to mitigate risk in your research, and if so, how?

The concluding section moves from research to considerations of the role of feminist policy creation around COVID-19, also looking at the limits of these efforts. It makes the case that feminist research is needed to produce accurate and nuanced data for implementing effective policy.

Section 4. COVID-19, feminist policy, and data disaggregation

Intersectional feminist insights are critical for policy research and implementation at the juncture of pandemics and the urban, working to hold governments accountable to the impacts of the pandemic on everyday life (PAHO and WHO, 2020). Given that pandemics are urban ecological phenomena, data and knowledge pertaining to the urban and the lives of those within it are indispensable for developing policy that is responsive to the actualities of pandemic life in cities. The pandemic is likely the single most-studied event of our lifetimes and has produced an enormous and ever-growing amount of information and data (Kabeer, Razavi and van der Meulen, 2021). The adequacy of this largely non–gender-disaggregated output for policy development is, however, questionable. Furthermore, while urban policy often suffers from a lack of feminist analysis, feminist policy rarely considers the urban (Wekerle, 2013). In consequence, a major gap exists between the lived experiences of those who shape policy and those it is intended to benefit (Centre for Feminist Foreign Policy, 2022). The insights of feminists are thus crucial to ensure that voices addressing the disproportionate social, health, and economic challenges faced by women and other marginalised groups in pandemics are heard in civil society arenas.

Feminist policy recommendations and responses to the pandemic vary, as evidenced by the Gender and Development Network (2019), which has compiled a comprehensive resource on feminist responses to the pandemic across the globe. These responses include calls for an intersectional feminist policy response to address social inequality effectively (Kelly, 2020), the need to account for transgender and nonbinary experiences in COVID-19 policy (Perez-Brumer and Silva-Santisteban, 2020), and for a global feminist COVID-19 policy, backed by 1,600 individuals and organisations demanding governments address inequalities related to education, social and economic inequality, water and sanitation, violence against women, access to information, and abuses of power (Feminist Alliance for Rights, 2020).

In the global North, there is evidence that mainstream, white feminist analyses overshadow intersectional ones. For example, in Canada, the Liberal government's Women and Gender Equality (WAGE) department prides itself on its use of a gender-based analysis

(GBA+) and used this framework to create dedicated emergency funding to women's shelters during COVID-19, based on evidence of sharply increasing violence against women. Other advocacy groups, however, expressed their concern with the limited scope of issues addressed by WAGE in its analysis and its failure to take other dimensions of women's experience into account, such as race, age, or socioeconomic status (Wright, 2020). The Canadian Women's Foundation, with the backing of many other NGOs in the social sector, argued that fitting policy responses must incorporate data that are disaggregated by factors other than gender alone to address the impacts of COVID-19 in a more rounded and appropriate way.

Pandemic policymakers rely on accurate, high-quality data to produce effective policy. Data disaggregation – the separation of collected data according to chosen variables – is a key factor in this effort, as disaggregated data shed light on 'the distribution of infection as well as the risks of worse outcomes for those infected' (Berkhout and Richardson, 2020, p. 4), accounting for the way that different axes of experience come together in different ways from group to group or person to person.

Given that data are collected in all fields across the natural and social sciences, the variables according to which data are disaggregated are context specific. For global pandemics, the United Nations' *Fundamental Principles of Official Statistics* suggests that the minimum set of indicators for disaggregation should be sex, age, education, income/economic status, ethnic origin, geographic location, and disability (Pan American Health Organization and World Health Organization, 2020). Even this minimum set of indicators goes above and beyond the level of detail that has been drawn on by major government and civil society policy bodies during the pandemic. The organisation Gender and COVID-19 cites a 'global data problem,' with only just over half of 192 data-producing nations providing pandemic data disaggregated for sex alone to the exclusion of any other indicators (Ogundele and Walton-Roberts, 2021). Disaggregated data are, however, crucial, contributing analytical granularity about the pandemic to reflect accurately the differential impacts that systems of power have on people occupying different social and geographical locations.

Table 5.1 lists sex-disaggregated data on the pandemic mapping onto various indicators, such as the UN Sustainable Development Goals (SDGs), public health measures, and more. It can be a helpful point of departure for researchers looking for differently disaggregated data (see Chapter 6 Feminist urban policy and the Sustainable Development Goals).

While data disaggregation is necessary, disaggregation alone is insufficient. Producing useful data for feminist research relies on several factors related to research design, such as who is producing the data, what questions are being asked, who is analysing the data, and

Table 5.1 Global sex-disaggregated COVID-19 data trackers*

Organisation	*Scale*	*Indicators*
World Health Organization Coronavirus dashboard	Global	Sex/gender, cases, deaths, vaccinations, public health safety measure impacts
UN Women's Women Count data hub	Global	Sex-disaggregated COVID-19 data mapped onto the United Nations SDGs
Global Health 50/50 data tracker	Global	Sex-disaggregated COVID-19 data about cases, deaths, vaccines, and other health-related indicators

Source: Adapted from Kabeer *et al.* (2021)

* Links to these organisations are available on the book's website www.routledge.com/9781032668680.

what the analysis process looks like (see Chapter 16, Section 3 Stages of qualitative data analysis). Even given the diversity of phenomena accounted for across the various datasets listed in Table 5.1, Naila Kabeer, Shahra Razavi, and Yana van der Meulen Rodgers (2021, p. 20) identify a lack of data on the 'intersectional dimensions of the crisis, particularly the losses to livelihoods and health by gender, race, class, disability, life course, and other markers of disadvantage'. In short, to be useful to feminist research, intersectional data are best produced through research processes informed by a feminist epistemological and ethical framework. Such a feminist intersectional approach to data is important for the disaggregation process because it can ensure that the variables that account for social difference (such as gender, ethnicity, age, etc.) are highlighted throughout the processes of data collection and analysis. Attending to intersectionality in research design through data disaggregation is a key area of concern that requires attention to all the stages engaging data as well as ensuring that the project objectives make reference to intersectionality. There is no 'one size fits all' model, but your ability to reflect and address intersectional analyses in your research will hinge on the data you collect.

Feminist approaches to research design are crucial in ensuring that the quality and content of data lend themselves to the creation of policy that can help to meet the needs of those whose experiences are often erased. What we do with data as researchers, moreover, can have deep impacts on the groups the data are supposed to represent (see Chapter 10, Section 2 Research data management).

Reflection exercise 5.3 is intended to introduce you to the way that data disaggregation can open or close down areas of inquiry.

Reflection exercise 5.3: Data limitations and feminist policy research

Explore the data resources available in Table 5.1. Identify the variables according to which data are disaggregated across these datasets.

- Based on your investigation, what are the more common variables that are included in data disaggregation across datasets?
- What are some of the less common variables that data are disaggregated by?
- What other variables might you need the data to be disaggregated by?
- What sorts of research questions would you be able to begin responding to using these data?

Section 5. Summary

This chapter has provided a brief overview of the COVID-19 pandemic as an urban ecological phenomenon, materialising pandemic geographies of social reproduction and carcerality to highlight the gendered and racialised relations and dynamics of pandemics and related crises that are crucially pertinent to feminist urban research. It further discussed the disruptions pandemics can bring to fieldwork and ways to adapt research to external the circumstances. The final section moved from research to policy addressing the role of feminist research in developing and improving pandemic-related policy. Offering examples of the range of feminist policy recommendations related to the pandemic, it has also discussed data disaggregation, a key aspect for producing thorough, effective, and intersectional analyses for policy development.

6 Feminist urban policy and the Sustainable Development Goals

Nasya S. Razavi and Linda Peake

Learning objectives

After completing this chapter readers should understand

- how a feminist approach to urban policy can lead to more gender-responsive and inclusive urban policy;
- the emergence and relevance of the SDGs;
- the synergies between SDG 5 'Gender equality' and SDG 11 'Sustainable cities and communities';
- and the possibilities and challenges associated with local implementation of global frameworks for urban governance.

Further, readers should be able to

- critique urban policy for gaps related to gender and other social differences;
- connect research to policy outcomes;
- and communicate policy-relevant research outputs.

This chapter provides an overview of feminist engagements with urban policy. It outlines how feminist approaches to gender in urban policy and planning can lead to more inclusive and gender-responsive urban policy. It then turns to feminist engagements with global frameworks for urban governance, focusing on the SDGs and specifically the synergies between SDG 5 'Gender equality' and SDG 11 'Sustainable cities and communities'. The chapter also delves into the challenges and opportunities associated with local adaptation of the SDGs, using GenUrb research to illustrate implementation at the city level. Finally, the need to communicate policy-relevant research is discussed.

Section 1. Feminist approaches to urban policy

In the early 21st century, just as it has been impossible to ignore the impact of COVID-19 on everyday life in cities, so too have the **Sustainable Development Goals (SDGs)** loomed large in urban planning and policy circuits, as they have for the first time ever introduced a global development goal relating to urban life: SDG 11 'Sustainable cities and communities'. While not a policy, this goal has initiated a global conversation about the role of the urban in sustainable development – a conversation in which many feminist academics, practitioners, and activists have participated. There is a long history of feminist engagement in **urban policy**, shaping the agendas, legislation, programmes, and planning

DOI: 10.4324/9781032668727-9

practices that determine the development of cities. Since the late 19th century, women – from revolutionary feminists in Russia (Alexandra Kollontai) to reformist feminists, like Jane Addams, involved in 'municipal housekeeping' in the United States (Hayden, 1981, p. 175) – have been a part of the quest for urban social justice. They have argued for and worked towards the redesign of cities through gendered urban programmes and policy, from 'women's right to the city' agendas to global frameworks for urban governance (Peake, 2024). Underlying all these interventions is the vexed question of grassroots practices and their incorporation into the state. The mainstreaming of radical ideas into formal policy arenas can lead, as Nancy Fraser (2009, p. 114) observes, to 'the discourse becom[ing] independent of the movement' and the rise of what Sophie Watson (1992, p. 186) calls 'femocratic feminisms' or what Janet Halley *et al.* (2018, p. ix) describes as 'governance feminism'.

This section briefly reviews the ways in which feminists have engaged with urban policy as a broad category of (usually geographically oriented) policies aimed at influencing the social, economic, and cultural development of urban areas and urban lives, promoting long-term, inclusive, and resilient urban development.

(i) Gender-responsive urban policy

With the women's rights movement of the late 1960s and early 1970s, Western feminists began entering the fields of urban planning and policy. In some places they gained sufficient strength in numbers to intervene in the design of urban environments. They sought to counter the patriarchal and capitalist processes that had led, from the Industrial Revolution onwards, to the structuring of cities according to the needs of men and of paid employment. This had resulted in an urban form that placed waged work at the centre of the city with radial channels of public transportation leading out to residential areas, facilitating the movement of male workers between home and work (McDowell, 1999; Peake, 2017a). It was a design that had failed to consider the different needs of women in relation to employment, childcare, shopping, and other daily care-based activities.

Feminists, whether activists, architects, planners, or policymakers, have challenged the gendered, classed, and racialised hierarchies of power that are evident in this approach to urban planning with the aim of producing built environments that are safe and sustainable and that address the everyday needs of all residents (see, for example, Matrix, 1984). Increasingly, groups of women – including housewives, lesbians, migrants, union members, and professional women – have challenged the public–private divide, putting so-called private issues, like childcare, onto municipal agendas (England, 1994a). From the 1980s onwards, and with the aim of making institutions more gender-aware, some feminist activists moved from lobbying city politicians into working in the institutions of government themselves, becoming **femocrats** and forming 'women's bureaus'. In England, for example, the first full standing committee for women was established in 1981 in the Greater London Council. Although relatively short-lived it worked on redistributing financial resources to women. In Toronto, the Toronto Women's City Alliance was formed in 2004 to address the invisibility of issues affecting women and girls on city agendas, particularly in relation to essential services. And Col·lectiu Punt 6 is a feminist cooperative of architects, sociologists, planners, and activists working in present-day Barcelona to rethink cities in order to eliminate gender discrimination.

In addition, since the early 21st century there has also been an increase in the number of female elected officials. Foremost in the public consciousness among these women are those who are mayors, elected to lead cities after decades-long movements for gender equality. They include, for example: Yvonne Aki-Sawyerr in Freetown, Sierra Leone; London Breed in San Francisco, USA; Olivia Chow in Toronto, Canada; Ada Colau in Barcelona, Spain; Iraćí Hassler in Santiago de Chile, Chile; Anne Hildago in Paris, France; Elia Jones in Ferguson, USA; Leila Mustafa, co-chair of the civil council of Raqqa, Afghanistan; Valerie Plante in Montreal, Canada; Claudia Scheinbaum Pardo in Mexico City, Mexico; and Asmaa Rhlalou in Rabat, Morocco. The current executive director of UN-Habitat, Maimunah Mohd Sharif, was previously mayor of the city council of Penang Island, Malaysia. Of course, being a feminist and being a mayor covers a wide range of political positions, from revolutionary to liberal, but also from being relatively safe to unsafe, some of them operating in contexts of **femicide**. It is now the norm for women in elected offices to receive misogynistic abuse, and for some, it is life threatening. For example, it has been impossible to find any mention of Leila Musphafa on the internet dating from after 2021. The silence could be interpreted in many ways, from her no longer being in post to her having been killed for the work she had been doing in reopening schools. Marielle Franco, the first Black openly bisexual woman to be elected as a city councillor in Rio de Janeiro, Brazil, was fatally shot in 2018, assassinated for her outspoken criticisms of police violence.

The issues that the mayors and councillors listed have addressed different aspects of gender inequality and inequity, such as reducing maternal mortality rates; developing basic income policies; introducing child care and taking measures to combat domestic violence; increasing employment opportunities for women and making small-business loans available to them; increasing women's access to essential services; increasing the availability of public transport and introducing 'pink transport' policies providing women-only transport; reducing homelessness and increasing the supply of public housing; improving public safety, with a focus on women's safety; promoting green power, reducing air pollution, planting trees, and providing recycling schemes, waste plants, and clean water; addressing COVID-19; increasing women's engagement in **urban governance**, democratic management of city affairs, and city planning; and introducing gender budgeting informed by gender-disaggregated data collection.

These programmes and policies speak to justice, sustainability, equity, and inclusivity, all of which are implicit in the feminist mandate of women's 'right to the city' (Fenster, 2005). The concept of the 'right to the city' was first introduced by Henri Lefebvre ([1967] 1996) as a revolutionary mode of resistance by urban citizens against capital, the reassertion of use value above exchange value. The right to the city has since been taken up as a slogan, a political programme, and an analytical category. Critiqued by some for its use and 'watering down' in the work of international institutions such as the United Nations, it has been taken up by feminists focusing on the interplay between urban policies and everyday urban life (Boudreau, Boucher and Liguori, 2009). Feminist charters on rights to the city – which are not policies per se but can include demands to be fed into policymaking – include a focus on feminist urban governance and increasing women's access to urban services, including employment. As the topics listed earlier show, they also address how processes of social exclusion and marginalisation, but also of political and social integration, are experienced through the spatialities of urban practices. See the book's e-resource website www.routledge.com/9781032668680 for links to women's-right-to-the-city charters.

Reflection exercise 6.1 asks you to think further about what you would include in a woman's charter for the right to the city.

Reflection exercise 6.1: A women's charter for the right to the city

Imagine you are the newly elected feminist mayor of City X (choose a city with which you are familiar). Your first task is to produce a women's rights charter for the city. Start by thinking about the following questions.

- Who would you involve in writing this charter?
- How would you envision the drafting process taking place?

Now think about what you would include in the charter.

- What are the guiding principles of the charter?
- What aspects of gender inequality and inequity would you address?

(ii) Feminist engagements with global frameworks for urban governance

Feminist engagement with urban policy has extended beyond the metropolis. Issues of **governance**, and specifically urban governance, have also been pursued at the global level, as exemplified in feminists' participation in the Habitat meetings, the New Urban Agenda (NUA), and the SDGs.

Susan Parnell (2016) argues that the global policy focus on the role of cities in the global system and in sustainability can be traced back to the Habitat conferences: I (1976, Vancouver), II (1996, Istanbul), and III (2016, Quito). In 1976 the World Bank's agenda dominated the debate, and hence the focus of both Habitat I and Habitat II was on ' "development in cities" instead of the role of "cities in development" ' (Parnell, 2016, p. 532). This was so even though more progressive voices addressed gender, urban poverty, and sustainability, including those of the feminists Barbara Ward, Harvard economist and founder of the International Institute for Environment and Development, and Caroline Moser, anthropologist and World Bank policy advisor. Feminists and other critical thinkers helped push Habitat agendas to address these pressing issues more critically. The focus of the Habitat conferences moved from 'slums' to shelter and the right to housing and, in Habitat III, to a much more capacious understanding of the right to the city and a concern not only with cities in the global South but with all cities (see Chapter 1, Section 1 (ii) The employment of 'global South' terminology).

It was at Habitat III that the right-to-the-city-based New Urban Agenda policy document was devised, incorporating long-term visions, priorities, and actions for cities. With many feminist interventions involved in its production, the New Urban Agenda has a focus on both poverty and inequality and provides for the mainstreaming of gender issues alongside the SDGs. However, at the time of writing, the extent to which the New Urban Agenda will change the way cities are addressed in global urban policy remains to be seen. While the New Urban Agenda makes cities more visible in development policy, Parnell (2016, p. 539) argues that it is unclear whether cities will come to be seen as

> key pathways in every aspect of sustainable development and not merely vehicles for the promotion of social, economic, or environmental objectives . . . . [The] central problem, working from the premise that [the] city's future is integral to achieving global sustainable development, is to define appropriate post 2015 actions that will lead the world into the urban [A]nthropocene.

Section 2 discusses the SDGs and feminist engagement with their formulation asking how women are addressed within them.

Section 2. The Sustainable Development Goals

In 2015, all 193 United Nations member states adopted the 2030 Agenda for Sustainable Development, comprising 17 SDGs and 169 targets intended to work towards the elimination of poverty and 'a better future for all' (United Nations, 2015). They adopted this agenda with an understanding that it necessitates achieving gender equality and empowering all women and girls: linking women's rights to the SDGs is crucial to the pledge that 'no one will be left behind' (United Nations, 2015) (see Table 6.1). This international framework promotes development goals that are global in their intention, national in their construction, and local in their implementation, and is intended to be an inclusive plan for a sustainable planet. In contrast to their precursors, the Millennium Development Goals (2000–2015), which were criticised as the products of a top-down exercise, the SDGs were developed with input from feminist organisations around the world, including the Women Environmental Programme, the International Women's Health Coalition, and the Women's Environment and Development Organisation (Gabizon, 2016). Feminists had critiqued the Millennium Development Goals not only for their lack of consultation with civil society in the formulation of the goals but also for their omission of gender issues such as violence against women, sexual and reproductive health, and the narrowness and inadequacy of indicators and targets (Antrobus, 2005). For the SDGs, however, feminists played an important role in securing a more inclusive design and reach. The current development goals also differ from the previous ones in that they have established global goals for all countries rather than targeting only global South countries, recognising that poverty and systemic inequalities exist among and within countries (see Chapter 1, Section 1 (ii) The employment of 'global South' terminology).

Although it is widely accepted that the SDGs improve upon their predecessors, their implementation is in tatters, as noted by this candid recognition from the United Nations (2023, p. 2) itself:

> Progress on more than 50 per cent of targets of the SDGs is weak and insufficient; on 30 per cent, it has stalled or gone into reverse. These include key targets on poverty, hunger and climate. Unless we act now, the 2030 Agenda could become an epitaph for a world that might have been.

Progress has been impeded by the climate crisis, COVID-19, new and continuing conflict and wars, a dysfunctional international financing system, a weak global economy, and

Table 6.1 The 17 Sustainable Development Goals

SDG 1 No poverty	End poverty in all its forms everywhere.
SDG 2 Zero hunger	End hunger, achieve food security, and improved nutrition and promote sustainable agriculture.
SDG 3 Good health and well-being	Ensure healthy lives and promote well-being for all at all ages.
SDG 4 Quality education	Ensure inclusive and equitable quality education and promote lifelong learning opportunities for all.
SDG 5 Gender equality	Achieve gender equality and empower all women and girls.
SDG 6 Clean water and sanitation	Ensure availability and sustainable management of water and sanitation for all.
SDG 7 Affordable and clean energy	Ensure access to affordable, reliable, sustainable, and modern energy for all.
SDG 8 Decent work and economic growth	Promote sustained, inclusive, and sustainable economic growth, full and productive employment, and decent work for all.
SDG 9 Industry, innovation, and infrastructure	Build resilient infrastructure, promote inclusive and sustainable industrialisation, and foster innovation.
SDG 10 Reduced inequalities	Reduce inequality within and among countries.
SDG 11 Sustainable cities and communities	Make cities and human settlements inclusive, safe, resilient, and sustainable.
SDG 12 Responsible consumption and production	Ensure sustainable consumption and production patterns.
SDG 13 Climate action	Take urgent action to combat climate change and its impacts.
SDG 14 Life below water	Conserve and sustainably use the oceans, seas, and marine resources for sustainable development.
SDG 15 Life on land	Protect, restore, and promote sustainable use of terrestrial ecosystems, sustainably manage forests, combat desertification, and halt and reverse land degradation and halt biodiversity loss.
SDG 16 Peace, justice, and strong institutions	Promote peaceful and inclusive societies for sustainable development, provide access to justice for all, and build effective, accountable, and inclusive institutions at all levels.
SDG 17 Partnerships for the goals	Strengthen the means of implementation and revitalise the Global Partnership for Sustainable Development.

Source: Adapted from United Nations (2015)

unsustainable debt in the countries of the global South (see Chapter 5, Section 2 The gendered and racialised social impacts of COVID-19 on everyday urban life).

Feminist critiques provide further insight into the stumbling blocks impeding the progress of the SDGs. Valeria Esquivel (2016) and Shahra Razavi (2016) have questioned how the SDG vision can be realised without substantive changes to the dominant economic model or structural power relations. Rather than challenging the dominant model of economic growth, the SDGs take the current system as a given without acknowledgment of structural inequalities or root causes of poverty (Esquivel and Sweetman, 2016). Hence, the SDGs do not aim to transform the power relations that maintain global inequalities, including colonialism's legacy of discrimination and disparity. The 2030 Agenda also encourages private-sector investment and innovation, particularly public–private partnerships, to achieve the SDGs. However, blended financing for the SDGs, a so-called innovative development policy that uses public development assistance funds to leverage private finance, leads to resources being shifted away from the poorest countries and sectors towards profitable investments in middle-income countries (Murray and Spronk, 2019).

As outlined in Table 6.2, each SDG has a set of ambitious targets and indicators, but these are not unproblematic. The goals on water (6), energy (7), infrastructure and industrialisation (9), sustainability (12), oceans (14), ecosystems (15), and the implementation of the goals have no gender-specific indicators despite women's centrality to many of them. Despite the relevance of gender across all SDGs, given that gender inequalities intersect with other inequalities, gender is reflected in only 52 of the 231 SDG unique indicators, just over 20 percent (UN Women and United Nations Department of Economic and Social Affairs, 2023). This omission is at odds with the recognition by the UN that for the SDGs to succeed gender considerations need to lie at their heart. Neither are the rights of Indigenous nor queer groups explicitly addressed in the SDGs. These omissions constitute a glaring exclusion contradicting the promise to 'leave no one behind': both Indigenous and queer people have repeatedly been left behind by national and international development initiatives. For example, recent work by queer organisations and activists in peace processes demonstrates how armed forces specifically targeted LGBTQIA2S+ people during internal armed conflicts as a tactic to assert dominance and maintain control in communities (Colombia Diversa, 2020).

Other problematic issues raised by the SDGs are addressed in the following sections. Reflection exercise 6.2 asks you to engage further with a few of the SDGs and to ask yourself how gender considerations are crucial for their ability to achieve their targets.

Reflection exercise 6.2: Gender and the Sustainable Development Goals

Review the complete list of SDGs in Table 6.1. You may also want to look at the accompanying targets and indicators (available on the UN website, which is linked on the book's website www.routledge.com/9781032668680). Consider two or three of the goals identified as failing to address gender.

- How is gender implicated in each of these goals, even when it is not explicitly mentioned?
- What are the consequences of overlooking the gender implications of the goals in their implementation?

Section 3 turns to an examination of the synergies between SDG 5 and SDG 11. Critically, neither officially addresses the other in its targets although their fortunes are inextricably linked (see also Mustafa *et al.*, 2024).

Section 3. The synergies between SDG 5 and SDG 11

The 2030 Agenda for Sustainable Development introduced the stand-alone goal SDG 5 'Achieve gender equality and empower all women and girls' to address gender and women's rights in response to feminist criticisms of the Millennium Development Goals (as noted earlier). Table 6.2 lists the SDG 5 targets. They are broad in scope and address the issues that were overlooked in the Millennium Development Goals. For example,

Table 6.2 SDG 5 'Achieve gender equality and empower all women and girls': targets*

5.1	End all forms of discrimination against all women and girls everywhere.
5.2	Eliminate all forms of violence against all women and girls in the public and private spheres, including trafficking and sexual and other types of exploitation.
5.3	Eliminate all harmful practices, such as child, early, and forced marriage and female genital mutilation.
5.4	Recognise and value unpaid care and domestic work through the provision of public services, infrastructure, and social protection policies and the promotion of shared responsibility within the household and the family as nationally appropriate.
5.5	Ensure women's full and effective participation and equal opportunities for leadership at all levels of decision-making in political, economic, and public life.
5.6	Ensure universal access to sexual and reproductive health and reproductive rights as agreed in accordance with the Programme of Action of the International Conference on Population and Development and the Beijing Platform for Action and the outcome documents of their review conferences.
5.a	Undertake reforms to give women equal rights to economic resources as well as access to ownership and control over land and other forms of property, financial services, inheritance, and natural resources in accordance with national laws.
5.b	Enhance the use of enabling technology, in particular information and communications technology, to promote the empowerment of women.
5.c	Adopt and strengthen sound policies and enforceable legislation for the promotion of gender equality and the empowerment of all women and girls at all levels.

Source: Adapted from United Nations (2015)

Note:

* Targets that address outcomes (circumstances to be attained) use numbers, whereas targets that address means of implementation (how targets will be achieved) use letters

they call for the elimination of violence against women and girls, the end of discriminatory gendered practices, and the provision of access to sexual and reproductive health services.

SDG 11 'Make cities and human settlements inclusive, safe, resilient and sustainable' is a significant milestone recognising not only the importance of cities in establishing sustainable development agendas but also the role of sub-national or urban governance in implementing the SDGs (Parnell, 2016). As urban scholars, practitioners and activists have long been arguing, urbanisation processes cannot be overlooked when addressing global problems (Kaika, 2017). The aim of SDG 11 is to focus more policy attention on the urban areas that are already the most inhabited places on earth, currently accommodating 55 percent of the world's population (United Nations, 2023). With over one billion urban residents living in 'slum'-like settlements and another two billion expected to join them over the next 30 years, urbanisation will exacerbate current levels of urban poverty and inequality (United Nations, 2023). Hence, 'slum' upgrading and reducing cities' environmental impacts lie at the heart of the SDG 11 targets (see Table 6.3).

The SDGs were intentionally designed as interconnected goals. The synergies between SDG 5 'Gender equality' and SDG 11 'Sustainable cities and communities' are particularly salient given that the growth of the urban population, stemming largely from natural increase but also rural-to-urban migration, leads to more women than men living in urban areas. Not only do women live longer than men, but women now also migrate in larger numbers than men. Gendered inequalities in urban areas are spatialised and present in divisions of labour, different levels of mobility and access to urban infrastructure,

Table 6.3 SDG 11 'Make cities and human settlements inclusive, safe, resilient and sustainable': targets

11.1	By 2030, ensure access for all to adequate, safe, and affordable housing and basic services and upgrade slums.
11.2	By 2030, provide access to safe, affordable, accessible, and sustainable transport systems for all, improving road safety, notably by expanding public transport, with special attention to the needs of those in vulnerable situations, women, children, persons with disabilities, and older persons.
11.3	By 2030, enhance inclusive and sustainable urbanisation and capacity for participatory, integrated, and sustainable human settlement planning and management in all countries.
11.4	Strengthen efforts to protect and safeguard the world's cultural and natural heritage.
11.5	By 2030, significantly reduce the number of deaths and the number of people affected and substantially decrease the direct economic losses relative to global gross domestic product caused by disasters, including water-related disasters, with a focus on protecting the poor and people in vulnerable situations.
11.6	By 2030, reduce the adverse per capita environmental impact of cities, including by paying special attention to air quality and municipal and other waste management.
11.7	By 2030, provide universal access to safe, inclusive, and accessible green and public spaces, in particular for women and children, older persons, and persons with disabilities.
11.a	Support positive economic, social, and environmental links between urban, peri-urban, and rural areas by strengthening national and regional development planning.
11.b	By 2030, substantially increase the number of cities and human settlements adopting and implementing integrated policies and plans towards inclusion, resource efficiency, mitigation and adaptation to climate change, [and] resilience to disasters, and develop and implement, in line with the Sendai Framework for Disaster Risk Reduction 2015–2030, holistic disaster risk management at all levels.
11.c	Support least developed countries, including through financial and technical assistance, in building sustainable and resilient buildings utilising local materials.

Source: Adapted from United Nations (2015)

exposure to violence, climate change vulnerabilities, and participation in governance. These gendered insecurities intersect with other forms of discrimination based on class, religion, race, age, sexuality, and identity, resulting in cities being gendered in specific ways (Varley, 2015; Chant and McIlwaine, 2016; Razavi, 2020b). Therefore, addressing these gendered insecurities is key to achieving gender equality in cities and to building sustainable cities.

The international commitments to the SDGs, in particular SDG 5 and SDG 11, have resulted in certain advances, such as the decline in female genital mutilation and child marriage, women's increased political representation, and strengthened efforts towards disaster risk reduction (United Nations, 2023). However, this limited progress does not hide the fact that achieving gender equality remains a very distant goal. In 2023, progress on SDG 5 is, according to the UN itself, 'way off track', with only two Goal 5 indicators being 'close to target' (UN Women and United Nations Department of Economic and Social Affairs, 2023, p. 26). And indications are that by 2030, over one billion women and girls will still live in slums or slum-like neighbourhoods (pp. 4–5).

GenUrb example 6.2 describes the research undertaken by the GenUrb project on SDG 5 and SDG 11.

GenUrb example 6.2: Research on SDG 5 and SDG 11

One objective of the GenUrb project was to understand what impact the new global framework of the SDGs is having on the lives of women living in neighbourhoods in the following cities: Cochabamba, Bolivia; Delhi, India; Georgetown, Guyana; Ibadan, Nigeria; Ramallah, Palestine; and Shanghai, China. Paying particular attention to the implementation of SDG 5 and SDG 11, members of GenUrb researched what was happening in each city in relation to the SDGs and the extent to which SDG 5 and SDG 11 are being addressed in practice, as well as their relevance or irrelevance to women's lives, and particularly how they mediate gender relations and women's rights.

Interviews were conducted over a five-year period to allow for a longitudinal study, with interviews undertaken in 2018–19 and in 2022–23 to explore how the implementation of SDG 5 and SDG 11 had developed over time and how policies had travelled between urban sites. A total of 143 interviews were conducted with policy shapers – the politicians, bureaucrats, practitioners, activists, and academics who influence and engage with urban policy – with the number of interviews fluctuating in each city due to challenges such as the difficulty of accessing interviewees due to elections, protests, strikes, and COVID-19 restrictions. Close attention was paid to the gendered implications of the SDGs' local adaptation process, to the historical specificities of coloniality, and to everyday micropolitics in each city.

Our objective was to explore what feminist comparative analysis can reveal about the transformative potential of the national and local adaptations of the 2030 Agenda for Sustainable Development. We wanted to understand whether and how urban policies incorporate the SDGs, how they might mediate gender relations and women's rights to the city, and how the latter are reconstituted and transformed in the context of urbanisation in the global South.

By Nasya S. Razavi

In Section 4 we explore some of the challenges to implementation of the SDGs at the city level.

Section 4. Local adaptation of the Sustainable Development Goals

To enable progress towards the 2030 Agenda and the SDGs to be monitored, a set of 231 indicators were linked to the 17 goals and their targets. The complex monitoring process involves different actors at global, national, and local levels (see Chapter 8, Section 4 Monitoring and evaluation). Reporting on the SDGs entails coordination among United Nations agencies, global consultancy firms, regional organisations, and member states, typically through Voluntary National Reviews. Given that cities are key to driving progress towards achieving the SDGs there is a need to adapt global frameworks for implementation at the local level in a process called 'localisation'. Sandra Valencia *et al.* (2019) and Gloria Novovic (2021, 2022) sum up the conditions that need to come together to make localisation successful:

- political buy-in and advocacy from politicians to facilitate cross-departmental government collaboration;
- policy coherence, aligning the SDGs to established national and local development priorities;
- robust coordination mechanisms to include a wide variety of actors, including civil society;
- effective communication and information flows to 'socialise' the SDGs;
- availability of data to support monitoring and evaluation, especially data disaggregated by gender, class, and race, for example, as well as by location (see Chapter 10, Section 1 Mainstream and feminist conceptions of data);
- and budgetary capacity to fund work towards achieving the SDGs and to track progress.

Recent studies, including GenUrb's own research, have revealed that localisation is a major challenge for cities (Croese and Duminy, 2023; Horn and Grugel, 2018; Valencia *et al.*, 2019; Novovic, 2021) (see also GenUrb example 6.2). As noted earlier, localisation depends on the local political context, levels of decentralisation, and national priorities; together, these explain why the SDGs have not been universally adopted and applied to development policies (Horn and Grugel, 2018). Technical roadblocks abound due to difficulties in producing data, disaggregating data, lack of experience measuring progress towards international targets, unclear methodologies, and inconsistent standards (Valencia *et al.*, 2019; Agencia Andaluza de Cooperación Internacional para el Desarrollo, 2021). Further, the quantitative nature of some indicators does not mean they are objective or straightforward; rather, they can be subject to contradictory interpretation at best, and, at worst, can fall prey to abuses of power (Sultana, 2018a). Indeed, as Valeria Esquivel (2016, p. 18) remarks:

> the goals and targets give the impression that progress is possible [on] all fronts – without tensions or inconsistencies between targets . . . . The underlying assumption behind the . . . template is that means of implementation (policies) are immune to power imbalances (politics).

Limited funding for Agenda 2030 and the SDGs has also significantly limited the process of localisation. Countries have adapted Agenda 2030 in a manner that matches the goals to their own established national priorities, thereby claiming ownership over the SDGs. However, the financial mechanisms tied to the existing development structure have allowed development agencies and donors to finance their own priorities (Novovic, 2021). This brings us back to the point that, in failing to address the dominant economic model, the SDGs risk failing to accomplish the structural changes needed to advance the goals.

GenUrb example 6.2: Understanding local adaptation of the Sustainable Development Goals

In GenUrb's research on the local adaptation of the SDGs one goal was to understand what policy shapers consider to be the main problems or priorities in the urban context and how these have changed in the period prior to and during the implementation of the SDGs. We were also interested in how policy shapers understand sustainability and inclusive urban policy and in how they view the links between sustainability, women's experiences, and gender equality in the city. Engaging both policy shapers and grassroots women helped us identify the gaps between urban policy rhetoric and the realities of women's everyday lives.

In our engagement with policy shapers on local adaptation of the SDGs, and specifically of SDG 5 and SDG 11, it became apparent that awareness of and knowledge about the SDGs was uneven. This unevenness stemmed partially from the fact that although it is national governments that agree on and sign up to the SDG framework, it is at sub-national levels that policies are implemented. In some cities there was neither a great awareness of the SDGs nor the finances or political will to implement them. In most instances, governments adapted the SDGs to existing frameworks and interpretations. Therefore, the policies that reached city-level institutions had different degrees of divergence from the SDGs. At the same time, non-governmental organisations in some cities received international funding to work on SDG-directed projects, and staff were very well versed in the goals, targets, and indicators. Localisation had not been a value-free process: significant flows of money and power relations were at play.

Overall, we found little evidence of local implementation of SDGs at the municipal level. We also found a disconnect between urban priorities and grassroots women's lived experiences. During the Cochabamba CRT policy-shaper interviews, for example, I engaged with Cochabamba's Department of Urban Governance. After explaining the GenUrb project, I was pointed to the only woman in the department and told to interview her, the assumption being that, as I was a woman researching gender and women's issues, the only person who would be able to answer my questions was another woman, even though she was not responsible for gender policy or programming specifically. This civil servant told me the story of the upgrading project for the city's marketplaces. Cochabamba is full of markets, open-air stalls, and street vendors. Market traders and street vendors are part of the informal sector, which forms a significant part of the Bolivian economy and is dominated by women. In the markets women fulfil several roles: they generate family networks with other vendors, resolve conflicts, form friendships, breastfeed, and take care of school-aged children out of school hours. The municipal government does not facilitate these activities. The markets have few washrooms; there are no facilities for children; and there are no official resources in case of an emergency. Instead, there is a drive to sanitise these spaces and move vendors off the streets as part of a city beautification process. The city ended up building a structure in a location away from existing markets to house a fruit-and-vegetable market (part of a pattern of building visible infrastructure projects to garner votes). But the new market sits half empty, mostly because there is not enough foot traffic to attract customers, and the vendors prefer to stay where they earn more. The city is not investing in or addressing any of these issues, even though by doing so they might alleviate the women vendors' burdens and anxieties, leading to a disconnect between urban policy and projects and the realities of women's everyday lives. Urban policy, whether connected to the SDGs or not, does not necessarily address women's specific needs.

By Nasya S. Razavi

In Section 5 we address the issue of connecting research on the SDGs to policy and why this is important.

Section 5. Connecting research and policy

There is often a disconnect between research and policy that can be framed as a communication problem: researchers are not effectively conveying their messages to policymakers, or policymakers fail to understand these messages (Cáceres *et al.*, 2016). However, this stance simplifies the policy creation and implementation process that involves a myriad of multi-sector actors and is shaped by complex political and social dynamics (Cáceres *et al.*, 2016). It is for this reason that national research agencies have been placing much more emphasis on 'research impact' or 'knowledge mobilisation' in research-grant applications, requiring researchers to outline concrete plans that go beyond project websites and academic conference presentations and that show how their research will reach with, connect, and have an impact upon their intended audiences (see Chapter 19, Section 1 Knowledge mobilisation). Even with such plans in place it can prove difficult to connect research with specific audiences, and this is certainly the case with connecting research to policy. The Canadian federal research funding agency, the Social Sciences and Humanities Research Council, even warns researchers of the potential need to scale down their ambitions about the extent to which their research will bring about policy change, given the barriers that exist to enabling this connection.

Different groups of knowledge makers and policy shapers – academics, activists, planners, policymakers, civil-society organisations, grassroots organisations, non-governmental organisations, and others – all too often work within their own circles, often lacking the connections needed to reach other groups. To counteract this situation researchers are increasingly involving these stakeholders in their grant applications from the very first stages of thinking about a proposal and ensuring ways of keeping these groups or individuals connected through all the stages of research. Still, as Farhana Sultana (2018a) points out, scholars' attempts to engage critically in development challenges can be blocked by development institutions that are either resistant or unresponsive to change. One obvious example is the choice to name the SDGs using the term 'sustainable development', although the term has been heavily criticised since its introduction in the 1980s, and its adoption as a buzzword encompasses multiple conflicting definitions (Sultana, 2018a). Other common challenges include working across different languages, including *disciplinary* languages, working to different timetables, and ensuring that links are being made to the most relevant stakeholders to enable research to make an impact. The time and resources needed to keep these links dynamic and workable should not be underestimated (see Chapter 9, Section 1 Research partnerships). Despite these challenges, critical research (and activism!) can influence and have a positive impact on policymaking, as was demonstrated, for example, through the improvement of the SDGs after the Millennium Development Goals were criticised for being too limited in the targets and indicators used to measure progress.

In planning your own research process, you may have considered the impact or contribution your research may have on shaping policy. Perhaps you have even encountered such a requirement through grant applications or discussions with peers on how your work might engage the broader public (see Chapter 19, Section 1 Knowledge mobilisation). A particular policy outcome may even be a condition of a grant or award. This requirement is not easily squared with the difficulty of measuring relevance in research. Further, the debates around relevance do not always account for academic pressures that can distort which research outcomes are most valued by institutions (Staeheli and Mitchell, 2005). For example, academic responsibilities often prevent sustained engagement with policymakers, and this engagement is not necessarily rewarded in job appointments or promotions in academia.

Thinking through the policy relevance of your research and practising your ability to communicate research findings to broader audiences that can drive meaningful action are useful skills to hone for a career in academia and beyond. Writing policy differs from academic writing because its main aim is to offer concise content on a particular issue and provide practical recommendations for decision-making purposes. Policy writing typically omits theoretical arguments and distils different viewpoints on a focused topic to inform a series of practical recommendations. Given that the audience rarely consists of subject-matter specialists, it is essential to communicate using simple and straightforward language that can be understood easily and quickly. Being succinct is imperative. Policy writing follows a logical structure, and each point must be clear and relevant to the issue at hand.

Policy documents most commonly take the form of proposals, strategy documents, consultation reports, briefs, and memos. Policy memos and briefs concisely outline the rationale for choosing a particular policy alternative or course of action in a current policy debate, providing objective summaries of relevant research, suggesting possible policy options, or arguing for particular courses of action or outcomes. They are written for decision-makers who influence and guide public policy. A policy memo is typically shorter than a brief, usually ranging from a single page to ten pages in length. Reflection exercise 6.3 prompts you to write a policy memo.

Reflection exercise 6.3: Crafting a policy memorandum

Pick one contemporary issue related to gender and urbanisation in one of the GenUrb cities or a city in which you conduct research that will be the focus of the policy memo. Your memo should be addressed to policymakers in a local or international organisation.

Explain the policy challenge and its importance. Provide some context on the policy issue, and make a case about why the issue is important and warrants action. Explain why the issue should matter to policymakers.

- What is the issue?
- Where does it take place?
- Who are the key actors involved?
- What are the implications and for whom?
- What is at stake?

Provide concrete advice on how to meet this challenge. The memo should analyse current policies designed to solve this problem and make specific recommendations. Provide two to three feasible policy options to meet the challenge described in your memo, and make the case for your recommended 'best option'. Explain the following:

- the benefits and opportunities for policy interventions;
- the implications of not addressing the issue at a policy level;
- the policy options that can be adopted.

Format

The policy memo should be written in plain and clear language and should be about two pages in length (a maximum of 750 words). Use 12-point Times New Roman font and 1-inch or 2.54-cm margins. Use a generous amount of white space, with clear headings and subheadings (if needed). The policy memo should include:

- a title and author;
- an executive statement providing a summary of the memo;
- keyword definitions;
- an objective that outlines the issue and explains its significance and urgency;
- contextual information that includes a summary of relevant facts, describes and analyses the issue using existing research, and highlights the benefits and opportunities of addressing the issue;
- a brief consideration of the implications, i.e. what could happen if the issue is *not* addressed;
- two to three policy recommendations, with an assessment of the challenges/constraints associated with each, and an indication of what, in your opinion, is the best option;
- works cited.

Section 6. Summary

This chapter has introduced the reader to feminist engagements with urban policy, providing a brief account of feminist efforts to address urban social justice through the gendering of urban policy and engagement with the SDGs. It has described the SDGs and feminist concerns about them, and addressed the synergies between SDG 5 on women and SDG 11 on cities. Local adaptation of the SDGs has been considered, and the chapter concludes with an overview of the challenges faced in engaging policy shapers with your research, focusing on how to write policy memos to gain their attention.

Part III

Feminist research standards

7 Feminist research ethics

Linda Peake and Wiley Sharp

Learning objectives
After completing this chapter, readers should understand

- the difference between philosophical ethics and professional ethics codes;
- the history of professionalised research ethics;
- the feminist conceptions of care ethics, intersectionality, positionality, and decolonisation;
- the challenges and importance of practising care across difference in feminist research;
- and the ways in which GenUrb has practised a feminist research ethics.

Further, readers should be able to

- anticipate ethical issues in their own research;
- develop strategies to mitigate ethical problems;
- and apply a care ethics to their own research.

This chapter addresses research ethics and the understanding of ethical research as that which values the well-being of participants and researchers alike. It begins by outlining the emergence of the principles and values on which research ethics are based. It then turns to a feminist ethics, introducing different schools of ethics, explaining why a feminist ethics has focused on the issue of care, and articulating a care ethics grounded in intersectionality, positionality, and decolonisation. Finally, it explores how this care ethics can be applied to feminist research, employing case studies from GenUrb to emphasise the nuances and challenges of conducting ethical feminist research.

Section 1. Ethics and research

Ethics is the lived process of answering the question: *how ought we be?* Also known as moral philosophy, ethics attempts to clarify the question by exploring the values, principles, and duties that help us live an ethical life. In the context of research, ethics takes a narrower scope: *how ought we create knowledge?* **Research ethics** can refer to individual approaches to the production of knowledge as well as the creation of guidelines, rules, or principles set by professional organisations. In other words, 'research ethics' can refer to both the practice of ethical deliberation and action during the research, which is dependent upon the particular ethical philosophy a researcher adopts, and the written or unwritten rules that govern the profession of knowledge creation (see Chapter 8, Section 2 Ethics policies). These diverging notions of research ethics underscore the fact

DOI: 10.4324/9781032668727-11

that conducting ethical research involves the negotiation of different, often competing, ethical values, such as the conflicts between professional research ethics guidelines and the researcher's ethical duties towards participants.

Research has not always been ethical in its practice. The main principles underlying professional paradigms of research ethics can be traced directly to the barbarous experiments conducted on thousands of prisoners by Nazi scientists during World War II (Bell, 2014). In 1947, during the Nuremberg War Crime Trials that followed the war, the Nuremberg Code was drafted as a set of standards for judging physicians and scientists who had conducted biomedical experiments on concentration camp prisoners. The code includes a range of principles such as the requirement of fully informed, voluntary consent, without any kind of coercion, risk–benefit evaluation, and avoidance of suffering. This code was followed in 1964 by the *Helsinki Declaration*, a set of ethical principles regarding human experimentation developed for the medical community by the World Medical Association (n.d.) and in 1979 by the *Belmont Report* for the USA (US Department of Health and Human Services, 2018). These provided the basis of many later codes intended to assure that research involving human subjects would be regulated and conducted in an ethical manner. From these codes and declarations, the ethics principles of various national research councils, such as the Social Sciences and Humanities Research Council of Canada (SSHRC) and the Economic and Social Research Council (ESRC) in the UK, evolved. A number of principles and values are common to these professional paradigms of research ethics (Carpenter, 2017; Ravitch and Carl, 2021):

- respecting the rights and dignity of individuals and groups, aiming to maximise benefit for individuals and society and to minimise risk and harm;
- ensuring that participation is voluntary and appropriately informed wherever possible;
- conducting research with integrity and transparency, with participants' data remaining private and confidential;
- clearly defining lines of responsibility and accountability;
- securing the independence of the research from vested interests and making explicit any conflicts of interest that cannot be avoided.

The legacy of unregulated and unethical research explains why the primary function of research ethics has been to protect research participants and researchers from harm. Taking these points into consideration, Reflection exercise 7.1 asks you to consider how 'harm' can be understood and how it can relate to research in the social sciences.

Reflection exercise 7.1: Values in feminist research

The Hippocratic maxim 'do no harm' is one of the most common expressions of human dignity in professional ethical conduct. In the biomedical model it is true that the participant can be at risk of harm from the researcher, but in the social sciences this asymmetrical power relation is not so stark and may even be reversed. Reflecting on this difference, answer the following questions.

- How do you define harm in relation to your own research?
- In terms of your research, to what extent could a focus on not doing harm lead to risk aversion (at the expense of risk awareness)?

- For how long after the official end of your research are you, the researcher, responsible for no harm being done to participants?
- Think of other values that you may consider relevant to your research (such as respect for cultural difference, collegiality, trust, and empathy). Would you prioritise these above causing harm? How would you formulate these values as ethical principles?

How are these ethical principles applied to research? While contemporary university ethics boards, often operating within a neoliberal liability framework, may frame research ethics as a technocratic exercise guided by professional ethical codes, the practice of ethics requires the researcher to be an active decision-maker, where what it means to 'do the right thing' may not always be obvious (see Chapter 8, Section 2, Ethics policies). Rather than assuming, as do university ethics processes, that issues about ethics can always be determined in advance, in practice guidance is needed to know how to deal with the complex relational and reflexive nature of involvement in ethical issues that occur throughout the research. As Rosalind Edwards and Melanie Mauthner (2012, p. 25) observe: 'Ethics is about *how* to deal with conflict, disagreement and ambivalence, rather than attempting to eliminate it'. This conflict occurs on multiple levels: between ethical paradigms, between actors, and within oneself. In this way, ethical questions underlie every part of the research process, from conceptualisation, fieldwork, and analysis to writing and dissemination of research results.

The following section takes up the ways in which feminist scholars have addressed questions of ethics and especially the attention they have paid to a feminist ethics of care.

Section 2. Towards a feminist ethics

Where do you begin in determining your ethical stance to your research? There are three major schools of an ethical philosophy, each resulting in a different approach to research ethics (see Edwards and Mauthner, 2012; see also Kvale, 2013).

- According to a *deontological* (or duty-based) approach, researchers have a categorical duty to follow principles such as honesty, justice, and respect. A deontological approach suggests that ethical research is that conducted in accordance with ethical duties, and it evaluates acts based on whether they are consistent with these duties. For example, if any act is not honest, just, or respectful, it is judged to be unethical, regardless of its intent or ends.
- A *utilitarian* approach judges ethical value based on the outcome of an act. Unlike the deontological approach, which values an array of duties, a utilitarian approach suggests that ethical research is that which maximises the total ethical good, even if it also causes harm. In this view, a dishonest, unjust, or disrespectful act, such as deception, may be considered ethically sound if it creates more good than harm.
- The third major ethical approach, *virtue ethics*, emphasises not the moral content of acts but the actor. By shifting the emphasis from evaluating acts to cultivating the capacities of the actor, it seeks to avoid the rigidity of the deontological approach and the muddiness of the utilitarian approach. This virtue ethics approach suggests that ethical research is that conducted by a researcher who has cultivated deep-seated

capacities such as honesty, solidarity, trustworthiness, and reflexivity. Accordingly, when a researcher lives sincerely in accordance with ethical virtues, they will conduct ethical research.

The deontological, utilitarian, and virtue-ethics models represent three ideal forms of ethical philosophy. Practising ethical research is rarely so clear-cut as following a single paradigm; rather, it necessitates understanding each approach and applying them with fidelity to the context of the research. As Sarah Jane Banks *et al.* (2015, p. 108) argue, there is a 'dialectical tension between impartial principles and rules and the responsibilities that arise from relationships of trust and care and a commitment to working for a better world'.

Feminist philosophers of ethics have drawn from these three schools to develop an ethical paradigm grounded in social reproduction and its virtuous counterpart, care: the everyday practices of attending to the emotional, material, social, and spiritual needs of other beings, usually – but not always – those close by. Edwards and Mauthner (2012) have outlined the characteristics of care within an ethical framing.

- Care is a moral activity. It is the 'activity of caring', as opposed to a set of principles.
- Care, rather than being formal and abstract, is a characteristic of specificity and context. Care thus puts an emphasis on contextual reasoning; it is contexts that define moral problems and resolutions.
- A focus on care puts an emphasis on responsibilities and relationships to others and ourselves (rather than rules and rights). The self is thus understood as self-in-relation i.e. a dialogic as opposed to an autonomous view of the self. Care not only values and attempts to maintain connections among individuals but also shows that there is an ethical importance to these personal connections.
- Care can be compromised. It can lead to the sacrifice or loss of self, a failure to recognise the autonomy of the other, or over-identification with the other. Hence, care requires acceptance of inevitable dependencies as well as a recognition that harm can be caused when caring connections are broken.

By positing care as a core ethical value, feminist philosophers seek to correct masculinist bias in ethical philosophy and assert that aiding other beings is an act with intrinsic moral value. The developmental psychologist Carol Gilligan and the political theorist Joan Tronto were some of the first to conceptualise these feminist ethics. For Gilligan (1982, p. 2), women deliberate in a 'different [ethical] voice' from men, and for Tronto (1993, quoted in Edwards and Mauthner, 2012, p. 22), 'caring work primarily undertaken by women is excluded from mainstream moral and political philosophy and theorizing because it is regarded as instinctual practice rather than willed action based on [ethical] rules'. These early interventions were important efforts in pushing back against masculinist notions of ethical deliberation and action, but their essentialist articulations of the virtue of care reified the very gendered division these theorists wished to critique.

In the last few decades, feminist theorists have built upon this early work and extended it beyond feminised care labour in the household (see, for example, Hill Collins, 2000; Held, 2005; Robinson, 2011; Engster and Hamington, 2015). These interventions have worked to de-essentialise the practice of care, emphasising that care is a collective duty that has been unevenly and unjustly distributed by the institutions and practices of cisheteropatriarchal oppression. One example is the work of feminist theorist Selma Sevenhuijsen (1998), who argued that an expanded ethics of care has the potential to transform our ideas of ethical deliberation. For Sevenhuijsen (1998, p. 57), ethical quandaries become 'no longer primarily conflicts between the legal claims of discrete, coherent persons; instead, the starting point for [ethical] deliberation lies in the experiencing of [ethical] dilemmas in individual and collective contexts'. In this way, a **feminist ethics of care** takes the historically feminised practice of care labour as the catalyst to reimagine ethical deliberation and action across all areas of human life. In other words, feminist care ethics articulates a feminist answer to the question of how to live – that is, a feminist way of life.

A feminist ethics of care dovetails with other central tenets of contemporary feminist thought, namely, positionality and intersectionality, as well as those of decolonisation. These ideas emerged from the dialectic between feminist theory and political action, particularly in response to the question of difference. Difference within the category of woman has long haunted feminist theory and politics because the mainstream feminist movement in the global North was built upon colonial notions of gender. Historical processes of capitalist colonisation rendered womanhood as exclusively white by imagining colonised peoples not just as inferior men and women but as altogether less-than-human beings (Lugones, 2007, 2010; See also Chapter 3, Section 3 Decolonising translation). In the 1970s and 1980s, Black and Latinx feminists took this history to task, grappling with their exclusion and erasure from the mainstream feminist movements in the global North and analysing how colonial axes of difference further marginalise women of colour (Combahee River Collective, [1979] 2017; Moraga and Anzaldúa, 1981; hooks, 1981; Davis, 1981; Lorde, 1984).

One extraordinarily influential concept that emerged from these debates was Kimberlé Crenshaw's (1991, p. 1245) notion of **intersectionality**: 'the ways in which the location of women of color at the intersection of race and gender makes our actual experience of domestic violence, rape, and remedial reform qualitatively different than that of white women'. Women-of-colour feminists and others have taken up intersectionality to illuminate how power functions across difference, creating unique forms of violence and vulnerability for those positioned at the intersection of axes of social difference (Hill Collins, 1991, 2000, 2019; McCall, 2005; Dill and Zambrana, 2009; Mollett and Faria, 2013, 2018; Nash, 2019).

Intersectionality builds upon the feminist concept of positionality: the notion that what can be known is shaped by the location of the knower. The idea that knowledge is differentially partial according to the location of its production emerged from feminist engagements with the relationship between difference and knowledge, particularly in the standpoint epistemology developed by feminist philosophers Sandra Harding (1986) and Donna Haraway (1991; see also Benhabib, 1992; Young, 1997; and Chapter 11, Section 4 (ii) Positionality). Standpoint theory was quickly taken up in the social sciences, including urban studies, by feminist researchers interested in examining the gendered exclusions of academic canons, research methodologies, and professional discourse (England, 1994b; Katz, 1994; Kobayashi and Peake, 1994). In short, as Gillian Rose (1997, p. 313) argues:

> The feminist task becomes less one of mapping difference – assuming a visible landscape of power with relations between positions ones of distance between distinctly separate agents – and more one of asking how difference is constituted, of tracing its destabilizing emergence during the research process itself.

In the last two decades, it has become increasingly common for feminist researchers to critically engage the constitution of social difference through the research process, and, in turn, address how this constitution delimits the knowledge we produce. By addressing our own positionalities, researchers can assemble our partial knowledges in acts of solidarity that can work towards dismantling the gendered and racialised exclusions of the academy (see Mollett and Faria, 2018).

In this way, intersectionality and positionality orient the question of difference, knowledge, and power towards *action*. One way that feminists have taken action to address the injustices of gendered and racialised difference is the praxis of decolonisation: the dismantling of the systems of power that distribute vulnerability and violence according to hierarchies of colonial difference (see Chapter 2, Section 1 Decolonising knowledge production in feminist urban research). Decolonising knowledge production sheds light on the enduring operation of racial and ethnocentric frameworks of colonial knowledge systems and their impacts on the organisation of the social world. It requires working towards research practices and relationships that incorporate concepts and methodologies from worldviews and intellectual traditions that have historically been excluded or marginalised in colonial knowledge frameworks and contributing to furthering the self-determination of colonised peoples.

Conceptions of decolonisation differ, especially between scholars working in 'postcolonial' nations and those working in states where settler colonialism is ongoing. Most agree, however, that decolonisation is constituted by the collective effort to undermine the material, social, and epistemological bases of colonisation, as well as ongoing work to revitalise Indigenous polities, languages, and cosmologies (Lugones, 2010; Tuck and Yang, 2012; Simpson, 2017; de Leeuw and Hunt, 2018; Tamale, 2020). As the feminist philosopher María Lugones (2010, p. 754) writes:

> One does not resist the coloniality of gender alone. One resists it from within a way of understanding the world and living in it that is shared . . . . Communities rather than individuals enable the doing; one does with someone else, not in individualist isolation.

Thus, decolonisation is not simply a *politics* but an *ethics*: a way of living with and caring for others that is grounded in an historical understanding of the relationships between difference, knowledge, and power.

A feminist research ethics informed by the concepts of positionality, intersectionality, and decolonisation grapples with the challenges of practising care across difference at the same time as it recognises care as an essential arena to challenge intersecting axes of oppression. It asserts that everyday relationships, emotions, and embodiment fall squarely within the domain of ethical deliberation and action. A feminist approach suggests that ethical research is research conducted with care for participants and researchers alike, emphasising the importance of good relations between all parties that account for differences in emotions, embodiments, ontologies, and histories. Banks *et al.* (2015, p. 100) point out that this means that we should understand:

> [a] moral agent not just as an impartial deliberator but also as an embedded participant with situated and partial relationships, responsibilities, values and commitments

> that frame and constrain ways of seeing, judging and acting in particular situations. Thus the 'ethical' is present in ways of being as well as acting, and in relationships and emotions, as well as conduct.

In the next section, we go on to explore the complexities of applying feminist ethics to research.

Section 3. Feminist ethics and research

When we consider the ethics of the practice of research, we can understand ethical research as that which values the well-being of participants and researchers alike, working to actualise human flourishing not just through research outcomes but through the very practice of research itself. Thus, a feminist research ethics centres the critical practice of care: engaging in reciprocal, accountable relationships across difference, recognising the limitations of our ways of knowing and the power of our actions at the same time as we work, carefully, not just to better understand intersectional structures of oppression but to actively dismantle them. Thus, applying care ethics in feminist research must highlight the importance of context, relationships, and power. To do so means not only entailing a formal commitment to the well-being and concerns of research participants and researchers, as discussed by Linda Bell (2014), but also accounting for the affective, or emotional, impact of research and the relationships formed during research on participants and researchers, recognising that these can result in harm. Feminist research ethics thus require addressing questions such as how research is conducted, who is involved, who is considered knowledgeable, whose interests are served, and what is the impact of research on participants, researchers, and others.

Some feminist researchers may not engage with research participants or research team members (they may, for example, be working alone on producing a discourse analysis or a literature review or be engaged in archival research). For those who do, however, there are phases during the research when ethical issues require consideration. Table 7.1 explores some of the ethical challenges that feminist researchers should consider both before and during the research process, specifically when conducting community-based research (it cannot possibly provide an overview of all the ethical issues you may encounter). Ethical issues must be taken into consideration when considering research design, and certain ethical issues will be considered in your institutional ethical review, with many such reviews having expanded their mandates to consider not only issues of consent and potential harm but also, for example, the question of where ultimate control (supposedly) lies. During research, various issues concerning how power differentials will affect teamwork, boundary formation, the cultivating of relationships, and the rights of groups versus the rights of individuals will invariably raise their heads.

GenUrb's multilayered organisation is both its strength and its Achilles heel when it comes to putting feminist research ethics into practice. In a research project that navigates several national and cultural contexts, conducted over a long period of time, shared understandings of ethical codes of conduct cannot always be assumed. In GenUrb example 7.1 we explore how ethical challenges have played out in the GenUrb project.

Table 7.1 Examples of common ethical challenges in community-based research

Research phases at which ethical issues arise	*Particular ethical challenges*
Pre-research phase	
Research design	• What are the specific social and personal locations and the needs of the people involved in your research (the participants and/or your research team) in relation to each other? • How will the research methods and practice avoid exploitation, extractivism, and exclusion? • What dilemmas may arise in terms of your research topic or in terms of the dynamics among your participants, among your research team, or between participants and researchers? What issues may these raise personally and socially for those involved?
Institutional ethical review process	• What modes of consent are appropriate? Is individual consent sufficient? • Will you maintain anonymity, privacy, and confidentiality of participants or community partners? • Are there clear divides between researchers and participants? • To what degree does control or responsibility lie with the principal investigator?
Research phase	
Teamwork, collaboration, and power	• How do the timelines and expectations of community organisations, funders, and researchers differ? • How will relations with partners be maintained? • How will partnership agreements be reviewed and adjusted as relationships change? • How do existing relationships, both professional and personal, affect your judgements? • How can you best communicate any ethical dilemmas to those involved, give them room to raise their views, negotiate with them, and respond to any difference of views between them (if possible)? • How will those involved understand any actions you may take, and are these actions in line with your judgement about your own practice?
Blurring of the boundaries between the researcher and researched, academics and activists	• Who are you identifying with and why? Who are you posing as 'other'? • How are appropriate boundaries between researchers and participants agreed upon and practised? • What are the challenges facing those who take up roles as both academic researchers and activists? • What are the challenges of those who are both researchers and participants? • How does the insider/outsider paradigm affect tensions between researchers, activists, and participants?
Community rights, conflict, and democratic representations	• What role does reciprocity, understood as the mutual negotiation of meaning and power, play in building trust and cultivating relationships? What are you taking from research participants as well as giving to them? • How do you extend rights beyond individuals to communities or groups? • How do you negotiate conflict within and between communities? • Who represents a community? How do you decide? • How will you challenge stereotypes or stigmatisation of community members?

Source: Adapted from Edwards and Mauthner (2012); Banks *et al.* (2015); Ravitch and Carl (2021)

GenUrb example 7.1: Examples of common ethical issues

Research phases at which ethical issues arise	*Particular ethical issues*
Pre-research phase	
Research design	• Different national contexts exerted varying degrees of control over the research design, limiting the extent to which researchers could help participants with their own issues and concerns that the research uncovered. • The circumstances of the daily lives of some participants had visceral effects on some researchers, leading not only to action on behalf of these participants but also to periods of withdrawal from the field of these researchers given the deleterious impact on their own mental health.
Institutional ethical review process	• Although all the partners in the research signed on to the institutional ethical review process it became clear over time that it was of little help in addressing the slow-burning ethical issues that we encountered during the research. Perhaps more importantly, the documentation it was based upon was not viewed as having the legitimacy needed to provide answers to the dilemmas we faced. • Securing agreement between all the City Research Teams (CRTs) and the participants on maintaining participant anonymity, privacy, and confidentiality was uncontroversial (see Chapter 14, Section 2, Planning research interviews). No problems were encountered in implementing these agreements. Somewhat unexpectedly, however, a few of the policy shapers interviewed did not want to remain anonymous.
Research phase	
Teamwork, collaboration, and power	• Relations between members of GenUrb were not always harmonious, and it was not always easy to see how our differences could be resolved. See, for example, GenUrb example 7.2 Muddling through ethical decision-making
Blurring of the boundaries between the researcher and researched, academics and activists	• While researchers in GenUrb took on different roles, with some being scholar-activists and others not engaging in overt activist activities, these differences did not give rise to conflict. It was understood that those not involved in activism were primarily operating in national contexts in which engaging in activism was a dangerous practice (see Chapter 13, GenUrb example 13.2 Security issues).

Community rights, conflict, and democratic representations	• Issues of conflict between community members engaged in the research and those who were not did not arise in the communities we worked with. We feel this is due to the care and diligence that was taken when introducing the research to potential participants. It may also be attributable to the fact that not all CRTs paid participants for their time and that some of those who did so deliberately delayed payments in order to be sure that participants were engaged because of their interest in the research rather than the remuneration alone. Hence, the research was not necessarily associated by other community members with the opportunity to secure monies. • One area in which we feel we fell short was in our failure to follow up on our original plans to put participants in each CRT in touch with each other through Zoom meetings. This proposal floundered over issues of time-zone differences, lack of time for everyone involved, and differing levels of interest in conducting this activity. This missed opportunity could have led to connections that could have taken the research and its outcomes in new directions.

By Linda Peake

In GenUrb example 7.2, we turn to the most serious ethical dilemma that GenUrb confronted as it worked to build trusting, accountable relationships across myriad axes of difference and to practise a feminist research ethics. GenUrb example 7.2 illustrates that boundaries, marginalisation, and silences in the negotiating of power are interwoven in ways that can make research based on teamwork and collaboration a messy and painful undertaking, fraught with difficulties. It also reveals how professional ethical guidelines often fail to speak to the nuances of complex ethical decision-making.

GenUrb example 7.2: Muddling through ethical decision-making

In GenUrb's grant application to SSHRC, it was necessary to write a section addressing ' Conflict resolution, accountability and decision-making authority'. We wrote:

> The partnership members have already established a respectful mode of working together across our cultural differences (an issue that was directly addressed at the August [2016] workshop). We are committed to reflecting on our actions, learning from our practices, and continually evaluating our capacity to achieve our objectives in line with our principles of accountability, co-creation of knowledge, integrity, inclusion, reflexivity, and respect.

> Issues that result in a specific conflict will be resolved through a Learning and Improvement Strategy but if this fails the Project Director will act as mediator according to an administrative review that will be established. . . . Decision making is the responsibility of the Executive Committee, governed by consensus, or a majority vote of members if consensus cannot be achieved.

In the six years it took to put the GenUrb network together, there were no apparent problems in working across our many differences. However, within a year of starting the project, it became increasingly apparent that cracks were emerging across our differences, although they were not always clearly defined or discernible. Moreover, not everyone was comfortable with bringing issues to the wider group. And just as Sara Ahmed (2013) refers to the 'brick wall' in her discussion about racism in the academy – 'You encounter a brick wall. . . . To those who do not come against it, the wall does not appear' – so were these cracks obvious to some but not to others. Hence, it was not always clear at any one point in time, given our widely dispersed locations across the globe, exactly what was happening in terms of our broad group dynamics.

These 'cracks' came to a head a few years into the project, putting the sustainability of the network into question. It was perhaps inevitable that it was at an in-person meeting, when we spent more time together than we had done before, that a picture began to emerge of the depth of feelings over issues. This was not a straightforward 'textbook' situation of white women versus women of colour or of women in the global South versus women in the global North. The ensuing discussion highlighted complexly drawn lines of division and raised uncomfortable questions around personalities and social locations, including distinct cultural understandings of the hierarchical relations of power between senior and junior faculty members, postdoctoral fellows and graduate students, as well as differences between academics and grassroots women, and issues of anti-blackness, and in turn, researchers' positionality. Comments on researcher positionality by Richa Nagar *et al.* (2016, pp. 510–511) speak pointedly to these tensions within GenUrb: 'the very trust and engagement that we invoke and celebrate can become hierarchical and exploitative if not interrogated, critiqued, revisited, and revised on an ongoing basis'.

An impasse followed the painful discussions from that in-person meeting: for months afterwards, we could not bring people together again to discuss the issue further. The aforementioned conflict-resolution strategy, so diligently written into our proposal, was dead in the water with the trust that had underlain the project shattered. It proved impossible to revive it. The outcome was that of some members being asked to leave the project. With time an ease eventually returned to GenUrb's daily work, although accompanied by a heightened sense of vigilance. Principles of accountability, integrity, inclusion, reflexivity, and respect are not merely terms in our grant application. They are the not so fragile, not so robust, infrastructure of an ethics of care that continue to hold us together.

As we have discussed earlier in this chapter, ethical decision-making is not always easy or straightforward: feminist reflexivity and being mindful of our individual and inter-personal praxis as researchers is an ongoing constant and continuous exercise. As Sharon Ravitch and Nicole

Mittenfelner Carl (2021, p. 222; see also Banks *et al.*, 2015) claim, 'healthy and enduring partnership requires a deep and ongoing reflective process that focuses on engagement in and exchange of constructively critical feedback and ideas'. And yet ways of seeing, acting, and decision-making in our large project were constrained by obligations to balance harm done with respect for confidentiality, and by geographical distances and infrequent in-person meetings enabling only imperfect and always-partial relationships with each other. What was important for GenUrb was to recognise the messiness of negotiating ethics in research, to pick up the pieces, and to find a way to continue.

Is this ethical dilemma behind us? Yes and no. The pain of that time has not fully dissipated. For several months, it was truly a dreadful time for many of us, and the consequences have reverberated in our personal and professional lives for a number of years. In determining whether to tell this story, I was aware of the power I held to dictate how and under what circumstances this story could be told. And yet what is written earlier can, and should, be understood as an inevitably partial account, one that would invariably be told differently by the others involved. In other words, the account I have given is a discursive exercise, and in all such exercises we make choices about the stories we tell and those we don't tell, what we reveal and what we conceal, and the ends we hope to serve in telling those stories. The partiality of this particular account leaves open the possibility of others in GenUrb disagreeing with my version. Its partiality also speaks to the realisation that awareness cannot be exhausted; that telling stories is not about reaching conclusions, but about opening yourself to the possibility that you might be wrong; that there's always more to learn.

By Linda Peake

Taking into account the points raised in GenUrb examples 7.1 and 7.2, turn to Reflection exercise 7.2 and consider hypothetical examples of addressing care ethics in research.

Reflection exercise 7.2: Enacting an ethics of care

Hypothetical example 1

Imagine that you are the principal investigator on a research project on which you hire a graduate student to work as a research assistant investigating violence against women in a small town. During the research, the graduate student confides that she was subjected to domestic violence in a previous relationship, and she discloses that she is being affected emotionally by the research.

- As principal investigator, how will you practise an ethic of care?

Hypothetical example 2

A group of researchers from a university social science department are working with an international development organisation to study the working conditions of informal workers, and they aim to recruit a group of women who work in waste-picking in a city neighbourhood to participate. They seek the assistance of a local women's organisation to recruit participants for their study.

- What ethical issues may arise from this scenario?
- How could you apply feminist research ethics to address them?

Section 4. Summary

This chapter has addressed how feminist researchers ought to conduct themselves. It has explored how the major paradigms of a philosophy of ethics shape research ethics, and explained the history of professional research ethics codes, which focus on shaping research design to prevent harm to both participants and researchers. Feminist researchers have developed a unique model of ethical deliberation and action that can be at odds with these professional paradigms. Feminist philosophers turned to an ethics of care, but early feminist conceptions of care ethics failed to account for racialised and classed difference, and Black and Latinx feminist critiques help us understand both the difficulty and the importance of practising care across difference. By placing an ethics of care in dialogue with the concepts of intersectionality, positionality, and decolonisation, we have outlined an ethical framework for feminist research. Throughout the chapter, we have discussed how GenUrb has worked to practise a feminist research ethics, demonstrating both the difficulty and the importance of enacting care across difference and of reconciling research-based institutional and feminist ethical procedures.

8 Professional standards in feminist research

Araby Smyth

Learning objectives
After completing this chapter, readers should understand

- the meaning of professionalism in the context of the academy and research;
- critical feminist responses to professional standards;
- and the evaluation of research projects according to professional standards of ethics, equity, diversity, and inclusion, monitoring and evaluation, and knowledge mobilisation.

Further, readers should be able to

- distinguish between professional practices of regulation and collaborative practices of accountability;
- and determine how to apply professional standards to their own work.

This chapter first outlines how professional standards function as a set of expected behaviours and a form of assessment in academic work and research. It discusses why feminists are critical of how professional standards have emerged and how they continue to be used to evaluate scholars in accordance with social hierarchies that valorise hegemonic whiteness and maleness. The chapter reviews professional standards in research in relation to: ethics policies; the application of equity, diversity, and inclusion recommendations to remove systematic barriers in research projects; requirements to engage with monitoring and evaluation; and with knowledge mobilisation across all stages of the research project. Links to the many resources in this chapter are available on the book's website www.routledge.com/9781032668680.

Section 1. Professional standards in the academy

Ethical review boards, research funding agencies, research institutes, and university departments commonly employ the notion of **professional standards**. Professional standards are a framework of practices, ethics, and behaviours that govern scholarship affiliated with universities. They ostensibly operate through the consent and self-regulation of affiliated scholars. This section outlines professional standards in academic work, including research, as well as feminist engagements with and critiques of professional standards.

DOI: 10.4324/9781032668727-12

(i) Professional standards in academic work

Students, teachers, and researchers in the academy are expected to exhibit certain behaviours and reproduce certain values. For graduate students and faculty, these behaviours and values are often called 'professionalism' or 'professional standards'. For example, the University of Michigan defines professional standards for faculty as an expectation that people 'engage each other in a professional manner, with civility and respect' as part of maintaining 'a community that enables all of its members to reach their full potential' (University of Michigan, 2022, no page).

How does one learn about professional expectations and behavioural norms as a graduate student, postdoctoral fellow, research associate, or faculty member working at a university? Many universities, especially but not only in North America, offer programmes, workshops, or courses that provide professional development guidance. For example, the National University of Singapore offers career workshops like 'How to project professionalism through image and etiquette' to undergraduate students pursuing a career in business. Many research institutes, including the City Institute at York University in Canada, offer professional training programs for graduate students interested in academic careers. These kinds of programmes, such as one at the University of Kentucky in the USA, focus on 'the dynamics of faculty work in the areas of research, teaching, and service' and 'introduce graduate students to the day-to-day realities of the professoriate' (University of Kentucky, n.d.). In addition to workshops, academic advisors and faculty members can be a useful resource for those learning to navigate professional demands in the academy.

While professional standards are often aimed at all the graduate students in a programme or university, not everyone experiences the expectations of and training about professional standards in the same way. Institutional and internalised barriers that affect people across intersections of disability, economic class, immigration status, gender, racial identity, religion, and sexual orientation result in expectations being different and resources being more accessible to some than others. It is therefore beneficial to learn about the professional expectations in academic environments through graduate student or professional organisations that serve underrepresented groups, such as the University of Washington's Office of Graduate Student Equity & Excellence, or through a student or labour union if there is one on your campus, or even outside these more formal venues – in conversations with peers, for instance.

Beyond the university, professional disciplinary organisations, such as the American Association of Geographers, offer workshops and webinars, such as their 2022 Summer Series, that provide guidance for students and recent graduates along their academic and professional paths (American Association of Geographers, 2022). Groups within these organisations, such as Black Geographers, the Black Geographies Specialty Group, and Queer Geographies Postgraduate Reading Group, along with organisations catering to specific groups, such as the Society of Woman Geographers, can also offer funding, links to resources, and workshops and speaker series on aspects of professional standards. Such groups foster inclusive and reflexive communities amongst scholars and work against exclusionary practices in university spaces and professional organisations.

A number of professional standards apply to academic research. Federal research funding agencies like the Canadian Social Sciences and Humanities Research Council (SSHRC), which funds the GenUrb project, evaluate researchers and their projects' adherence to the use of professional standards. Upholding professional standards is seen as an indicator of a researcher's ability and the overall potential of the research project (SSHRC, 2021b).

Next, we turn to recent scholarly critiques of and feminist strategies for navigating professional standards in academic work.

(ii) Feminist engagements with and critiques of professional standards

Feminist scholars challenge regimes of evaluation that demand a compromise on ethics of care and those that seek to discipline behaviours that deviate from the mainstream norms of professional standards (van den Hoonaard, 2017). University statements about what constitutes professional standards in academic work, like those mentioned, can be elusive. The feminist urban geographer Emily Kaufman (2021) writes about the discursive construction of professionalism in academia in the USA. She found that professional standards are often vague and that they consequently "reinforce social hierarchies that value white maleness above all" and create intentional discomfort for everyone else' (Rios, 2015, quoted in Kaufman, 2021, p. 1740). Since regimes of professional standards emerged through institutions already divided along social axes such as race and gender, actions that challenge racism, sexism, ableism, and other structures of oppression in the academy can be vulnerable to being portrayed as unprofessional (James and Tynes, n.d.; Ahmed, 2012, 2021; Lau, 2019; Gilmore, 2022). Feminist critiques of professional standards may differ in their theoretical paradigms and strategies for action, but they all call on people to re-evaluate how they work and change how they relate to each other in the university.

In addition to offering critical insight into the power relations that structure the academy, feminist collectives explore how scholars can navigate and resist the norms of professionalisation. For example, the Women in Geography group at the University of Wisconsin–Madison held workshops that addressed 'how professionalisation itself is tied to exclusionary legacies and continuities in academia' and that focused on 'broader systemic issues, such as racism and sexism, and the ways these manifest and are resisted in everyday settings' (Al-Saleh and Noterman, 2021, p. 461). Other feminist geography collectives in North America, such as the Feminist Geographies Specialty Group, the Feminist Intersectional Solidarity Group, the GeoBrujas, the Society of Woman Geographers, and Supporting Women in Geography, have brought women together to discuss geographical research related to women and women's experience in the discipline of geography. They attempt 'to unsettle the relations of power that reproduce precarity and collectively hold space for alternative futures' (Pascoe *et al.*, 2020). Making spaces in which people can be supported and protected when they challenge the racism, sexism, and structural violence of academia is a crucial part of their work (see Linz *et al.*, 2021). Recently, some of these groups have responded to calls from scholars of feminist and queer theory to think beyond cis women's experiences and include non-binary and trans experiences in the discipline (Brice, 2023; Gieseking, 2023; Kinkaid, 2023; Rosenberg, 2023).

Feminists are trying to bring about change in professional standards in many ways. The Great Lakes Feminist Geography Collective calls for a slow scholarship movement grounded in a feminist praxis of care (Mountz *et al.*, 2015; Loyd *et al.*, 2022). They suggest collectively talking openly about the challenges of academic work, recognising the types of work that have historically been marginalised, making space for communal modes of scholarship, taking care of each other, limiting time on email, and making the space and time to think and write differently in order to shift the social and material conditions of academic labour. Their calls are reflected in recent publications which stress that slowing down knowledge production would not only improve the quality of scholarship but also have a positive impact on the mental health of its producers (Peake and Mullings, 2016). Eli Meyerhoff and Elsa Noterman (2019) build on these ideas, arguing that while collective action and more caring relations are necessary, producing scholarship slowly is a privilege only available to few, such as tenured faculty. Instead, they call on people working in universities to take action on urgent political issues both inside and outside the university. To undo the exclusionary relations that underlie professional standards, scholars must build solidarity within the university and, beyond it, with social movements seeking to dismantle oppressive power relations.

Reflection exercise 8.1 encourages you to engage with issues of professional standards at your own university.

Reflection exercise 8.1: Professional standards at your university

Identify ways professional standards work in your department and university.

- Does the department handbook or university website include language about professional standards?
- Does your university offer professional development workshops or courses? Note the expectations for behaviours and assessment that they promote.

Consider feminist engagements with and critiques of professional standards.

- Are there internalised or institutional barriers that have a negative impact on you or your classmates?

Now identify groups such as organisations and unions that exist at your university or beyond and that speak to intersectional differences to work against exclusionary practices.

- How might you participate in and contribute to these groups (see Chapter 4, Section 3 Defending space for feminist activism in the neoliberal academy)?

It is widely accepted that responsible research adheres to ethical and professional standards. These standards apply not only to the conduct of research itself but also to the processes of 'applying for funding, financial management, the process of [knowledge] dissemination, monitoring for potential conflicts of interest or commitment, [and] maintaining a fair and equitable work environment' (Simon Fraser University, 2022). In the remainder of this chapter, we turn to these standards to see how feminist scholars have engaged with them. In the following sections, we address how professional standards apply to conducting research in four main areas: ethics policies; equity, diversity, and inclusion; monitoring and evaluation; and knowledge mobilisation.

Section 2. Ethics policies

States, institutions, and professional associations regulate academic research via their ethics policies. **Research ethics policies** are sets of rules and regulations that govern the conduct of research primarily to mitigate possible harm to participants, researchers, and communities. These policies respond to the fact that harm can result from the research process, even unintentionally, and they work to prevent or minimise harm to everyone involved. Research ethics policies require specific practices to protect the privacy and autonomy of participants, such as avoiding harm, obtaining the voluntary and informed consent of participants, anonymising research data, and ensuring fair treatment of participants in publication of results (see also Dingwall *et al.*, 2017) (see Chapter 7, Section 1 Ethics and research).

There is general agreement that the public, research sponsors, and end users expect ethics to be taken seriously. However, the various institutional approaches to research ethics across countries reveal differing understandings of ethical principles and their applicability. In Canada, where the GenUrb project is based, all research with human participants is subject to the *Tri-agency framework: responsible conduct of research* (CIHR, NSERC and SSHRC, 2021a). Over 30 years ago, the three major federal granting agencies in Canada (SSHRC, Natural Sciences and Engineering Research Council of Canada, and Canadian Institutes of Health Research) decided to take a unified approach to policies concerned with research practice in the belief that fundamental ethics principles apply across the remit of all three bodies. In 1998, they circulated a joint statement on the ethics of human research, which was updated in 2018 and again in 2022 as the *Tri-council policy statement: ethical conduct for research involving humans* (CIHR, NSERC, and SSHRC, 2022b). Though social and biomedical research and specific approaches to ethics differ in many respects, the core ethical principle in conducting research with people is respect for human dignity, which has three aspects: respect for persons, concern for welfare, and justice. Researchers have a responsibility to meet these ethical standards. While the three agencies have no regulatory authority to impose their ethical guidelines, compliance with these policies is required to receive their research monies. The Canadian approach is considered 'regulation with a light touch', with an emphasis on 'education and compliance, rather than on misconduct and discipline' (Zimmerman, 2017, p. 108).

At the university level, ethics review boards provide 'support, guidance and oversight' to researchers on how to maintain professional standards about how people are treated in research (York University, n.d.-b). At York University, the Office of Research Ethics states that 'all research involving humans, animals or biological agents must be reviewed by the relevant ethics review committee *prior to the start of any research*' (York University, n.d.-b). The research ethics office at your university may have a slightly different name (such as, for example, Office of Research Ethics and Integrity, Institutional Review Board, University Research Ethics Committee, Office of Research Integrity, Research Ethics Board, Human Research Protection Programme, etc.), but your university likely has an ethics committee that must approve research affiliated with the institution.

Implementing research ethics policies requires more than simply following university protocols and knowing what [not] to do. It is a commitment to continuous ethical engagement that cultivates respectful, reciprocal relationships with research participants (Kingsolver *et al.*, 2003). Rather than assuming, as many university ethics forms do, that issues about ethics can always be determined in advance of the research, in practice, consideration needs to be given to how to deal with the reflexive and continuing nature of involvement in ethical issues on an everyday basis. In addition, the ethical obligations of research are not just limited to the jurisdiction of your home institution; it is essential that you become familiar not only with university ethics policies but also with the ethics policies and practices that apply at your field site and that you learn about how the communities you will be working with engage researchers and research ethics.

Academic researchers must take seriously the ethics boards in their universities and deal with them in good faith, knowing that the filling in of an ethics form is only a preliminary step of what can be a difficult process of decision-making that never ends and the consequences of which can be felt long after the research is 'finished'.

Reflection exercise 8.2 encourages you to engage with the research ethics policies at your own institution. While for some these will be easy to discover and follow, for others it may not be so easy. Such policies may even not exist, and you will then need to consider how you will draw up your own guidelines.

Reflection exercise 8.2: Research ethics policies at your university

Do a search to find out whether there are national and/or institutional research ethics policies in your country and your university. Find out whether you must undergo an ethics review process before you are permitted to begin your research project.

- What are the requirements of the research ethics review process?
- If there are no ethics policies where you are located, determine whether you can use existing ethics policies of governments or universities in other locations to draw up an ethics strategy for your own project.

Section 3. Equity, diversity, and inclusion

Equity, diversity, and inclusion (EDI), also known as DEI (diversity, equity, and inclusion), and other variations such as JEDI (justice, equity, diversity, and inclusion) and DEDI (decolonisation, equity, diversity, and inclusion), refers to practices dedicated towards identifying, understanding, and transforming structural inequalities in the places where we learn and work. SSHRC (2021a, no page) defines equity as the 'removal of systemic barriers . . . enabling all individuals to have equitable opportunity'; diversity as 'the variety of unique dimensions, identities, qualities and characteristics individuals possess along with other identity factors'; and inclusion as 'the practice of ensuring that all individuals are valued and respected for their contributions and are supported equitably in a culturally safe environment'. From these definitions, it is clear that equity is about more than providing all individuals with the same opportunities; diversity is about more than bringing multicultural groups of people together; and inclusion is not just a welcoming environment. Effective EDI requires identifying and removing systemic barriers that have historically excluded groups of people; undoing biases, harassment, and discrimination; and cultivating environments in which difference is acknowledged and embraced and all people can thrive in unique ways.

Anti-racist, decolonial, and feminist scholars have described the ways that EDI practices are resisted and undermined in university settings. Sara Ahmed (2012) has written extensively about the gap between the commitments to diversity that universities make and the lived experiences of people in the university who embody diversity. Universities often present diverse faculty, staff, and students as evidence that there is not a problem with racism at their institution, but critical scholars tell a different story. They describe traumatising microaggressions and dismissal (Joshi *et al.*, 2015). For example, Aretina Hamilton (2020) writes about how Black scholars recruited for diversity positions have been 'used to fill classes, for their emotional labour, and or as props in department political games'. Writing about a racist attack on a Black student by a white student at a university in the USA, Jordan McCray (2023) argues that while EDI practices are the norm, not enough is being done by universities to stop the racist violence that is happening on campuses and disproportionately harming racialised faculty, staff, and students. Removing barriers, undoing discrimination, and embracing difference is a start. But much more is needed. Scholars, students, and administrators all need to be committed to anti-oppressive praxis and building solidarities across difference (Oswin, 2020; Kinkaid, Parikh and Ranjbar, 2022).

Reflection exercise 8.3 asks you to look at the EDI priorities at your university and identify at least one strategic practice to implement in your research.

Reflection exercise 8.3: EDI at your university

Research EDI at your university or place of work. For example, look for a strategy or strategic plan for the university or college or an EDI overview from your department. If you can't find such a document at your institution, search SSHRC's website, www.sshrc.crsh.gc.ca, for ideas about addressing equity, diversity, and inclusion in applications.

- What are the priorities for identifying, understanding, and transforming social inequalities at your university?
- Identify at least one concrete practice you might put in place in your research project to ensure a more equitable, diverse, and inclusive research environment.

Section 4. Monitoring and evaluation

Monitoring and evaluation processes are those through which the progress, goals, and impact of a research project are assessed. They are employed to determine the ways in which research design as well as the impact of research can be improved, especially in relation to knowledge mobilisation activities. Although both are employed throughout the research, they differ in intent and timing. Monitoring includes continuous assessment of the research project, from start to finish, based on the collection of information about its progress or delay; it is often conducted with the aim of improving research design. Evaluation involves examination of the degree of success of the research project in terms of its relevance, effectiveness, efficiency, and impact, with relation to short- and long-term objectives.

It is important to know whether and how your research may be subject to monitoring and evaluation. Some national research bodies, including SSHRC in Canada and the USA National Science Foundation, require researchers to implement monitoring and evaluation practices for proposed evidence-based research projects. They require evaluation of the short- and long-term results, most commonly in terms of outputs, outcomes, and impacts.

- *Outputs* are short-term results that comprise items or 'products' that result from activities undertaken during the research, including publications, presentations, and research partnerships.
- *Outcomes* are short- or medium-term effects resulting from research, including but not limited to the number of people in target audiences that use research findings, the number of people trained, and policies and business strategies developed.
- *Impacts* are long-term outcomes that change thinking and behaviours and could be evaluated using indicators related to economic performance, quality of life, or social inclusion. Impacts may include, for example, long-term behavioural changes, network development, or long-term relationship building.

Monitoring and evaluation involve a somewhat mechanistic measurement of impact, often reduced to short-term quantitative measures of outputs and outcomes, with less emphasis on impacts, which are themselves fraught with measurement difficulties and are consequently much harder to evaluate. Funders commonly expect researchers to establish quantitative indicators that can be used to measure outputs and outcomes. GenUrb example 8.1 shows how GenUrb had to indicate in our SSHRC application how we would apply monitoring and evaluation indicators to our knowledge mobilisation outputs for the different groups of knowledge producers and users with whom we engaged.

GenUrb example 8.1: Monitoring and evaluation indicators used to measure how knowledge mobilisation is applied to different groups of knowledge producers and users

Knowledge producers and users	*Monitoring and evaluation indicators*
Grassroots women research participants	• Measures to ensure dissemination, connectivity, and accessibility through informing, translating, and active knowledge brokering (e.g. number of neighbourhood fora, plays, exhibits, partner workshops, conferences, and meetings held) • Measures to ensure partnership growth, long-term collaboration, and research capacity building (e.g. number of new partners; number or type of capacity-building efforts; number of training opportunities; experience in research protocols, analytical frameworks, and new methods)

Policy-shapers engaged with SDGs 5 and 11	• Engagement of policymaking bodies in partnership activities and in addressing issues of relevance to the partnership and SDGs (number of invitations to participate in policy forums; policy consultations; number of keynote addresses before policy bodies; number of policy interventions; participation of key policymakers in GenUrb network activities)
Academic researchers	• Measures of influence on academic and policy debates (e.g. number of books and refereed journal articles, policy papers, or reports; number of citations; number of downloads; number of curriculum and teaching materials produced)
General public	• Measures to ensure dissemination, connectivity, and accessibility to a wide general audience (e.g. number of videos, graphic pamphlets, and other visual means of dissemination; social exposure, via number of media interviews, website hits, op-eds, plain-language summaries, policy briefs, blogs, presence on social media; and reach on X [formerly Twitter], Instagram, Facebook, YouTube, and WeChat)

By Linda Peake (based on GenUrb knowledge mobilisation plan, SSHRC application)

GenUrb example 8.2 indicates some of the challenges this quantitative approach to monitoring and evaluation raised for the GenUrb project manager.

GenUrb example 8.2: Monitoring and evaluation challenges

As the GenUrb project manager, it wasn't until I had to write a report for SSHRC that asked me to address the success of GenUrb's knowledge-mobilisation activities in a numerical fashion that I really became aware of the level of detail we had to engage in for monitoring and evaluation of knowledge mobilisation. For instance, I had to report (using Google Analytics) on the number of visitors to the GenUrb website as well as the number of page views per month. In addition, I had to report on the number of followers and impressions on our X (formerly Twitter) profile. I also consulted our GenUrb social media expert about developing a social media plan, down to minutiae such as the best times of the day to post to attract more viewers and how to keep on adding new followers to our profile. These were not things that I had previously thought to do or to keep track of (or had thought were important). Once I became aware of the level of quantitative detail required by SSHRC, I became more systematic in my approach.

Engaging in monitoring and evaluation has served the purpose of showing SSHRC that while we were hitting our targets in relation to some knowledge users, for others we were less successful. However, such numerical reporting cannot account for the underlying reasons for our uneven results. For example, keeping policy-shapers informed of our work has been very uneven across the City Research Teams. In Delhi, no attempt was made to inform policy-shapers about our work, as this could have attracted unwanted attention to the feminist nature of our project. In contrast, in Ramallah, where the research team lead knew the mayor and where the SDGs are driven by a donor agenda, it has been much easier to keep engaged with policy-shapers.

Finally, COVID-19 severely affected our ability to engage in knowledge-mobilisation activities such as hosting public events and meetings. This significantly reduced our monitoring and evaluation obligations, although they continued in relation to other outputs such as publications, conferences, and other presentations.

By Leeann Bennett

Monitoring and evaluation practices have sparked concerns that certain kinds of knowledge are being recognised and privileged over others. These concerns speak to a tension between monitoring and evaluation as dictated by funding agencies and how feminists tend to engage in evaluating research outcomes, emphasising the creation of useful knowledge and the nurturing of knowledge-making relations as opposed to numerical measurement (or the provision of solutions). In Shawna Wakefield and Daniela Koerppen's (2017) OXFAM discussion paper exploring the application of feminist principles to monitoring and evaluation, the authors stress the need to recognise that understanding and measuring changes in gender and power is neither straightforward nor simple. It is not unusual for there to be no 'clear pathways from program activities to program results' (Wakefield and Koerppen, 2017, p. 3). Change is non-linear and complex, and yet research funders expect there to be a straightforward process of change that can be numerically measured.

For monitoring and evaluation processes to be conducted according to feminist principles there needs to be a commitment to feminist values and research ethics. For example, this requires participation in the design and implementation of monitoring and evaluation by all programme partners and participants, and this requires time to build trust in and understanding of the process. The process needs to be 'flexible, adaptable and responsive to context' (Wakefield and Koerppen, 2017, p. 20). It must be understood as a learning process and not merely a set of data outputs. Of course, it is not only feminists who have these concerns. They are shared by other critical scholars, especially those who engage in participatory research methods. For example, development geographer Glyn Williams (2015, p. 60) assesses monitoring and evaluation in research projects based in the UK that focus on the global South and finds that:

> impact-evaluation practices . . . contain a particular view of 'relevant' research which is itself framed around explicit separation of 'creators' and 'users' of knowledge and of research from education. This directly contradicts the ethics of knowledge co-production and mutual learning that have been central to 'alternative development' practice.

These worries about monitoring and evaluation and similar feminist concerns are connected to broader conversations within and beyond the academy about whose knowledge matters, the role of research, the ways that it is conducted, and the ways that researchers are or are not held accountable.

Reflection exercise 8.4 asks you to consider how you might implement a feminist version of monitoring and evaluation in your own research.

Reflection exercise 8.4: Engaging in monitoring and evaluation in your research

While there may be no formal requirement to address monitoring and evaluation in your research project it is useful to consider how you may want to engage in a monitoring and evaluation process informed by feminist principles. Make a list of the aspects of your research that you could usefully monitor to improve your research design.

- Can these all be known in advance of the research starting?
- How often will you assess your progress?

Make a list of what it is that you want to evaluate.

- How are you measuring change?
- How are you measuring success? What are potential positive and negative effects of your research for the participants?
- To what extent will the length of your project limit your evaluation to measuring outputs and outcomes?
- How realistic is it to assume that your research project will lead to any kind of change?
- How are you valuing different ways of knowing and representing different voices?
- Reflect on your own biases as a researcher and how these evolve over the research process.

Section 5. Knowledge mobilisation

A knowledge-mobilisation strategy links research to broader issues and ensures that results have an impact in the real world. According to SSHRC (2012a), knowledge mobilisation 'encompass[es] a wide range of activities relating to the production and use of research results, including knowledge synthesis, dissemination, transfer, exchange, and co-creation or co-production by researchers and knowledge users'. In other words, knowledge mobilisation is about the practices that connect the research process and its results to other people, often outside the academy. Knowledge mobilisation has been given an increasingly prominent role by research funders, with much more attention being paid to how research results reach diverse audiences (see Chapter 19 for a detailed account of knowledge mobilisation in feminist research).

Section 6. Summary

Professional standards encompass ethics policies, EDI, monitoring and evaluation and knowledge mobilisation. Although these standards are ubiquitous in academic workplaces, feminists are critical of how they have emerged and how they continue to be used to evaluate scholars. This chapter has explored how, as you begin to plan a research project, you should: learn about the ethics policies at your university, in your country, and in the places where you plan to do research; apply EDI recommendations and constantly evaluate your research project for how to remove systemic barriers and biases; and engage in monitoring and evaluation and knowledge mobilisation throughout the research process to ensure that you connect the research and its findings to diverse audiences and that you can assess the success of such activities.

9 Partnerships and teamwork in feminist collaborations

Araby Smyth

Learning objectives
After completing this chapter, readers should understand

- the value brought to research by partnerships and why they are increasingly being employed;
- the relevant management and leadership practices;
- the challenges that research partnerships bring;
- and the importance of feminist praxis in teamwork for challenging hierarchical power relations.

Further, readers should be able to

- think creatively about the value of research partnerships;
- practice teamwork critically and respect difference;
- and implement a feminist ethics of care during the research process.

This chapter discusses the value of research partnerships. It describes how to establish, manage, and lead research partnerships while also addressing the challenges that research partnerships pose. It responds critically to mainstream practices of partnership by putting these in dialogue with a feminist politics of knowledge production and the challenging of hierarchical power relations. In particular it addresses the need to investigate partnerships both internally and externally, most commonly through transnational feminist research and praxis.

Section 1. Research partnerships

Research partnerships, both within and beyond the academy, have the potential to add value to research processes. They can bring together people from different areas of expertise to conduct research that addresses challenges that are uniquely relevant to communities, policymakers, academic, and other end users. Partners may include, for example, think-tanks, government bodies (e.g. municipal authorities, planning departments), corporations, businesses, or other private institutions, community, not-for-profit or other grassroots or advocacy organisations, multi-lateral organisations (e.g. United Nations bodies and specialised agencies), non-governmental organisations, and other groups and individuals (see GenUrb example 9.1, which lists GenUrb's partners).

DOI: 10.4324/9781032668727-13

Such partnerships between academic and non-academic stakeholders are increasingly leading to collaborative and action-oriented research that includes the co-production of knowledge between researchers and practitioners. Within the context of the academy, bringing together financial and intellectual resources across disciplines within partnerships also provides an array of opportunities to train students through embedding them in research ecosystems, facilitating their access to 'laboratories', equipment, scholarly networks, knowledge, and expertise. Partnerships have the ability not only to shape academic agendas and debates but also to build (inter)national research capacity, influence policy and community development, and play an ambassadorial role (Flinders, 2020; Oldfield, 2023).

GenUrb example 9.1: The formal GenUrb partnership

GenUrb's formal partners include:

- two women's organisations: Red Thread in Georgetown, Guyana, and Centro de Estudios y Trabajo de la Mujer in Cochabamba, Bolivia;
- two civil-society organisations: the Centre for Human Development in Ibadan, Nigeria, and UR°BANA in Berlin, Germany;
- two policy organisations: the UN-Habitat Gender Hub in Nairobi, Kenya and Women in Cities International in Ottawa, Canada;
- and two academic institutions: York University in Toronto, Canada (the host University), and Queen's University in Kingston, Canada.

Other partners have joined the partnership for limited periods of time. They include:

- a lesbian organisation in Shanghai, China (name withheld);
- various universities including the New York University Shanghai; the University of St. Andrews in Scotland; Trinity College in Hartford, Connecticut, USA; the American University in Cairo, Egypt; and the Tata Institute for Social Sciences, Mumbai, India.

This list presents, however, only a partial picture of the teams engaged in the research. Some teams did not bring a formal partner into the research. Some universities in the global South could not join the partnership given their university's requirement for research overheads to be covered and the policy of the Social Sciences and Humanities Research Council (which funded GenUrb) not to provide overheads. Others preferred not to involve their universities given the length of time it would take to process funds. In the daily work of GenUrb, our contacts are with specific individuals, some of whom are attached to partners and others who are not.

By Linda Peake

Whether it is a formal research partnership, collaborating with a colleague, or working on a team, working together allows us – possibly – to get more done in a shorter amount of time and to benefit from multiple perspectives on the collection, interpretation, and analysis of data. GenUrb's partnership approach allows for collaborative dialogues and alliances that seek to challenge dominant epistemologies and to interrupt the researcher's authority by incorporating multiple voices. GenUrb example 9.2 describes the multifield value of our partnership approach.

GenUrb example 9.2: The value of the GenUrb partnership

In aspiring to a global reach and working across social, cultural, and urban differences, the geographical range of GenUrb transcends the capacity of any one scholar, institution, or discipline and requires a multilingual, transnational, multi-disciplinary, and multi-sector team-based coordinated approach that only a partnership can provide. The value added by GenUrb's partnership approach is manifold and speaks to:

- the ability to influence academic agendas in urban studies through the utilisation of a comparative feminist methodology based on transnational feminist praxis and partnerships;
- the ability to increase the variety of ideas and the breadth of perspectives in urban studies agendas;
- bringing grassroots and civil-society knowledge into academic conversations and evidence-based research into policy conversations;
- increasing research capacity through the training of a future generation of urban feminist post-colonial scholars, including undergraduates, graduate students, postdoctoral fellows, and early-career scholars;
- building a global network of urban feminist post-colonial scholars through the engagement of other scholars, activists, and policy-shapers in GenUrb knowledge-mobilisation activities.

By Linda Peake

In what follows we discuss the work that goes into establishing, managing, and leading research partnerships.

(i) Establishing partnerships

A formal research partnership is different from ordinary collaboration with a colleague or working on a team project. Research partnerships are efforts of knowledge co-production between academic researchers and academics at other universities, government agencies and departments, or other kinds of organisation across the public, private, and not-for-profit sectors. Your university and relevant granting agencies will likely define and provide guidelines for establishing research partnerships. Many universities have multiple partnerships with industry, government, non-profits, the private sector, and other universities that scholars can draw upon. National funding agencies, such as SSHRC, have different funding opportunities for projects that connect academic research to domains beyond the academy, such as businesses or local communities. Other research foundations also allocate funding for urban research, and one of these, the Urban Studies Foundation, recently added partnered research to its remit.

Research partnerships have become increasingly common as the need for interdisciplinary knowledge and research ecosystems to address crisis-based issues (such as climate change, migration, and gender violence) has been recognised. Partnerships tend to be large; are often international; can involve more than one funder; and take seriously equity, diversity, and inclusion (EDI) and collaborative knowledge mobilisation. Such large and cross-sectoral partnerships take time to evolve. It is common that years of experience and networking are needed before a partnership can develop, though some organisations, like the Urban Studies Foundation, also allow PhD students to apply for small research grants involving research partnerships.

Various factors must be considered when establishing research partnerships (Brown *et al.*, 2022; Connolly, Henriquez and Hursthouse, 2022; O'Connell and Brannen, 2022).

- Personnel will be needed, including:
    - a principal investigator(s) who is willing to take on leadership; has the time and determination to devote to the project; can implement equitable and transparent budgeting and governance decisions; is aware of the different needs and expectations of project members; can work closely with the project manager; and is adaptable to changing circumstances and open to inclusivity;
    - and a project manager who understands the importance of communication; has financial and administrative experience; can organise intellectual property agreements, timelines for research, outputs, and knowledge-mobilisation strategies; can monitor the partnership's progress, including its EDI strategy; and connect with funders.
- Resources and infrastructure are needed including:
    - a budget sufficient to provide adequate administrative support;
    - the commitment of host partners to support the project and to assist in overcoming any financial and legal issues that may arise;
    - a realistic understanding of the requirements of funders, especially of the organisational and financial barriers faced by international partners;
    - and the development of strategic plans to deal with loss of resources (financial, personnel, researchers) likely to occur in partnerships lasting several years.
- Researchers are needed who are:
    - familiar with and interested in the research topic and with a range of appropriate methods;
    - trained in cognate disciplines with the skill sets, knowledge, and experience of working across cultural, national, and disciplinary backgrounds;
    - proficient in the dominant language of the project;
    - and highly motivated to engage in constructive and productive relationships.

- Time and effort must be devoted to:
  - building relationships with researchers, research teams and networks prior to applications for funding;
  - establishing a feminist ethics of care; building trust, consensus, and new relationships; developing new research skills; managing career progression, especially of graduate students and early-career scholars; and bringing in new members;
  - and determining strategies for 'navigating issues around identity crises, external disruptions, potential academic credibility issues and disciplinary boundaries' (Connolly *et al.*, 2022, p. 72).

(ii) Managing partnerships

The work required to manage a research partnership should not be underestimated. Managing a research partnership is a large-scale undertaking that involves coordinating administrative tasks; managing budgets and financial contributions; keeping the project on schedule while also juggling team members' commitments, timelines, and schedules; completing regular project evaluations and reports; supporting knowledge-mobilisation activities; and providing training and mentoring opportunities for team members. Fortunately, there are resources that provide information on best practices. For example, the SSHRC website provides tips for managing a research partnership (SSHRC, 2020).

As this section makes clear, the work of managing a research partnership is no small task. This is why large, funded projects tend to employ a project manager. And even with a project manager in place, all the administrative and managerial components of the project are not necessarily undertaken by the manager. Project members, including the principal investigator, need to work alongside the project manager on a range of issues throughout the duration of the project. In GenUrb example 9.3 the project manager, Leeann Bennett, outlines the many activities of her average workday to illustrate how the work of managing a research partnership is much more than coordinating schedules and budgets and adhering to university protocols.

GenUrb example 9.3: A day in the life of a project manager

As GenUrb's project manager, I am responsible for everything administrative and logistical. An average day might find me scheduling meetings, including our regular steering committee meetings; writing annual reports for GenUrb partners; booking travel; maintaining our project website and social media accounts; keeping project records; making sure that we all follow various SSHRC policies, including ensuring, if possible, that our publications are open access; helping to hire students, postdocs, research associates, and other staff; acting as our project's internal HR person; and supervising undergraduate students working on the project. Every day is busy!

Many of my responsibilities are financial: I administer our budget, which means I process payroll for York University staff or international research assistants; send money to our CRTs (all have their own annual budgets), either by institutional transfers or invoice payments; process claims for reimbursement, either on paper or online; complete monthly financial reconciliations; complete annual partner-contribution statements for SSHRC (and ensuring that we meet our 35 percent partner-contribution requirement); and make sure we are not over- (or under-) spending.

In addition, I assist project members with various other tasks, such as providing support for new grant applications or editorial assistance. I'm also always available to answer anyone's questions or emails. I also offer support to other staff at York University who are employed in similar roles, helping them figure out how to do these tasks and navigate the university bureaucracy.

In addition to adhering to SSHRC protocols, I also have to be mindful of the professional expectations set out by York University, particularly by the Vice President Research and Innovation Office, the Office of Research Services, the Office of Research Ethics, and Finance Services, as well as the protocols of those units in which the project is located, namely the Faculty of Environmental and Urban Change, the Faculty of Graduate Studies, and the City Institute.

However, my role is not just about office professionalism. My role in GenUrb is also about providing a feminist ethics of care to the project. I see my role as providing support to everyone working on the project to make their work easier (most project members have other full-time jobs). I also see my job as keeping GenUrb members connected (those based at York University and those located internationally), and I meet weekly with project members for coffee, either in person or virtually, where issues of project members' physical and mental health may arise. I also care about the safety and security of our members. For instance, I once had to liaise with university lawyers and embassies in a case where we needed to help a researcher quickly leave a particular country.

In addition, I take my position seriously, as a white woman, located in the global North, who happens to manage the project finances, who has the security of a full-time job on a research project about women's placemaking in the global South, and the possible power dynamics that could result from such a role. For example, I have had to address issues of structural inequalities and post-colonial realities related to GenUrb members' participation in academic conferences and SSHRC rules (for instance, being able to reimburse 'collaborators' only for travel-related expenses, which has served to exclude payment for some of our researchers who are neither formally designated co-applicants or collaborators).

By Leeann Bennett

(iii) Partnerships and leadership

Working together as a team requires communication, coordination, flexibility, and a sustained practice of reflexivity. It also requires leadership. We refer to research leadership here not in the dominant hierarchical sense (which conjures up an image of the lone scholar 'leading' their field) but in relation to the contribution of a leader or leaders of a research project in supporting and nurturing the careers of other project members and that resonates with a feminist ethics of care (see Chapter 7, Section 3 Feminist ethics and research). This understanding of research leadership recognises that there is more than one leadership role in any project, 'which resonates with existing theories of "leading from the back" with [their] emphasis on nurturing, facilitating and supporting others'

(Flinders, 2020, p. 13). While this model does not deny the individual development of a 'leader' it relates foremost to supporting and facilitating the production of research in an inclusive manner that maximises the academic quality of that research. Matthew Flinders (2020) suggests the following ways to develop an awareness of leadership:

- learning about different leadership styles;
- nurturing self-leadership and self-awareness;
- developing the capacity to work in inter-disciplinary contexts;
- developing networks;
- being innovative in relation to knowledge mobilisation;
- developing an intellectual vision;
- assessing and managing risks;
- and learning from failure.

In GenUrb there are many leaders. Each CRT has its own lead or co-leads, and each committee organises itself. There is, however, one person who is recognised by SSHRC as the principal investigator, namely Linda Peake. GenUrb example 9.4 shares how Linda thinks about her leadership style.

GenUrb example 9.4: Developing a feminist leadership style

Dominant understandings of leaders and leadership are problematic from a feminist perspective, tied up as they are with domination, oppression, and the suppression of difference (see, however, Liu, 2020). In the GenUrb project, with its alliances across racialised, classed, monied, language, institutional, and global North/South differences, my own understanding of leadership has been defined by learning from the anti-racist feminist praxis of the grassroots organisation Red Thread (with whom I have worked for over 30 years). Contesting my own whiteness and the vagaries of the project have revealed that leadership is not a fixed position but a political process, my own experience of which has shown me that much as I may want to believe that my approach will be stable and knowable in advance of events, it can be difficult to know what may happen in the face of crises or fatigue.

As principal investigator of the GenUrb project, I would want to think of my leadership style as rooted in an anti-racist feminist ethics of care but also 'tidal' in its practice. In certain times and places my leadership has been up front and defining, as when assembling the original members and partners to form the project or when organising steering committee meetings or working with the project manager. At other times, I have taken much more of a backseat role, working with individual members, especially the postdoctoral fellows and early-career scholars, building up trust among members and attempting to set the pace for the project. Even after six years, it can be difficult to know *how* to set the pace, given the tension between the competing demands on people's time and the obstacles that have slowed us down (such as COVID-19, maternity leaves, elections, and so on) versus the finite deadlines of the temporary members of the project such as students and postdoctoral fellows. It has not always been an easy set of considerations to balance. I find I equivocate, sometimes consumed by worry that we will not be able to get everything done on time, at others trying to adopt a more *laissez-faire* approach – *que sera sera* – given that I know that none of us can work any harder than we are already doing.

My leadership is mostly about communication, co-ordination, and administration, enabling collaboration, teamwork, and decision-making to take place. Although we have been working across twelve time zones and dealing with COVID-19, we have ensured that decision-making has largely taken place at regular team meetings of the steering committee and of all other teams, meeting with differing degrees of frequency. These meetings take place mostly over Zoom – although prior to COVID-19, our annual general meetings took place face to face, and we would also meet in person at conferences. I also meet very regularly with our project manager, sometimes two or three times a week. Like our project manager, I attempt to be in contact as much as possible with all project members, touching base not only about research and publications but also about mental and physical health issues.

Leadership has meant addressing head-on the fact that the project operates within the strictures of a Canadian government-funded project that determines in advance how monies can and cannot be disseminated and who gets to be in decision-making positions. This situation was met by most with weary acceptance and by a few with reluctance. The structural inequalities of the project, imbued with colonialism and whiteness, can make decision-making difficult. An urban scholar once said to me that when we have research monies we have to write, write, write, but in the feminist context of GenUrb, I find we have to talk, talk, talk. And having to make difficult decisions is par for the course in such a long-term and transnational project: I have tried not to shirk them while still operating within a feminist ethics of care that has solidarity and justice as guiding principles. Doubtless, there are times when project members have found me lacking or not stepping up as much as they may want: for example, in relation to intervening in problematic relationships or issues. I know that not all decisions I have made have been agreed with by some project members, and I have found myself often reflecting (even years later) as to whether I made the right decision at the time. And still, we have nearly all of us made compromises and grown in different ways. At times, the weight of the project – its large budget, its length, its global reach, its employment of staff and postdoctoral fellows – can be heavy and time-consuming, but when all goes well, when there is trust and solidarity and laughter, GenUrb is a good place to practise leadership.

By Linda Peake

In Reflection exercise 9.1 we ask you to think about how you might navigate the power dynamics of a research partnership through a hypothetical research project.

Reflection exercise 9.1: Research partnerships across difference

As a feminist academic researcher, you are conducting a comparative research project on women who work in grassroots organisations and other informal advocacy channels in support of labour rights for domestic workers in a city with which you are familiar. Answer the following questions in relation to a specific city.

- Which partners would you invite to join your project and why? Which parts of the project would they be involved with?
- What skills or capacities do these organisations offer, and what are your expectations in terms of research outputs from these partnerships?
- What logistical challenges may come up in this collaboration? How can you work together to resolve them?

In the following section, we briefly discuss some of the challenges facing partnered research.

Section 2. Partnership challenges

Notwithstanding calls for reciprocity and intentional collaboration between project partners, there are no standard practices for conducting research in partnership with participants or organisations. The lack of clarity is amplified by the contingent and negotiated nature of research partnerships. Further, research partnerships present an array of unique challenges for academic researchers. These include uneven power dynamics, differences in organisational timelines, working on different timescales, a lack of time, and unequal access to resources, especially money (Bayley and Phipps, 2017). Examples of uneven power dynamics include different rates of pay for research assistants and graduate students in different countries and team members with different politics and cultures (see Chapter 2, Section 4 Circulations of labour in feminist knowledge production). Practical challenges also abound, including the fact that not all universities have the same academic calendar; working across time zones; and translating materials across two or more languages (see Chapter 15, Section 3 Planning for translation and interpreting throughout the research process).

One common set of issues in a partnership with academic and non-academic partners is the differing emphases placed on co-authorship (see Chapter 4, Section 4 (ii) Co-authorship as feminist alliance work). Alison Bain and William Payne (2016, p. 338), who conducted participatory action research with Toronto residents volunteering with a queer youth programme, found that the academics were more invested in seeking academic publication because it was beneficial for their careers, whereas the participants found it challenging ‘to feel or see or value themselves as scholarly knowledge producers with significant enough meaning-making expertise to respond to the judgments and criticism of peer reviewers’. This situation can be somewhat mitigated by the production of outputs that speak more directly to the needs and circumstances of participants. Further, working across differences in power dynamics and through practical roadblocks can be generative (see Chapter 3, Section 3 Decolonising translation).

One factor that can help reduce the emergence of potential challenges is the specification of the roles and responsibilities of all participants at the outset and sharing this information with all partners. However, even with written documents and agreements, challenges can arise as partnerships change over time. These problems can sometimes be mitigated by consistent and transparent management and leadership practices.

The following GenUrb example 9.5 by the GenUrb project manager discusses some of GenUrb's challenges with managing partnerships over the years.

GenUrb example 9.5: Challenges to managing partnerships

Throughout the project, we have had many challenges in managing our partnerships.

One of the most significant challenges to managing these partnerships is time. All the organisations that we are affiliated with have their own work agendas. While GenUrb is a project that they are generously supporting and interested in, we cannot ask our partners to do more than they already do. Most partners provide a mix of administrative support (if GenUrb funds are held at their universities), dissemination of GenUrb news on their networks, and participation in GenUrb conferences (including our 2019 Feminist Explorations of Urban Futures conference) or other events, while our partners working in the field (the Centre for Human Development, Centro de Estudios y Trabajo de la Mujer, Red Thread, and UR°BANA) are all involved in the research, selecting participants, gathering data, and undertaking analysis.

Another major issue we have with managing partners is financial. Before officially joining GenUrb, some partners hesitated, asking if there was an expectation that they provide monetary funds to GenUrb. It was not a requirement that they do so. Partners can make in-kind contributions to which we apply a monetary value given that we need to leverage 35 percent of our total grant funds from our partners. Given that most of our partners lack funds, they make in-kind contributions, with our largest cash contribution coming from our host partner, York University. Perhaps the financial issue with which we have had the most problems is the SSHRC stipulation in the *Tri-agency guide for financial administration* (CIHR, NSERC and SSHRC, 2022a) that research funds cannot be used towards overheads. Some partners in the global South have needed overheads to be covered, and this has stopped us from officially partnering with several groups or institutions. Hence, our official list of partners provides only a partial list of organisations engaged in the GenUrb project.

It has also been difficult to remain equally connected to all our partners given the different roles they play in the project. Some partners have had to leave the project for political reasons (for example, when it became too dangerous for a partner to be formally active in a feminist project) or when a researcher has moved to a different academic institution or a partner organisation has ceased to exist after financial collapse.

Another aspect of maintaining contact has been the impact of COVID-19, which prevented any in-person meetings, reducing the level of intimacy among members. Even when in-person meetings have been possible it has been complicated and expensive to coordinate everyone's schedule, and so since COVID-19 we have turned to Zoom meetings.

In addition to managing our partnerships, it is also challenging to engage them with broader publics. Our partners working in the field have been the most successful, as they work directly with the women living in marginalised neighbourhoods and have been able to host workshops to talk about the research findings and themes, as well as hosting radio shows, film screenings, art installations, and photo exhibitions, among other activities.

By Leeann Bennett

In the following section we build upon a feminist understanding of partnership, looking outwards to transnational feminist research and praxis and inwards to teamwork, and the importance for both of challenging hierarchical power relations.

Section 3. Transnational feminist praxis and teamwork

Research partnerships are valued in part because they can bring together academics and non-academics (such as grassroots community activists and policymakers) to investigate commonly recognised challenges and devise meaningful interventions. In addition, they are 'a valuable site for engaging with feminisms as a reflective political praxis' (Houston *et al.*, 2010, p. 75). In this section we engage in a feminist exploration of partnerships. First, we discuss the ways in which feminist research partnerships that cross national boundaries have the potential to engage in transnational and decolonial feminist praxis. Second, we discuss the politics of the teams that make up partnerships, although it should be noted that while feminist research is often undertaken as teamwork, this is not necessarily in the context of formal institutionalised partnerships.

(i) Transnational feminist praxis

A central tenet of feminist theory is that knowledge is situated (Haraway, 1988, 1991). What we come to know is rooted in our embodied, geographically located, historically and culturally specific perspectives (Rose, 1997; Hill Collins, 2000). Thus, the carrying out of a research project, from its inception to completion, is shaped by power dynamics around social axes of difference such as age, disability, class, gender, immigration status, race, and sexuality. Feminist scholars have long grappled with these questions of power and representation as they work across borders and build transnational alliances that address issues in women's everyday lives. The decolonial feminist scholars most prominent in this field of working across difference have worked primarily to validate the everyday knowledge production of subaltern groups (Grewal and Kaplan, 1994; Nagar, 2014, 2019; Sangin Writers Collective and Nagar, 2006). Richa Nagar and Amanda Lock Swarr (2010, p. 5) define these kinds of intersectional transnational feminist praxis as three-pronged insofar as they:

> (a) attend to racialized, classed, masculinized, and heteronormative logics and practices of globalization and capitalist patriarchies, and the multiple ways in which they

(re)structure colonial and neocolonial relations of domination and subordination; (b) grapple with the complex and contradictory ways these processes both inform and are shaped by a range of subjectivities and understandings of individual and collective agency; and (c) interweave critiques, actions, and self-reflexivity so as to resist a priori predictions of what might constitute feminist politics in a given place and time.

These three points indicate that the gaze of feminist scholars engaged in transnational research must be focused not only outwards to their research and its participants but also inwards to its own daily practices of communication, co-learning, peer-mentoring, training, and more. Transnational feminist praxis requires that members of the research team constantly grapple in both spheres with issues of accountability, collaboration, and decolonising knowledge production (see Chapter 2 Decolonising feminist knowledge production).

(ii) Teamwork

One major difference between mainstream and feminist paradigms of research partnership relates to understandings of collaboration and flows of knowledge production and dissemination. The former commonly frame collaboration in terms of a hierarchical relationship between researchers and partners, suggesting that knowledge is transferred from 'expert partners' in universities to community members without 'expert' credentials. For feminist scholars, however, decolonising knowledge through 'knowledge co-generation' is important (Klein *et al.*, 2011, p. 427). Knowledge co-generation requires cultivating relationships in which team member contributions are equitably valued and resources are equitably distributed so that all can contribute (see Chapter 2, Section 2 Decoding affinity and difference in a feminist project). Successful co-generation of knowledge requires partners to outline expectations of how groups will be governed; how they will communicate, manage the budget, share data, engage in knowledge mobilisation and monitoring and evaluation; and whether they will do all this or whether different groups will take on different responsibilities. Managing co-generation with differing responsibilities requires constant vigilance and reflexivity, as hierarchical assumptions ('the academic knows best') can easily creep in.

While much attention has been paid to exploring the power dynamics between researchers and research participants, less attention has been paid to stressing the importance of exercising critical reflexivity with regards to research partnerships and teamwork. However, a few feminist geographers (Mountz *et al.*, 2003; Houston *et al.*, 2010) have argued that teamwork is a reflective political praxis that requires ongoing collaboration with all members of the research team and with research participants (see also Chapter 2, Section 2 Decoding affinity and difference in a feminist project and Chapter 4, Section 4 The challenges of feminist scholar-activist research in transnational contexts).

In relation to the research team, Alison Mountz *et al.* (2003, pp. 41–43) argue that:

> feminist team management means working productively with differences among team members, accentuating positive contributions, creating adequate modes of communication and decision-making, and enabling equal platforms for voices . . . . Though equal participation may not be possible, forms of collaboration that destabilise power relations are. Team organisation requires collaborative tending to communication,

expectations, divisions of labour, networks of support, distribution of finances, geographical location, language, personality, position, and so on, none of which can be taken for granted.

It is important, therefore, to decide together, at all stages of the research, how the project will be carried out. Embedding EDI in research design and the daily work of the research partnership is important for feminist projects that wish to remove barriers to participation that some members of the research team may face, respect the perspectives and lived experiences of members of the research team, and ensure that all team members feel supported (Wakefield and Koerppen, 2017; SSHRC, 2021a). Open communication is also central to building trust across research teams and with research participants. While 'communication takes up much needed time . . . infrequent and ineffective communication is likely to waste everyone's time' (Klein *et al.*, 2011, pp. 437–438). Well-outlined plans and strong communication with EDI at the centre of the partnership lay solid groundwork for addressing the challenges that will inevitably arise. But even with team organisation, collaborative praxis and effective communication, teambuilding is difficult. Not all team members are equally invested or interested in teambuilding and racism, sexism, and other structures of oppression, even when under constant vigilance, will raise their heads, even in feminist projects that are committed to antiracist and decolonial practices (see Chapter 7, GenUrb example 7.2 Muddling through ethical decision-making).

Reflection exercise 9.2, using another hypothetical research scenario, asks you to think about the hierarchical power relations that can arise in the research process.

Reflection exercise 9.2: Issues in feminist team research

Imagine you are working in a research project in which you are partnered with a volunteer group of lawyers and a local activist organisation collaborating in political struggle to stop the evictions of people living in encampments in a city park.

- How will you coordinate working across this partnership to communicate important aspects of the research project, including funding and resources, data collection and analysis, and sharing research outputs?
- What hierarchical power relations may arise in this research partnership?
- How might research ethics policies, EDI, and feminist praxis help you create frameworks for conducting this research in a way that attends to and seeks to break down some of these hierarchies?

Section 4. Summary

Partnerships are valuable because they bring together people from diverse backgrounds with different expertise to collaborate, but establishing a formal partnership is also a significant undertaking that requires administrative, budgetary, and communication skills. Partnerships also require management and leadership. Feminists have thought strategically about how to resist norms of professionalisation that continue exclusionary practices, and they stress non-hierarchical, decolonial praxis in coordinating and leading research partnerships and in addressing the challenges that partnerships pose. Feminist scholar-activists have created partnerships across national borders engaging in transnational feminist praxis. Feminist engagement with such partnerships requires that researchers must be focused not only outward on the project but also inward on the daily practices of their teams, constantly engaging in self-reflexivity and solidarity with individuals and political movements within and outside the university seeking to dismantle oppressive power relations.

10 Data management in feminist research projects

Mel Mikhail

Learning objectives
After completing this chapter, readers should understand

- the meaning of data in the context of research;
- the basic elements of data management;
- the critiques of data as static objects;
- and decolonial, anti-racist, and feminist practices of data management.

Further, readers should be able to

- implement a data management plan;
- address a project's particular data security needs;
- and anticipate the practical and ethical challenges of digital data collection and storage.

This chapter draws upon mainstream and feminist theories of data, science, and the internet to open the data management process to questions of ethics and politics and especially to questions of data justice. It explains the increasingly formalised process of research data management in university research, including the data management plan, as well as questions of data ownership for feminist researchers. It introduces the field of feminist data studies, which engages with issues raised by machine learning and open access data practices as well as data archiving. Links to the many resources in this chapter are available on the book's website www.routledge.com/9781032668680.

Section 1. Mainstream and feminist conceptions of data

Data are the engine of every information system. Data, the basic units or building blocks of empirical research, are 'observations about the social world' (Garwood, 2006, p. 57), and, as such, there is a vast range of data – quantitative and qualitative, textual, and visual. The data you collect during research depends on the nature of your methodology, methods, and your research question(s). Primary data, for example, can come from field notes, research diaries, or the transcripts of interviews. And secondary data sources can come from outside the field: for instance, from archives, censuses, newspapers, and popular culture, from both digital and analogue sources (see Chapter 11, Section 5 Qualitative methods in feminist research).

Mainstream research paradigms differentiate data from information and knowledge, the former being raw 'bits' of information that have no meaning on their own (Tuomi,

DOI: 10.4324/9781032668727-14

2000, p. 112). Researchers analyse data to produce information about a phenomenon, event, or object of study, which forms the basis of purportedly impartial scientific knowledge (Tuomi, 2000). Feminist researchers, however, view the research environment as a dynamic and value-laden epistemological field that also implicates their own social locations differently. This means that data cannot be disentangled from value and 'bias'. For feminist sociologist Ann Oakley (1981), 'raw' data are already structured by ideology, or a framework of ideas that make meaning of the world, since research methods are created and employed by humans from particular social and political positions. Thus, feminists understand data as being always already value laden. Unlike mainstream accounts that view such 'bias' negatively, feminist approaches to research see bias as unavoidable, although researchers can take steps to reduce levels of bias by ensuring the rigour of the research process at all stages (see Chapter 7, Section 3 Feminist ethics and research). Consequently, feminist methodologies work to cultivate awareness of the value-ladenness of data and grapple with the ethical and political consequences of this fact.

Feminist approaches to data draw upon feminist theories of knowledge which recognise that all knowledge claims are located, partial, and embodied, produced through (and productive of) gender, race, class, and other axes of difference (see Chapter 11, Section 3 Feminist epistemologies). For feminist geographers, it has also been important to account for how space mediates and produces social difference and thus knowledge about the world (see D'Ignazio and Klein, 2020). Such considerations matter in the digital register as well, since a lack of attention to socio-spatial considerations misses the extent to which knowledge produced online still reflects Western, masculinist ontologies. Recent initiatives by feminist anti-colonial scholars in the global South, such as the global campaign 'Whose Knowledge?', point out the need to 'decolonise the Internet' – and, specifically, its data structures and languages – to begin building an internet that reflects the fact that 75 percent of internet users are from the global South, while 45 percent of all women in the world are online (Whose Knowledge?, 2019).

Bringing these feminist, anti-racist, and decolonising considerations to bear on research data management allows us to recognise that collecting and working with data from the field – be it in person or online – is always charged and full of meaning. In turn, the question becomes: to what ends will data be used? In what follows, we investigate the practical and technical aspects of implementing an appropriate research **data management plan**. Reflection exercise 10.1 can help you to start thinking about the theoretical implications of research data.

Reflection exercise 10.1: Types of data and their implicit values

Consider the elements of your own research project(s) or research interests – the places/locations, people, structures, systems you might be engaging with.

- What types of data would be the most suitable to collect (e.g. quantitative, qualitative, textual, visual)?
- What methods do you think would be the most appropriate for collecting the necessary data?
- Having decided what methods you think you might use, what kinds of biases and ideological beliefs might you expect to inform your data?

In the following section we turn to issues of research data management and, specifically, the data management plan. We consider how to devise your own plan and discuss why questions of data ownership are not normally addressed in such plans.

Section 2. Research data management

In its most basic form, research data management refers to the plans you make for the collection, analysis, securing, preservation, archiving, dissemination, and disposal of data procured for a research project. Proper research-data management ensures that data are organised, classified, preserved, and secured in ways that facilitate the smooth operations of the research process (Surkis and Read, 2015). Data should be managed according to relevant standards (for example, that they be interoperable (the ability of a product or system to work with other products or systems) or proprietary; accessible and widely available or private; archived indefinitely or disposable). For social groups that have been historically dispossessed of bodily, spatial, and discursive autonomy, such as people with disabilities, the stakes of data management can be particularly high. Thus, it is crucial that feminist researchers develop mindful data management strategies that reduce the likelihood of misuse. Feminist data management strategies should protect the privacy, rights, and wishes of communities that are very often subject to a politics of 'big data' (i.e., data that are so large, fast, or complex that they require new systems to process) that understands their lives and experiences as alienable sites from which to mine content.

Figure 10.1 illustrates a common data management workflow. After planning for and launching the project, the 'data life cycle' identifies the different phases of data collection, analysis, storage, preservation, and dissemination that you may pass through before the project is terminated and data are archived or appropriately disposed of (if necessary).

Universities and national research governance bodies have begun to establish protocols for data management. For example, the Canadian government has published a

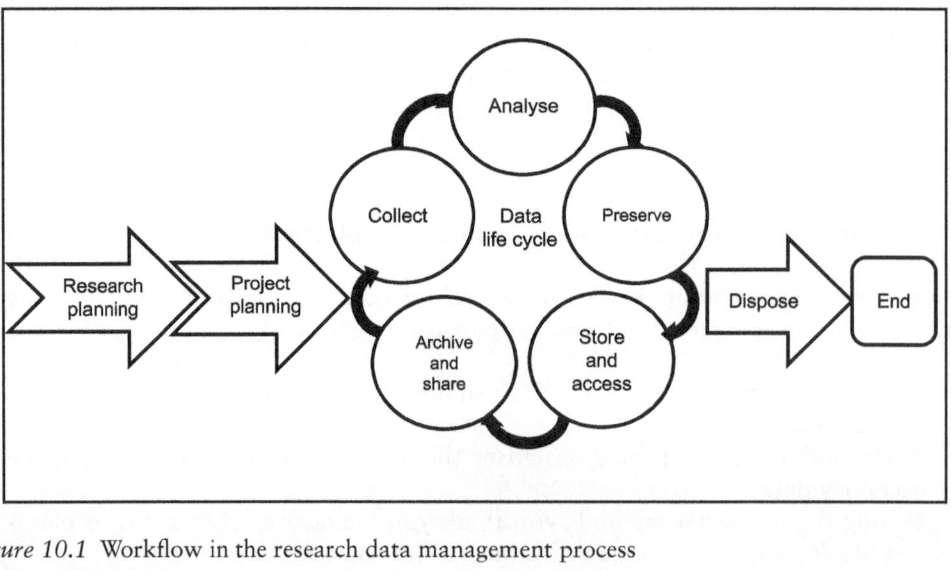

Figure 10.1 Workflow in the research data management process
Source: Mel Mikhail

Tri-agency research data management policy that outlines guidelines for institutions and individual researchers (CIHR, NSERC and SSHRC, 2022c). The Canadian Social Sciences and Humanities Research Council also runs a 'Research data management capacity building initiative', which aims to help researchers better understand data management and incorporate data management considerations into research practices. Increasingly, universities' research departments offer resources on data management, including workshops, online guidelines, or on-site data scientists and librarians. Be sure to check with your university about what resources are available to you.

These protocols for data management may differ between fields, research organisations, or universities, especially when working across national contexts, but what they all do is require information from researchers about their intended approach to data management. This usually takes the form of a specific plan, which we turn to next. We will also take a look at devising your own plan.

(i) The data management plan

Researchers often now find themselves having to complete a data management plan to satisfy institutional or funder requirements for more stringent data management (see Document 10.1 GenUrb's *Data Management Plan* on the book's website www.routledge.com/9781032668680). While not all funding bodies or research stakeholders yet require one, composing data management plan. A clear data management plan at the beginning of a project can nonetheless help you to ensure transparency around team roles and responsibilities, promote efficiency, and help eliminate logistical problems in reporting, data sharing, and dissemination phases.

These plans must articulate the various attributes of the 'data life cycle', including:

- data collection – the process of gathering data;
- data analysis – the application of techniques to evaluate data;
- data storage and access – the keeping of data in secure locations and the process by which team members can access the data;
- data preservation – the protection of data security and integrity;
- and data dissemination and archiving – the process of sharing and distributing data.

Data management plans are documents that outline the approach you and/or the research team will take to research data management – how you will collect, analyse, preserve, store, and archive/share your data – and the practices you will use to achieve your ends. Such plans also contain information pertaining to accountability and management – who is responsible for the data and other relevant team roles – as well as any relevant data-related constraints (ethical, legal, or otherwise). Data management plans are 'living documents' that can be updated and amended throughout a project life cycle.

During the early phases of research design, there are four areas of your data management plan requiring particular attention:

- financial sustainability – you should establish that you have a sufficient budget for all aspects of your data plan, including what resources, such as software or hardware, you will need;
- laws and regulations – indicate your understanding of and willingness to comply with intellectual property laws and regulations governing your data, e.g. in relation to

informed consent and data sharing, including the laws and regulations that apply at the research sites;

- research team organisation – specifiy how you will communicate about data collection, storage, access, and security to ensure all team members follow the same protocols as well as how you will resolve conflict over data use and care;
- an account of access – pay attention to how you and/or the research team will address the growing demand that research data be open access (see Section 3 Feminist perspectives on data management).

Most funding agencies will require some combination, if not all, of these institutionally formalised considerations to be present in a data management plan.

GenUrb example 10.1 provides a look into the complex dynamics around data management protocols that can arise in a transnational feminist research project, suggesting that regular engagement with the data management plan as a living document can help alleviate or mitigate potential disagreements.

GenUrb example 10.1: Data management issues

At the time we submitted the grant application for the GenUrb project, SSHRC encouraged but did not require a data management plan. The grant application did not have a separate section on data management; space permitting, it had to be included in the main section of the application. Given the lack of space, we were unable to include any such plan; however, as principal investigator I wrote a separate data management plan that was circulated to GenUrb members. Each City Research Team (CRT) used this document to agree upon certain protocols and procedures (e.g. file naming). However, after we had disseminated the data management plan and secured agreement on its use, we found that it was not discussed regularly within the project. It was perhaps not surprising, then, that differences among the CRTs about approaches to data started to appear.

Some teams, for example, had to be reminded about the data management plan, as it became clear that file-naming protocols, for example, were being ignored. A major indication of the different attitudes to data, and by implication the data management plan, came to light through a methodological disagreement when one CRT commented that they did not intend to participate in the agreed method of collecting data through interviews, preferring instead to conduct an ethnographic study based on field notebooks. That CRTs might diverge from the methods that they had initially amicably agreed to, and had signed the SSHRC agreement that they would undertake, had not been anticipated. While the team lead in question came to agree that interviews were necessary to ensure that some basic, comparable data were collected across all sites, it could no longer be assumed that agreeing on the contents of, or signing, a document, including the data management plan, would automatically be followed.

These differing approaches as to how we would proceed caused frustration and tensions between the CRTs and highlighted that the data plans were being interpreted in differing ways. This illustrates the structural limitations of a cross-cultural project: for some CRTs, the data management plan was understood as having been finalised and for others it was still open to interrogation. We realised that workshops had to be held with each CRT to remind them of the data management protocols. With hindsight, paying more regular attention to the data management plan might have alleviated some of the problems we faced, although it would not have resolved underlying differences between the CRTs. The significance of a data management plan in our transnational research context only came to light through its imperfect implementation.

By Linda Peake

(ii) Devising your own data management plan

In this section, we present a road map for producing your own data management plan. Table 10.1 poses relevant questions for each moment of the research process that should aid the process of drafting your plan. Some questions also invite you to consider the process of planning data management from a feminist, anti-racist, and decolonial perspective. The second column provides resources that may help you address these questions in the context of your own research. Researchers should inquire with their university's research centres and libraries about institutional access to proprietary data management tools. Remember that most universities have dedicated staff with expertise in data management; consider consulting them when drafting your plan. Also note that a data management plan requires close attention to detail and an ability to keep larger project goals and trajectories in mind.

Table 10.1 Data management considerations for each stage in the data life cycle

Questions to consider	*Resources (if applicable)*
Data collection, documentation, and management	
What is your data documentation plan?	See, for example, the University of Helsinki's *Guide for data documentation in the social sciences* (Fuchs and Kuusniemi, 2018). The Data Documentation Initiative offers an international standard for data description that provides valuable resources and information.
What power dynamics exist in the collection, documentation, and management phases, and what steps are you taking to ensure a horizontal and collaborative process?	The *Centering racial equity throughout data integration toolkit* (Actionable Intelligence for Social Policy, 2022) provides helpful resources on ensuring racial equity throughout the data life cycle.
How will you approach data management practices in cross-cultural research?	Numerous resources exist for interpreting and translating across languages. See Chapter 15 Feminist practices of translation and interpreting.
What type of data will you be collecting (images, text, geospatial, or numerical data, etc.)?	Resources will vary depending on the research methods you are employing. See Chapter 11 Feminist methodologies and methods.

(*Continued*)

Table 10.1 (Continued)

Questions to consider	*Resources (if applicable)*
What file-naming convention will you use? Clarifying such conventions early on will help to establish consistency and organisation.	Bulk Rename Utility (2023) provides a free online tool you can use to rename a large set of files all at once.
Do you need to anonymise your data? The answer to this will depend on your research goals for knowledge mobilisation.	The *Anonymisation decision making framework* (Mackey, Elliot and O'Hara, 2016) can be used to share data without breaching confidentiality, although it primarily concerns large datasets in a corporate context.
What is your data backup plan? This will help to ensure you do not lose your data irretrievably.	One popular backup method is the 3–2–1 rule: keep three copies of all data; store data on two different media; keep one copy off-site.
Data processing and analysis	
What format(s) will you use to store your data? NB: Certain file formats are better suited to file types (e.g. the JPEG format is a more versatile format for images than the PDF), and over time, operating systems and technologies will change.	The UK Data Service (2023) provides a guide to recommended file formats which can help you make this decision.
What kind of software will you need to purchase and/or download to analyse your data? This will depend on your research design.	Examples of relevant software include Dedoose and NVivo (see Chapter 17 Software-aided analysis for feminist research and Chapter 18 Using NVivo in feminist research).
Data storage and access	
What is your plan for data storage? How long will you need to store your data for the purposes of your own research project, and what hardware or software will you need?	Cloud-based services are useful for projects involving collaboration but may not be the most secure. If your research is hosted at a university, long-term storage options may be provided through your institution.
You may need to keep copies of de-anonymised data that contain sensitive information or find ways to prevent others from de-anonymising data. How will you secure and/or encrypt sensitive data?	The Massachusetts Institute of Technology Poverty Action Lab has developed security procedures for researchers (see O'Toole *et al.*, 2018).
Data preservation and archiving	
Do you know which archives and repositories are available to you for your data preservation needs?	Most academic institutions have their own data repositories. Independent researchers may use an open data repository, such as Figshare.
Do you have specific preservation and archival needs, for example, relating to the culturally appropriate preservation and storage of traditional Indigenous knowledge? Would your project or the broader community benefit from your data being uploaded to a culturally and/or community-specific archive?	Mukurtu (2023) offers a content management and preservation protocol developed in conjunction with Indigenous communities and researchers. The South Asian American Digital Archive (SAADA, 2023) is another digital archive, managed by members of the South Asian American diaspora.

Table 10.1 (Continued)

Questions to consider	*Resources (if applicable)*
Data sharing and dissemination	
Who can use your archived data? This question requires addressing access permissions, privileges, and restrictions.	If you and your team wish to make the data freely and widely accessible, the *Open data handbook* (Open Knowledge, 2023) is a valuable resource.
Can the data from this project contribute to any long-term, collective memory-building projects? Would such dissemination require active and ongoing consent from any persons or communities, and can you obtain such consent?	See, for example, the Chicana por mi Raza (2023) feminist digital memory archive.
Data 'destruction' and disposal	
Note any university-specific guidelines with respect to data destruction (such as how long you can keep data), including destruction certificates (if applicable).	Data disposal procedures will be determined on a case-by-case basis and/or by institutional requirements.

Source: Adapted from information provided by the UK Data Service, 2023

Remember that a data management plan is a living document. Unforeseen circumstances will arise and may require changes in the plan, so it should be open-ended enough to account for this unpredictability. You should also think through how you will ensure that standards are applied consistently through all these phases. Reflection exercise 10.2 can help you think through how to devise your own data management plan.

Reflection exercise 10.2: Brainstorming a data management plan

Imagine you are the lead investigator on a one-year research project that will investigate the impacts of urban expansion (for example, urban sprawl, infrastructure development) on residents of a rural area adjacent to a city with which you are familiar. Imagine you will use a mixed-methods approach, conducting interviews (offline), analysing census data (online), and studying historical records (online and offline).

Using this information, take time to fill out the sample data management template that follows. You may want to adapt this step's context to better reflect your own research priorities.

Data description. Provide a brief description of the data you will gather and the scales at which the data will be collected.

Ownership and responsibility. Demonstrate knowledge of university or research centre, funding bodies, or government policies around data ownership and any agreements that might need to be entered into between research participants and institutional bodies. This includes addressing questions around intellectual property rights and copyright.

Sharing. Outline your plans for sharing data among members of the research team and how shared data will remain easily retrievable.

Data access and metadata. Outline the metadata (the contextual information that describes and explains the data including, for example, author, date created, and file size), documentation, and any other material needed to interpret the data, as well as your strategy for making the data open access if applicable.

Data analysis. Detail your plan for analysis, including the software or hardware you will use.

Consent. Address any issues of consent, confidentiality, and anonymity.

Storage, archiving, and preservation. Outline short- and long-term storage plans. Specify the archives you will use for data storage, if any, and how access will be negotiated.

Source: Adapted from the Digital Curation Centre UK (n.d.) and the Inter-University Consortium for Political and Social Research (2023)

(iii) Addressing data ownership in feminist research

Data management plans often fail to address data ownership because mainstream research paradigms assume that data are owned by the researchers and their institutions. Given that university researchers are bound by the data ownership and intellectual property policies of their respective institutions, such policies towards data access may differ accordingly. Moreover, ownership of data can create unique challenges for research projects that work across multiple institutions or countries. In these situations, data may be managed across teams that are bound by different legal, political, cultural, and disciplinary norms. The result can be that researchers in different places come to the research project with different expectations about practices related to data sharing (see GenUrb example 10.1 Data management issues).

A range of possible paradigms – including, for example, possession, accountability, execution, production, curation – can be used to claim data ownership based on the type and degree of contribution involved in the research. In GenUrb example 10.2, we consider the questions we used to address data ownership in GenUrb. The formal data management plan that we established did not include any statement on data ownership. Hence, the CRTs discussed data ownership before data collection began and agreed that each CRT should own its own data. Most participants had no interest in having any engagement with their data, so we decided not to transfer ownership of the data to participants themselves. We believed that the project met a duty of care towards the research participants and their data through our commitment to an interview protocol based on informed consent that told participants about their rights in relation to the process of data collection (see Chapter 14, Section 2 (iv) Obtaining consent).

GenUrb example 10.2: Questions of data ownership

Questions of data ownership	*Who in GenUrb owns and has access to data*
Possession	
Who is creating the data?	The research participants and the CRTs
Who is responsible for producing the data?	The CRTs
Who can use the data?	The CRTs and GenUrb postdoctoral fellows
Who can access the data?	The CRTs and GenUrb postdoctoral fellows
Accountability	
Who understands the purpose/goal of the data?	Each CRT, the principal investigator, the project manager and GenUrb postdoctoral fellows
Who can verify that the data are correct?	Each CRT
Who is responsible for data loss?	Each CRT
Who is liable if data are corrupted?	Each CRT
Who fixes problems with the data?	The CRTs, the project manager, and the *NVivo* specialist
Execution	
Who decides what data should be collected?	The CRTs
Who is responsible for collecting the data?	The CRTs
Who can add data to the system, remove data, or change the data on the system?	The CRTs and the NVivo specialist
Who decides who can access the data?	The CRTs
Production	
Who can benefit from the data?	Academics and the general public
Who can publish the data?	The CRTs and postdoctoral fellows
Who can share the data?	The data can only be shared within GenUrb
Curation	
Who owns the data after the project has finished?	The CRTs
Who determines whether or not to archive the data?	The CRTs (who have all agreed not to archive the data for reasons of confidentiality)
Who can choose the data repository?	The CRTs

By Linda Peake

As GenUrb examples 10.1 and 10.2 suggest, the embedded power geometries of transnational research projects can create ethical and political tensions that require a great degree of work and care to navigate.

In addition to data ownership questions relating to project members, feminist scholars must also address data ownership in relation to other organisations. Data ownership implies not only legal possession of but also responsibility for information. This responsibility includes defining the use and distribution policy of the data. In other words, data ownership is primarily a data governance process: the data owner can create, edit, share, and restrict access to the data. As mentioned, determining data ownership is not merely a technical exercise, and you may have to consider not only whether your institution owns your research data if they funded the research but also whether any other institution, such as public- and private-sector sponsors that funded your research may also make claims to ownership. You should also consider your relationship to publishers, who may want you to sign a copyright document that covers the material, including data, in the publication. Before signing you should correspond with your publisher and clarify the policies involved. It is important to delineate the rights, obligations, expectations, and roles played by all interested parties. Compromises to data integrity can occur when you are not aware of data ownership and thus fail to clearly describe rights and obligations regarding data ownership.

Reflection exercise 10.3 gives you an opportunity to apply these ideas about data ownership to your own research.

Reflection exercise 10.3: Problematising ownership of and access to data as feminist researchers

Today, data and information are forms of currency, and having ownership of the data and information we produce or which are produced about us is a key area of political struggle. As a feminist researcher, you will likely encounter situations regarding the ownership of and access to data about research participants that require you to make complex ethical decisions.

Consider your own research – a project of which you are currently or have been a part or a potential project.

- Who was, is, or would be involved?
- To whom do you believe you would be responsible when negotiating questions of data ownership?
- Would you seek out open access opportunities to share the data? If so, how would you go about doing so?
- In the context of your research, how would you begin to bridge the divide between university and national research funding bodies' understanding of data as objective, unembodied, and proprietary versus feminist understandings of data as relational?

In the following section, we turn to further feminist considerations of data management, namely feminist data studies, feminist engagement with machine learning and open access data practices, and decisions about archiving data.

Section 3. Feminist perspectives on data management

Institutionalised data management principles and best practices (such as efficiency, transparency, and increased access) risk reproducing harmful power relations, especially in work characterised by cross-cultural differences. To explore how we can reduce this risk, this section places the data management planning process in dialogue with data feminism: a growing field of study that critiques mainstream data cultures and articulates feminist alternatives. It then turns to associated questions of data injustice and also turns the spotlight on the implications of machine learning and data management. Finally, it addresses feminist engagements with archives and the archival process, in which data management questions inevitably arise.

(i) Feminist data studies

Data-driven research in the social sciences and humanities is increasingly subject to standardised digital management infrastructures that naturalise an extractive paradigm of empirical research inspired by big data companies (Leurs, 2017). This paradigm incentivises the collection of gargantuan datasets and the statistical processing of data into 'objective' knowledge about population groups. This mainstream archetype views the global South as an 'untapped' resource that is ripe for harvesting. Without critical reflection, transnational research risks reproducing the neocolonial power relations of such extractivist tendencies.

Feminist data studies is a growing interdisciplinary field that demonstrates that 'big data is both ontologically and epistemologically part of the social structures from which it emerges' (Uprichard and Carrigan, 2015). In other words, the philosophy of big data emerges as a logical outgrowth of contemporary global capitalism's search for profit. Feminist data scientists therefore ask: 'how can we mobilise feminist and postcolonial philosophies of science to develop a situated, reflexive and more ethically-sound epistemology and praxis of data-driven research?' (Leurs, 2017, p. 136).

Table 10.2 responds to this provocation by synthesising key principles for a feminist internet (that is, a digital realm that is universally inclusive and non-oppressive to women, gender-diverse, and queer people) and tenets from feminist data studies. In doing

Table 10.2 Feminist principles of the internet for more equitable cross-cultural data management

Principle	*Summary*
Privacy	Data creators (i.e. internet users) have a right not to be subject to surveillance through the data they produce.
Embodiment, embeddedness, and difference	All data about individuals are only representations of embodied realities embedded within singular, everyday contexts and should be understood this way. As such, digital security issues do not affect everyone equally.
Building power	Instead of appealing to institutions to change, feminists can commit to build power with people and communities by harnessing the liberatory aspects of digital ecosystems (knowledge and skill sharing, bottom-up community-building, etc.) to refuse data logics that replicate colonial desires for capture and discipline.
Consent and equitable participation	Consent to use someone's data should be willingly given, reversible, and well informed; data ought to be used in the interest of self-determination and to promote the self-representation of the people they concern.
Restorative and transformative data	Data made with and for vulnerable subjects must be oriented towards a data methodology of the oppressed and avoid data-extractivism practices.

Source: Adapted from Feminist Principles of the Internet, n.d., and Cifor *et al.*, 2019

so, it highlights the disjuncture between institutionalised data management protocols and feminist research ethics. No-strings-attached **data justice** and related rejection of market and governmental discipline via data capture is a feminist issue, requiring networks of care, long-term relationship-building, and stewardship.

Reflection exercise 10.4 offers an opportunity to apply insights from Tables 10.1 and 10.2. It invites you to think through a hypothetical exercise related to ethics and security when sharing sensitive data with team members.

Reflection exercise 10.4: Navigating the feminist ethics of sharing sensitive data

Read the following hypothetical scenario regarding the management of sensitive data about the lives of working women in an Indian city and respond to the questions.

Part of your research process involves gathering and analysing sensitive data about the communication habits of a group of women micro-entrepreneurs who operate in a precarious and informal context in a low-income urban neighbourhood in Chennai, India. These entrepreneurs have given you their consent to use information about their 'grey' business practices for your research project. The rest of your research team is in different parts of the world, so you need to share the data with them using digital and web-based applications. Additionally, the organisation funding the research requires that you make your data publicly available to fulfil national demands for knowledge production. This stipulation was discussed when recruiting research participants. Although the research participants appear to have no concerns about the sharing of the data, some team members have expressed reservations about this practice.

- How will you resolve this issue?
- How will you share the data with your team?
- How will you reflect your plans to store and/or share and/or disseminate sensitive data in your data management plan?
- What is your rationale in each case?

(ii) Feminist engagement with machine learning and open access data practices

Negotiating data management practices that respond to the specific ethical and social arrangements of a project while also satisfying institutional expectations, which are often indifferent to such concerns, can be difficult. The difficulty is particularly apparent in the context of the rapid emergence of artificial intelligence, AI, through machine learning and digital automation, in both education and research. The recent push towards 'findable, accessible, interoperable, and reusable', or FAIR, data, is a good example of newly formalised data priorities to promote machine learning that do not align with feminist approaches to data.

In 2016, a consortium of scientists and other stakeholders 'representing academia, industry, funding agencies, and scholarly publishers' published 'The FAIR Guiding Principles for scientific data management and stewardship' (Wilkinson *et al.*, 2016, p. 1). They argue that good data management practices serve the function of making research data widely available to and re-usable by a broader scientific community in the name of 'extracting the maximum benefit from our research investments' (Wilkinson *et al.*, 2016, p. 1). For this consortium, ensuring that research data can be easily accessed and adapted to the needs of machines is central to maximising the benefits of FAIR data practices. They call for:

- *findability* – standardised metadata practices to allow ease of search by machines;
- *accessibility* – metadata protocol to be free and open and metadata itself to remain available even after the original data have been destroyed;
- *interoperability* – metadata and data production should follow protocols that are interchangeable with many different systems;
- *reusability* – data and metadata can be easily interpreted and understood for use in many different applications.

As feminist researchers committed to anti-racist and decolonial research practices highlight different values, however, understandings of findability, accessibility, interoperability, and reusability are likely to differ. A commitment to a feminist research ethic that prioritises the needs of research participants and the broader goal of sustained social change means that data extraction and use priorities are more likely to reflect those of the communities we work with rather than those of the broader scientific community.

On the other hand, open data practices (a publishing model that makes research data available to readers at no cost) can be a crucial step in the process of creating equitable data paradigms in line with feminist research practices (Craig, Turcott and Coombe, 2011). Not all feminists agree with making their data open access (increasingly a requirement of funding bodies), given that doing so can be at odds with promises given to research participants about anonymity and privacy but also that allowing open access to data decontextualises them from their relational meaning, from their creation in the relationship between the researched and the researcher. Open access does not automatically mean increasing transparency, accountability, and FAIR access in all contexts, especially cross-cultural ones where users may be affected by the digital divide (whereby gaps in access to and usage of internet and other advanced information communication technology can be observed affecting populations in the global South more than those in the global North). To encourage truly collaborative cross-cultural data partnerships, relations of power and knowledge around technology across contexts must be identified and capacities developed (Dalton, Taylor and Thatcher, 2016; Johnson, 2016; Heeks and Renken, 2018; D'Ignazio and Klein, 2020).

Open access data have specific advantages. They are universally accessible to all internet users at no cost to the user; they allow for the copying and distribution of the data; and they are available in international open access data repositories, such as OpenDOAR, arXiv, or Zenodo, among many others. Yet the adoption of open access protocols alone does not guarantee equitable practice, which depends on a community of feminist and other critical researchers actively working to encourage ethical open-data practices. Considering open access data practices in the context of the 'open science' paradigm (a move towards transparency and accessibility of information in the natural and social sciences),

Mary Brabeck (2021) identifies five key themes for feminists working towards ethical open access practices:

- encouraging the participation of women, people of diverse gender identities, and other social groups who have been historically left out of the project of knowledge production in the development of open science practices;
- addressing implicit biases in readily accessible scientific data owing to researchers' subject positions;
- addressing the limits of transparency in data access and sharing – what they are and what they ought to be;
- confronting institutional power dynamics and the lack of community access and participation in decisions around open access;
- and taking action to adopt a feminist ethics in open science projects.

(iii) Archiving data

Archiving concerns the life that data take on after a research project has concluded. Feminist and decolonial scholars demonstrate that archives are not neutral sites of information storage: rather, they privilege some forms of knowledge and viewpoints over others (Spivak, 1985; Stoler, 2002; Arondekar, 2005). Indeed, these institutions are defined by a 'useful in/stability': a precariousness that shapes knowledge through removal, destruction, and exclusion as much as through care, storage, and inclusion (Gieseking, 2015, p. 25). Feminist researchers should anticipate the consequences of archiving data: Whose stories are being recorded? Is participant anonymity being respected? How could the data affect future scholarship? How could access to this information affect the participants' future lives? Through a critical reflection upon data afterlives, feminist scholars can work to challenge extractive research paradigms and put the archive to work to different ends.

Feminist archival practices can take different forms. For example, archiving in feminist research can function as a 'small data' counter-practice that challenges big data paradigms of representation (Strobel, 1999; Heeks and Renken, 2018). When researchers are working with grassroots organisations, archiving can be used to preserve the memories of groups that harbour subjugated or situated knowledges (Strobel, 1999; Arondekar, 2005) (see Table 10.2). When they are working with official archival institutions, archiving can work to subvert the hegemony of official narratives by preserving the self-representations of groups who are marginalised through mainstream archival paradigms. How might these insights about the useful in/stability of community archives translate to digital archival practices?

GenUrb example 10.3 details the decision taken in GenUrb not to archive data.

GenUrb example 10.3: Archiving data

The *Tri-Agency statement of principles on digital data management* states:

> The ability to store, access, reuse and build upon digital research data has become critical to the advancement of science and scholarship, supports innovative solutions to economic and social challenges, and holds tremendous potential for Canada's productivity, competitiveness and quality of life.
>
> (CIHR, NSERC and SSHRC, 2021b)

The sharing and reuse of data is not new in the social sciences, but it is still not the norm, particularly as regards social science and humanities data. There is still uncertainty about being able to identify 'which data might be shared, by whom, with whom, under what conditions, why, and to what effects' (Borgman, 2012, p. 1072). Hence, despite the arguments outlined earlier for sharing and archiving data, the GenUrb project decided against doing so for several reasons.

- The data contain confidential information about individual women. Even with the metadata removed, there is a risk that women could be identified individually, particularly in the smaller cities of Georgetown, Ramallah, and Cochabamba, constituting a breach of participant anonymity.
- We did not inform our participants that we would archive the data and thus did not request their permission to do so.
- Without knowledge and understanding of the geographical and political context from which the data emerged, the data could serve little purpose in terms of further secondary analysis. Indeed, the secondary value of such free-floating data is of a dubious nature.
- One reason for sharing data is for replication of data and verification of findings. In our research, this is not feasible because the conditions in which the data were collected cannot be replicated by other researchers.
- There is no guarantee that secondary users of the research data will apply stringent legal and ethical standards to their use of the data.
- As Charlotte Borgerud and Erik Borglund (2020, p. 285) state, another challenge is that of 'choosing persistent file formats for storage, as well as ensuring that they remain compatible over time, as operating systems and technologies change on a regular basis'.

By Linda Peake

To decide against archiving your data does not mean that you do not have any more obligations in relation to the data. You may need to have control over or ready access to original research materials if you are required by funding agencies to keep data for a statutory period of time; if you are the person against whom allegations of misconduct or fraud would be made as a result of being responsible for a particular output (e.g. publication); or if you would be the person contacted by other researchers if you have agreed in principle to make data available for purposes of replication or reanalysis.

Section 4. Summary

Data management is a technical, highly digitised set of practices of collecting, analysing, storing, archiving, disseminating, and destroying data. It is increasingly becoming a core element of any research project such that data management plans are often required by research funding agencies. These articulate the type of data to be gathered, uses of the data, data governance rules, methods of data analysis, data storage, and data access practices. Feminist scholarship not only critiques mainstream data management practices, while also identifying data management as an important site within the larger research project at which to apply the principles of a decolonial, anti-racist, and feminist liberatory politics. Feminist data management practices advance alternative data management and data archiving models based on the values of justice and care. These models protect participant privacy and safety, respect participants' data sovereignty, and build alternative archives.

Part IV

Feminist methodologies and research methods

11 Feminist methodologies and methods

Linda Peake and Mel Mikhail

Learning objectives

After completing this chapter, readers should understand

- the distinctions between the three fields of a philosophical approach: ontology, epistemology, and methodology;
- the relations between feminist ontology, epistemology, and methodology;
- the issues raised by methodologies in the research process i.e. reflexivity, positionality, the relational nature of research, research ethics, and accountability and social change;
- and the relationship of feminist epistemologies and methodologies to methods.

Further, readers should be able to

- determine their ontological and epistemological stances to their own research;
- apply feminist methodological principles to their own research;
- and apply feminist methods in their own research.

This chapter introduces you to the role of feminist methodologies in knowledge production. It starts with a discussion of the connections between data and knowledge production and distinguishes between the use of naturalist and anti-naturalist philosophies in urban studies. It then turns to a description of the building blocks of philosophy – ontology, epistemology, and methodology – while also addressing the same constituent elements of a feminist philosophy. The discussion of feminist methodologies engages with the issues of reflexivity, the politics of representation, positionality, the relational nature of research, research ethics, and accountability and social change. Finally, the chapter turns to the ways in which feminist methodologies and epistemologies and other factors influence the choice of research methods in feminist research.

Confusion abounds over the terms commonly employed in relation to the research process. Hence, we provide a brief overview of the most common of these terms on the book's website www.routledge.com/9781032668680 (see Document 11.1 *Introduction to the terminology associated with research*).

Section 1. Introduction to the philosophical fields underlying research

What purpose do data serve? Data comprise the material that researchers use to produce knowledge, and knowledge can help to change the world. In this way, data can, in both small and significant ways, contribute to social change. But how do we make data into

DOI: 10.4324/9781032668727-16

knowledge? How do we know what counts as knowledge? For a claim about the world to count as knowledge, it needs to be justified, which, for knowledge produced through quantitative data hinges on the reliability and validity of the data, with the equivalent for qualitative data being its trustworthiness. This is not a straightforward proposition: the procedures underlying trust and belief in the knowledge we produce vary depending on our philosophical and political positionings. Feminists, for example, do not necessarily always agree with Marxists or humanists about the ways we justify knowledge production. There is also disagreement among feminists, as feminism is characterised by diverse approaches to philosophical debates and methodologies. However, researchers of all political stripes agree that there must be ways of guaranteeing that the knowledge we produce is reliable, valid, or trustworthy.

Increasing the trustworthiness of the research is especially important for qualitative research, because unlike quantitative research, it is unlikely to be replicable or generalisable. Tom Clark *et al.* (2021) claim that trustworthiness in qualitative research should be established through credibility, transferability, dependability, and confirmability. Credibility relates to how compatible the findings are with the researcher's conceptualisation of reality, requiring that the research has been conducted according to ethical and political guidelines and that the social world defined by the research is recognisable to those participating in the research. Transferability, on the other hand, is about whether the results would make sense in other milieux. Dependability relates to the consistency between the findings and the data collected. And confirmability refers to the researcher having acted in 'good faith', not allowing their personal values or politics to sway the results achieved. This is not to say, however, that personal values and politics will not influence your approach to research; indeed, they are inherent to the philosophical approach that you adopt.

There is a plethora of philosophies employed in urban studies that can be broadly divided into two camps: naturalist approaches that use the same **methods** as natural sciences, such as **positivism** and empiricism; and anti-naturalist approaches that eschew the scientific method and turn to **hermeneutics** that understand human behaviour as meaningful and interpretable. The latter includes feminism, humanism, Marxism, postmodernism, post-colonialism, and poststructuralism (to attempt to explain these terms in the space of this chapter runs the risk of simplifying complex subjects; see Richardson *et al.*, 2017 for a discussion of these and other 'isms'). While these philosophical approaches move in and out of fashion, it is rare that they die off. It is also not uncommon that a researcher may draw on more than one philosophical approach (although you cannot take both a naturalist and anti-naturalist position in a consistent and coherent manner) (Graham, 1997).

Despite their differences, if we were to examine the building blocks of these approaches, we would find that they comprise three fields pertinent to research (see Figure 11.1). These concern our understanding:

- of the reality of the world, what makes it what it is, i.e. an **ontology**, which comprises theories of being and existence;
- of how we, as human beings, can understand and learn about that world and especially about the social world, i.e. an **epistemology**, which comprises theories of knowledge and addresses how and what we know;
- of how we do research, i.e. a **methodology**, a framework or rule book of practices and processes that govern the ways in which we do research.

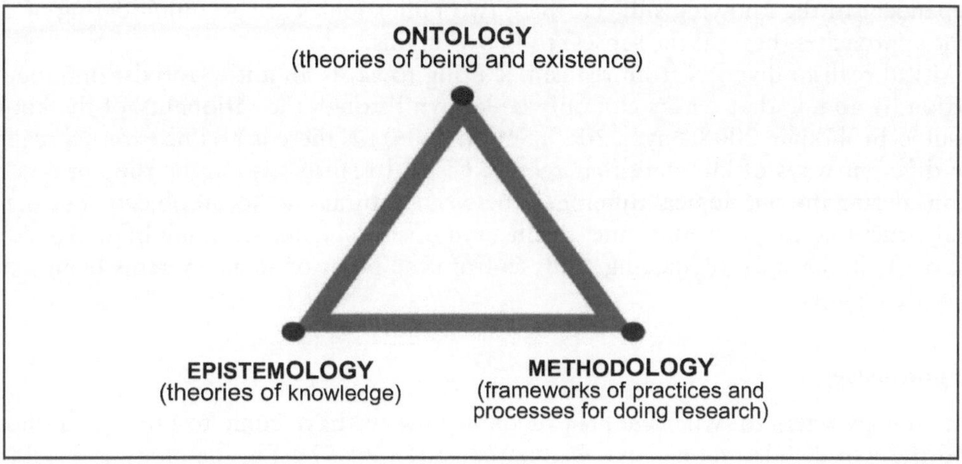

Figure 11.1 Three philosophical fields underlying research
Source: Linda Peake

The ontology, epistemology, and methodology of the philosophical approach that you adopt for research lays out the foundations for analysis. Thus, an appreciation of philosophical approaches is needed to make us aware of the linkages between theory and practice. It is important, therefore, that we discuss these in some more detail. Note that although we separate their discussion, ontology and epistemology are necessarily connected. And while ontological and epistemological beliefs may seem far from your own research interests, they will structure the way you ask research questions and the concepts you use (Hammond and Wellington, 2013).

(i) Ontology

Ontology concerns beliefs about the nature of reality, bodies, being, and existence. In the history of Western philosophy, ontology has aimed both to articulate the 'essence', the most irreducible and unchangeable elements, of existence – although this 'essentialist' impulse is widely critiqued by feminist and other critical philosophers – and to conceptualise the conditions of possibility for perceptive self-consciousness of the world.

A divide exists between those who believe in an objective reality that exists independently of the knower – an objectivist ontology that characterises empiricism, positivism, and realism – and those for whom reality is subjective and contingent, constantly being constructed, the meanings of which are dependent on the perceptions and assumptions of the knower – an anti-positivist ontology informed by constructivism or interpretivism. While strict objectivist positions are no longer widely held in the social sciences, this is not the case across all intellectual and geographical contexts, and you should be aware that the extent to which positivism holds sway varies across countries and disciplines.

The ontological position of empiricism posits that the external world is an empirical reality known through our senses. The philosophies of positivism and realism both also hold that the external world is hidden from us, exists beyond the world of appearances, and is independent of our knowledge. While accepting that a world exists that is independent of the knowing subject, these two philosophies are accommodating of different approaches (beyond the senses) to knowing this.

Critical realism diverges from realism, shifting towards an anti-positivist ontological position by adding that reality can only be known through the rationality of the knowing subject (Bhaskar, 2008; Sayer, 2000; Smith, 2006) i.e. there is no one external reality, only different ways of knowing that reality. Critical realists also argue the importance of considering the ontological differences between natural and social objects. For them, social structures are place and time specific and open to change through human agency. Hence, the importance of starting analysis from the point of social systems being open rather than closed.

(ii) Epistemology

Epistemology refers to 'what we believe about how we have come to know and understand the world' (Hammond and Wellington, 2013, p. 57). Epistemology is a field of philosophy concerned with whether or how we can have knowledge of reality. As such, epistemology determines how beliefs and information might obtain the status of knowledge, establishing rules about what constitutes valid knowledge i.e. knowledge that represents reality or the truth and the justification for this.

A strong divide exists in terms of philosophical approaches to epistemology, namely empiricism, positivism, and realism versus interpretive epistemologies. The former share the epistemological principle that the natural and social worlds can be studied in the same way. Empiricism is based upon 'facts' derived from systematic observations of the world; only knowledge gained through experience and the senses is acceptable. Positivism similarly holds that knowledge is derived from the external world, but this knowledge is then subjected to procedures established in natural science, mainly the collection of data, the testing of hypotheses, and the establishment of laws. This is a set of procedures commonly referred to as the logico-deductive method (see Chapter 16, Section 1 Data analysis). Although positivism is most frequently associated with deductivism, it is also amenable to inductivism, i.e. knowledge arrived at through the collection of facts that provide the basis for laws.

Interpretivism's primary epistemological principle is that the natural world and social world require different ways of knowing (and different logics of research) (Clark *et al.*, 2021). It sees social research as requiring a different logic from research on the natural world, concerned with 'uncovering the meaning associated with social activity' (Hammond and Wellington, 2013, p. 57). It has spawned a variety of approaches, including hermeneutics, phenomenology (the study of how individuals make sense of the world), and critical realism. Critical realism accepts epistemological relativism (that knowledge is always located and partial) and that inquiry goes beyond observation, although this is still important, to engage with the deeper underlying structures in society that shape reality and which need to be understood to transform the status quo.

(iii) Methodology

Methodology, or the philosophy of methods, refers to the framework of research principles, rules, and procedures that guide the research process. While this understanding may appear to be straightforward the line dividing methodology from epistemology is not always clear, as methodology can be defined in an overly broad way, covering, for example, 'the nature (or theory) of knowledge; the approach to empirical research; and the specific methods used' (Hammond and Wellington, 2013, p. 109). For instance, the feminist researcher Faye Harrison, in her articulation of a methodology and how it intersects with epistemology, states (Harrison, 2007, p. 25):

> methodologies articulate conceptual, theoretical, and ethical perspectives on the whats, whys and hows of research and the production of knowledge . . . . Methodologies provide the philosophical or logical rationale for the links researchers make among theory, pragmatic research strategies, evidence, and the empirical world . . . . A feminist methodology clues us in on which combination of methods is likely to be most suitable for meeting the pragmatic and ethical objectives of a feminist research project.

For some researchers, this broad definition is more akin to that of a conceptual framework (this might be a good point to turn to the book's website and look at the *Introduction to the terminology associated with the building blocks of research*). While the dividing line between methodology and epistemology can be blurry and can differ between philosophical approaches and disciplines, the distinction between methodology and methods, which are the particular techniques used to collect data and produce knowledge, should be clear. A methodology provides the rationale for the use of particular methods in that it configures particular methods into a research practice that is analytically commensurate with understanding the theoretical and empirical objects of knowledge constructed by the ontological and epistemological presuppositions of the research. It does not, however, determine the use of a particular method or methods: in other words: a methodology can accommodate many methods. This indeterminate relationship between methodology and methods results in a gap between theory and research practice, which can be mediated by well-thought-out research questions.

According to Elspeth Graham (1997), there are two aspects to this gap. The first aspect refers to the way that a research question (or the aim of your research) is articulated will contain

> assumptions that tend to direct researchers towards certain methods . . . . Thus deciding what to study and how you are going to study it are not separate dimensions of research design but are closely intertwined. This means that in formulating your research aim (or aims) you are to some extent prejudging the issue of what methods are appropriate for achieving your objective/s.
>
> (Graham, 1997, p. 30)

In other words, the choice of appropriate methods is inextricably entwined with the formulation of your research question.

The second aspect speaks to how the specific framing of the research question will also hint at how your epistemology can inform your methodology. 'For example, the positivist may ask "cause and effect"-type questions', such as, 'How does class background affect experience of domestic violence?', while the (feminist) interpretivist may ask, 'What different meanings have been ascribed to concepts of class and domestic violence?' and 'What type of explanation has been put forward to argue that class influences experience of domestic violence?' (questions adapted from Hammond and Wellington, 2013, p. 115). These questions point not only to different epistemological interests (facts versus meanings) but also to different methods (quantitative versus qualitative). The first question is best answered using a large-scale survey (preferably based on a random sample), while the second question requires the use of qualitative methods, such as in-depth interviews, and the third needs an extensive review of secondary sources and data. Another point to keep in mind is that research questions may well change over the life of a research project, meaning the methodology employed may also need to be reconsidered.

The following sections build upon these constituent fields of a philosophy (ontology, epistemology, and methodology), turning to their articulation in feminist research.

Section 2. Feminist ontologies

Anti-essentialist feminist philosophers refute the 'purity' of ontology, tending to blur its boundaries with epistemologies. Anti-essentialist feminist ontologies are anti-hierarchical, disrupting the gendered dualisms of mind/body, object/subject, reason/emotion, thought/sensation, knowledge/experience, culture/nature, public/private, and universal/particular that have been prevalent in the dualistic (and masculinist) ontologies of the Western philosophical tradition. Dualistic ontologies are based on a logic of domination – 'the belief that men were superior to women, and that knowledge and mind were superior to experience and body' (Barbour, 2018, p. 210). The exclusion of women from philosophical thought enabled the belief that knowers were neutral, objective, and independent, requiring feminists not only to add women's embodied understandings into knowledge but also to argue that there are multiple knowledges regarding the nature of reality.

Feminist ontology emphasises relationality and is open to revising categories. Feminist philosophers have made the case that sex and gender – two categories of thought that are taken for granted by 'male-stream' philosophers as undeserving of examination – concern the structure of conscious existence and being and are thus ontological concerns: they are significant in the process of becoming a subject and a knower. Gender as a subject position, moreover, is understood as negotiated, contingent, and dynamic. Ontological questions for feminist researchers concern uncovering and understanding the existence and effects of systems of gendered interlocking oppressions, namely cisheterosexism, patriarchy, racism, classism, colonialism, and capitalism.

Section 3. Feminist epistemologies

Epistemology has been a central focus of feminist theory, although there is no one feminist epistemology, with feminist scholars traditionally adopting a range of epistemologies, including those of empiricism, standpointism, structuralism, and postmodernism (Harding, 1991). Standpointism, for example, has received much attention from feminist philosophers. Its epistemological tenet is that starting research with women's lives and voices, with those who are oppressed, gives a less distorted view of the social world. It argues that women's experience provides a more valid basis for knowledge production because women as an oppressed group see not only their own experience but also that of their oppressors; it emphasises that women's knowledge emerges through struggle against oppression (Harding, 1993). While the early accounts of standpoint theory – such as those by Sandra Harding (1986) and Nancy Hartsock (1983) – valorised the epistemic position of women, they did so by relying upon universal claims about gendered oppression. Other work (Haraway, 1991; Hill Collins, 1991) challenged these essentialising feminist tendencies. Donna Haraway (1988, 1991), for example, argued for 'situated knowledge', which acknowledges not only the partial and situated nature of all knowledge production but also its embodied nature. She argued that situated knowledge also 'allows us to become answerable for what we learn how to see', thus adding **accountability** to the feminist remit for the knowledge we produce (Haraway, 1988, p. 583).

As scholars of **Black feminism** have pointed out, the notion of 'women's standpoint' is problematic: 'women' is not a unitary category, and women's experiences differ across race, class, sexuality, and other social differences (Hill Collins, 2000; Mullings, 2000; Porter, Sulé and Croom, 2023). Furthermore, if knowledge is produced from standpoints, then why should women's knowledge be revered above that of others? Researching the experience of women still has epistemic value, however, if it is recognised that there are multiple standpoints that constitute points of entry into research to understand women's lived experiences of social difference.

More recently, feminist, trans, and queer scholarship has also led to questioning of the gender binary, of the categories of 'woman' and 'man'. An epistemological focus on **embodiment**, on gendered identities coming into being through performativity, has led to the recognition that gender exists not in terms of a binary but along a continuum. The concept of embodied difference is of major concern to feminist epistemological questions of knowledge production: bodies, like minds, have 'explanatory power' (Grosz, 1994, p. vii). Subjectivities create knowledge not only through reasoning but also through the practice of their different bodily, situated, and affective experiences. Like phenomenologists, feminists claim that 'embodiment is the existential condition of being a person' (Barbour, 2018, p. 220, discussing Merleau-Ponty, 1962). Karen Barbour (2018, p. 221) argues that embodied knowledge points the way to knowledge that is 'liveable':

> For an individual woman using an embodied knowledge strategy, living with alternative understandings to dominant knowledge will likely create challenges and tensions that she will have to resolve throughout her life. Resolutions will not come only through rationalization or intuition, but through embodying and living out the possibilities. In living out the possibilities, she will necessarily come to discard knowledge that is not liveable.

What unites feminist epistemological approaches is that they address the ways in which embodied experiences of difference influence knowledge production. Asking questions – such as 'how can we know and who can be a knower?' – has allowed feminist epistemology to challenge masculinist knowledge production (Hartsock, 1983; Code, 1991). It makes women's lives and voices visible. Scientific approaches that were seen as objective and disembodied ('the view from nowhere') were critiqued in favour of the view that all knowledge is situated, embodied knowledge able to reveal only partial truths (Haraway, 1988). Feminist geographers have added substantially to this understanding of knowledge production as a geographically placed process (Kobayashi, 1994; Massey, 1994; Rose, 1997). For instance, Linda Peake (2017b, p. 2334) explains Cindi Katz's contribution to the inherent material spatiality of knowledge production as follows:

> Haraway's concept of situated knowledge is often taken to refer to knowledge from a single site and a knowing subject, and, while situatedness may imply locale, it is most commonly constituted as a subject position – as Katz (2001, 1230) puts it

'a space of zero dimensions', located nowhere specifically. Situatedness, Katz suggests, implies location in abstract location to others, but not any specific geography leading to a 'politics of "sites" and "spaces" from which materiality is largely vacuated' (2001, p. 1230), erasing the effect of specific historical geographies and the difference that space makes . . . . Turning from the abstract topological representations of space inherent to Haraway's formulations she gives ontological priority to topographical (territorial) space. Katz's intention is to reinsert materiality into feminist theorizing through an insistence on studying the processes of global capitalism in particular places, each with its own specificity but also with its connections to other places.

Section 4. Feminist methodologies

Feminist methodology addresses the rules for how to engage in research practices that can be considered feminist (see also Moss, 2002). It is characterised by the following issues: **reflexivity, positionality,** relational approaches to research, research ethics, and a desire for social change (see Chapter 9, Section 3 Transnational feminist praxis and teamwork).

(i) Reflexivity

Reflexivity is important in two respects: not only do you need to be reflexive of yourself i.e. of your influence as researcher (self-reflexivity) on your research, but you also need to be reflexive about your research design (your research questions, analytical framework, methodological approach, and plans for analysis and knowledge mobilisation (see *Introduction to the terminology associated with research* on the book's website www.routledge.com/9781032668680).

Reflexivity is based upon the understanding that the researcher is not an objective entity but an embodied interlocutor who influences the research process and the knowledge produced in myriad ways. It involves the researcher becoming the object of scrutiny in a process of self-reflexivity. As Annette Markham (2017) states:

> the basic position of reflexivity is analyzing the self recursively and critically in relation to the object, context, and process of inquiry. It's more than just reflection, which is what we get when we look in a mirror. Rather, it's like trying to look at yourself looking in the mirror.

This is a useful way of understanding that you can never step outside your research.

Rooted in feminist epistemology, reflexivity involves the researcher continuously evaluating how their subjectivity – their beliefs, judgements, and behaviours – influences the research process, identifying the power relationships inherent in research, and making those power dynamics explicit. The exercise of self-reflexivity is thus key to becoming aware of your own positionality as a researcher. Reflexivity can therefore be thought of as a method that enables you to become more aware of the cultural, historical, and geographical contexts of your research, developing a sense of rigour in your research, and developing a sense of yourself as researcher.

The advantages of engaging in reflexivity include:

- increasing the authenticity of the relationship between the researcher and participants, which lies at the heart of a reflexive approach to research (Ravitch and Carl, 2021) (see what follows in (iii) A relational approach to research);
- improving the quality of the research by identifying researcher biases and beliefs that can affect the research at all its various stages;
- ensuring that the researcher is not perpetuating oppressive structures within the research more generally and specifically in relation to participants;
- holding researchers accountable through the publishing of the reflexivity practices the researcher has engaged in;
- and increasing the trustworthiness of the research.

Reflexivity is not, however, a complete solution to the problem of hierarchical relations. It is always partial and in process. As Sarah de Leeuw and Sarah Hunt (2018, p. 3) point out 'unfettered self-reflexivity' may serve to maintain the discursive and material power of the researcher (see also Coddington, 2017), while an overemphasis on the exercise of reflexivity can result in centring the concerns of the researcher at the expense of research participants. And yet, as noted, being reflexive can improve research quality and make researchers better at what they do.

Being reflexive is not necessarily easy or straightforward; it is a sensibility that you learn over time (Markham, 2017). It requires the development of procedures that can recognise and account for the harm biases may have caused and the extent to which they have affected the credibility and trustworthiness of the research. For example, being reflexive should lead to your reflecting on the politics of representation and interpretation i.e. on how effectively the research has represented participants' lived experience. In this sense, reflexivity is the first step towards making ethical decisions (Markham, 2017). You can start the process of being reflexive by keeping a written record of your feelings, assumptions, experiences, and decisions. Reflexivity also entails the responsibility to make any changes to your research design clear to others, meaning that you communicate, write up, and present your work in a way that is sensible and meaningful, being transparent about the choices you make throughout the research process. Reflection exercise 11.1 provides a set of questions that can help guide you through the practice of reflexivity.

Reflection exercise 11.1: Practising reflexivity

This exercise aims to help you articulate and reflect on reflexivity as an exercise that requires you to address your own positionality and how your privileges but also experiences of oppression affect how you conduct research, including the dynamics of your interactions with collaborators, communities, research participants, and the academy. The following questions can be used as part of a reflective practice.

- What role does my positionality play in gaining access to funding, networks, and communities?
- What role does my positionality play in the way I interact with my collaborators in research projects? What are my/their emotional/visceral reactions to my/their actions?

- Is my choice of methods and practices emotionally triggering for the participants in ways that were not predicted by the research ethics process? Why is that happening? What can be done?
- If working with a specific community, what is my relation to this community? In what ways is my daily life different from and similar to their daily life? How does this matter, and does anything need to be done?
- If working with a specific community on a specific issue, do I have a direct lived experience of the issue? How does this matter, and does anything need to be done?
- What is the direct impact of my research on the life of individuals/communities and collaborators?
- How should my relationships with individuals and communities I conduct research with evolve beyond the research project?

Being able to engage in reflexivity is a vital part of feminist scholar-activist and decolonial praxis, not as a standalone practice but as an on-going process, exercised on an individual and collective level, attuned to the challenges of feminist collaboration across difference.

(ii) Positionality

Positionality is a process of becoming critically aware of the social, cultural, political, and economic aspects of your own background, experience, education, and embodied presence in the world and how these context-specific factors have shaped your identity, values, and beliefs as well as your intellectual orientation and worldview (England, 1994b; Sultana, 2007; Ackerly and True, 2008; Faria and Mollett, 2016). Your positionality as researcher, for instance, can introduce bias and enhance power imbalances through racist, classed, heteronormative, and insider/outsider assumptions and behaviours. Positionality is always a concern given how it is tied up with power, privilege, and control (Moewaka Barnes and McCreanor, 2022). The greater the differences between the researcher and the researched, the more issues of positionality give rise to a need to address how it can limit understanding of a particular context.

The importance of positionality is rooted in feminist methodology. As we have noted, opposed to universal or objective claims to truth, feminist philosophers have shown that all knowledge is shaped by the specific contexts or circumstances in which it is situated and produced, rooted in bodies as objects of knowledge (Haraway, 1988, 1991). Bringing embodied subjectivity, voice, and presence into knowledge production makes positionality an important methodological concern for

Audio clip 11.1: On being reflexive about positionality

Audio clips recorded by the GenUrb CRTs in Delhi and Ibadan are available on the book's website www.routledge.com/9781032668680.

By Grace Adeniyi-Ogunyankin, Anindita Datta, and Swagata Basu feminist researchers, shaping your research agenda, presence in the field, data analysis, and research findings.

Audio clip 11.1 presents three members of two GenUrb City Research Teams (CRTs) – Grace Adeniyi-Ogunyankin from the Ibadan CRT and Anindita Datta and Swagata Basu from the Delhi CRT – reflexively discussing aspects of their own positionality relevant to the research they are conducting, namely their educational, economic, and class privilege.

(iii) A relational approach to research

A relational approach to research requires being aware of the relational dynamics between researcher and participants, which can vary depending upon whether the participants are socially, economically or politically more or less powerful than the researcher. This means making yourself aware of these dynamics and engaging in research with a 'receptive sensibility, meaning that you are open to changing your opinions, approach to the research, and even critical aspects of the research as you learn with and from the research' (Ravitch and Carl, 2021, p. 194).

In community-based research, feminist researchers aim to break down the divide between the researcher as 'expert' and the participant as 'non-expert', engaging in research relationally within an epistemic community of knowers. They employ methodologies that are, as much as possible, non-hierarchical, aiming to reduce the distance between the researcher and participants and often adapting collaborative approaches to knowledge production in which there is fluidity and a sharing of meaning.

Researchers who are from the same or similar cultural, age, gender, ethnic, racial, or class background as their participants may use their status as 'insiders' to gain access and establish trust with participants. Insiders can understand participants' experiences at a level of complexity that it would not be possible for other researchers to grasp in a short time frame (Mullings, 1999; Cuomo and Massaro, 2016; Jamil, 2017). However, being an 'insider' may prevent researchers from fully understanding the experiences of their participants. They may find it difficult to separate their own cultural assumptions from the narratives of their participants, and as Dána-Ain Davis and Christa Craven (2022, p. 71) point out: 'shared membership in a group does not automatically mean there is complete sameness within that group'. The assumption of being an 'insider' can also serve to obscure lines of difference between researchers and participants, for example, along the lines of education or class (Gunaratnam, 2003).

Researchers who are 'outsiders' to the community being researched need to be aware of the ways in which the differences between them and their participants affect their engagement with that community. They need to establish trust with research participants, and they may need to negotiate access in ways that are accountable for the power and privilege they exercise. At the same time, outsider researchers may, in some cases, be better positioned to deal with topics without preconceived assumptions that 'insider' researchers may bring to the research.

Of course, a researcher's status as an insider or outsider may be fluid and shift at different points. Indeed, binary notions of inside and outside can obscure the effects of the social position of researcher. For example, Ghazala Jamil (2017) reflects on the way that her identity as a Muslim woman positioned her as an insider in the communities she was researching. At the same time, she had to differentiate herself as an outsider, a researcher, to project herself as 'worthy of initiating and facilitating a discussion about Muslims and

Muslim localities in Delhi' (Jamil, 2017, p. 104). Jamil concludes that the 'intuitive methodological tightrope act of balancing insider–outsider roles afforded to me a situated view that threw up insights and helped me access the complex and embedded knowledge of people about their own lives' (p. 105).

Reflection exercise 11.2 asks you to consider some questions you can ask yourself about relational aspects of your own research.

Reflection exercise 11.2: Engaging with others

Building on *Reflection exercise 4.1: Situating the self in feminist activist praxis* we now ask you to address how you situate yourself in relation to research participants and other people you may be conducting research with. How does difference, and specifically, your positioning as insider/outsider, affect your relationships? Deliberate on the following questions in relation to your own research.

- How are your relationships with your research participants affected by your positionality as a feminist researcher?
- How is your relationship to the field (site) you are working in affected by your positionality as a feminist researcher?
- If there are other people engaged in your research, such as faculty, research assistants, and/or other students, do you stand in different epistemic relations to them? If so, what are the potential issues you need to be aware of?
- To whom are you accountable for the knowledge you produce?
- How will you cultivate relationships with the people you are working with so that trust is established and that dynamics of power and privilege are addressed?

(iv) Ethical issues

Ethical issues, such as those of care, arise from the relational nature of feminist research and the unequal relations between researchers and participants. Chapter 7 on research ethics discusses in detail these ethics in practice and the decision-making that has to take place as researchers deliberate on and respond to different ethical situations. A key element of the relational ethical dynamic is concerned with how researchers strike a balance between caring for participants, conducting the research, and addressing other responsibilities they may have (Hallowell, Lawton and Gregory, 2005; Jackson, 2021). This division of labour may result in ethical issues falling under the radar as the time for reflection becomes squeezed. A feminist methodological approach can take this into account through creating the time and space during projects and within research teams to 'discuss ethical concerns, make changes to procedures, and . . . enable researchers to develop their ethical competencies' (Jackson, 2021).

(v) Accountability and social change

Social change is important to feminist scholar-activists, given their desire not only to understand oppression but also to challenge it (see Chapter 4, Section 2 Being a feminist scholar-activist). While not always achievable, there is a desire among feminist scholar-activists to conduct research ensuring that researchers can be held accountable for the knowledge they produce and that participants can exercise agency to work towards social change. In GenUrb, our practices of accountability speak not only to the knowledge we produce – we are committed to methodologies that engage in the production of anti-oppressive, anti-racist, and decolonial knowledges – but also to our practices as researchers, to each other, and to our participants. This has involved engaging with the ways we continue to participate in racist, capitalist, and colonial regimes of knowledge and power and also with revealing the social, geographical, and historical specificity of knowledges and how particular knowledges reproduce structural relations of inequality and oppression. GenUrb example 11.1 discusses the difficulties experienced by the City Research Teams (CRTs) in undertaking such research.

GenUrb example 11.1: Approaches to research and social change

The ability to engage in social change through research – that is, conducting research that aims to improve individuals' or communities' capacity for dealing with their lives – varied across the GenUrb CRTs. This was not only a result of structural factors, such as conducting feminist research in countries with governments antagonistic to feminism, which meant operating under the radar of local officials and reducing the amount of time spent in the field, or of the emergence of COVID-19, which prevented outreach and other activities from taking place. One of the most significant factors affecting the ability to engage in any activities that could work towards social change was the ties of CRTs to grassroots women's organisations and the local communities they worked with. Two of the CRTs were directly connected to grassroots women: indeed, they were run by women's organisations as opposed to university-based teams. One of these organisations, Red Thread, only agreed to participate in the research on the understanding that they would work towards 'betterment' in the neighbourhood of Sophia, Georgetown, in which they were conducting research. At the time of writing, for example, they were investigating the possibility of organising a 24-hour day care centre in Sophia because the research highlighted the high number of female single parents from Sophia who worked as security guards and who are often forced into 24-hour shifts, obliging them to leave their children at home alone.

By Linda Peake

To conclude, research can be considered feminist when it is informed by a feminist methodology, engages with feminist ontologies and epistemologies, and considers feminist politics such as having the intention of producing knowledge that may also lead to engagement in social change.

In the last section, we move from methodologies to a discussion of methods including the most common methods used in feminist research.

Section 5. Qualitative methods in feminist research

As we discuss in Section 1 (iii) Methodology, methodologies do not determine the research methods you use, since a methodology will provide a rationale for the use of methods but can usually accommodate a variety of methods. Methods are the specific procedures, operations, techniques, or research practices through which research is conducted for the collection of the data needed to answer research questions. In and of themselves, methods are neither feminist nor non-feminist, although qualitative research methods have largely been seen as the preference of feminist researchers.

Methods are usually divided into two camps. Quantitative methods produce numerical data that aim to measure or quantify aspects of the social world, while qualitative methods elicit the meaning behind data through detailed examination and interpretation of social phenomena, which usually requires some kind of descriptive and interpretive coding in order to be analysed. Quantitative methods include the use of surveys and experiments, while qualitative methods include interviews, field notes, focus groups, and observations (see Chapter 12, Section 2 (ii) Methods in the field and Chapter 14, Section 1 Interviews in feminist research). Some social scientists also include within methods the means by which data are analysed. According to this understanding methods could include, for example, content analysis and inferential statistics (Hammond and Wellington, 2013). It is common for qualitative methods to be associated with an inductive approach, which seeks to explore data and generate theoretical insights, and quantitative methods with a deductive approach, which has a focus on hypothesis testing and the use of descriptive and inferential statistics (see Chapter 16, Section 1 Data analysis).

The qualitative/quantitative divide is not always useful: it conflates methods with philosophical positions and obscures the relationalities and similarities between qualitative and quantitative researchers. Quantitative methods, for example, can be used outside a positivist approach. Indeed, feminist researchers who consider themselves qualitative researchers may use a mixed-methods approach, conducting, for example, focus groups or interviews that can collect data on attitudes. Attitudinal data can be used to rank participants' responses, allowing for the formation of crude numerical data (called ordinal data) that can be entered into statistical analyses. Alternatively, a researcher may start with a questionnaire survey, collecting quantitative information, and choose a small subset of those interviewed for a qualitative focus group.

The choice of methods in a feminist research project (as in any research project) will be influenced by a variety of factors. As mentioned, epistemological assumptions and methodological approaches guide the choice of appropriate methods. Feminist researchers want to adopt methods that respect the embodied experiences of participants as well as giving them a voice; that reduce the hierarchical relationship between researchers and participants; and that allow researchers to remain reflexive about their own experiences as part of the research process. Hence, the traditional methods of feminist researchers have been qualitative, suited to in-depth explorations of the multifarious dimensions of women's lives. In particular, the everyday lives of women open themselves up to qualitative methods that can address not only the materialities of work and home but also the embodiment of senses and emotions and the mobilities and stasis that mark the multiscalar spatialities and temporalties in which the experiences and relationalities of women's lives are enmeshed.

Institutional, cultural, and political contexts also play roles in determining choice of methods. For example, ethics boards are increasingly broadening their remits to have a say in the appropriateness of the methods the researcher wishes to employ (see Chapter 8, Section 2 Ethics policies). In places where gatekeepers control access to research sites and participants, they may also want to have a say in the vetoing of certain methods (see Chapter 14, Section 2 (iii) Gatekeepers). For example, in places where feminist research can only be conducted discreetly, without the awareness of the authorities, it would not be possible to engage in large-scale survey-based research. Samples of participants may have to be restricted to small numbers and methods that can be employed in a short period of time may have to be prioritised.

The stock qualitative methods that have been traditionally employed in the social sciences – interviews, focus groups, observations, and field notes – now sit alongside a broad range of newer creative and experimental practices, including storytelling, mapping, **visual methods** (including photographs, videos, and digital methods) as well as new takes on interviews (e.g. mobile interviews). The most common types of qualitative research methods that have been employed in feminist urban research include:

- interviews, including structured interviews, semi-structured interviews, unstructured interviews, mobile interviews, and focus groups;
- participatory methods such as participant observation or those employed in participatory action research (PAR), ethnographies, and geo-ethnographies;
- biographical methods such as autobiography and biography employed in oral and life histories;
- textual methods used in the analysis of archives or creative texts (e.g. maps, diaries, newspapers, poems, music, literature, exhibits) employed in content analysis, discourse analysis, and narrative analysis;
- and visual methods such as theatrical productions, photo-voice, and video and film production.

In some research projects, a mixed-methods approach, including both quantitative and qualitative methods, may be used to provide a more complex picture across multiple scales and dimensions of inquiry (Hesse-Biber, 2014). Mixed methods may also be used as a form of triangulation (i.e. observing the research question from different points of view). This can lead to the generation of additional knowledge to validate (or question) the research findings.

In GenUrb, it was important to employ the most appropriate methods to address our various needs and purposes. All of them were qualitative. They included interviews (see Chapter 14 Feminist interviews), life histories (see Chapter 14, Section 1 (i) Types of interview), and observation and field notes (see Chapter 12, Section 2 (ii) Methods in the field). Less central but still called upon in GenUrb have been focus groups (see Chapter 14, GenUrb example 14.2 Deciding (not) to use focus groups), the use of ethnographic fiction (Adeniyi-Ogunyankin and Peake, 2021b), photo diaries, and storytelling (see Chapter 19, Section 3 (iii) Post-research phase).

Section 6. Summary

This chapter has introduced you to the philosophical fields of ontology, epistemology, and methodology that underlie how we ask research questions, organise the research process, and engage in analysis. It has outlined philosophical fields in relation to feminist approaches to conducting research, with a particular emphasis on methodologies, which have been shown to require engagement with reflexivity, positionality, the relational nature of research, ethics, and social change. Finally, the chapter has addressed the ways feminist methodologies relate to research methods, as well as other factors that influence the choice of methods by feminist researchers.

12 Feminist approaches to fieldwork

Araby Smyth, Elsa Koleth, and Linda Peake

Learning objectives

After completing this chapter, readers should understand

- the meaning of 'the field' in the context of research;
- the colonial history of fieldwork;
- the process of conducting fieldwork;
- and the challenges of conducting fieldwork.

Further, readers should be able to

- anticipate the power relations embedded in fieldwork;
- build accountable relationships while conducting fieldwork;
- and address colonial or extractivist tendencies in their own fieldwork.

This chapter introduces fieldwork: the stage of the research project in which the researcher collects primary data from the field site. Drawing on feminist and critical geography, it begins by explaining what fieldwork and 'the field' constitute and giving practical suggestions for disrupting colonial practices in fieldwork. It then outlines how to prepare for fieldwork and how to use methods such as observation, participant observation, and field notes. Finally, the chapter turns to the importance of cultivating a feminist ethics of care when working with research participants and of navigating disruptions to fieldwork and unplanned events in the field.

Section 1. Fieldwork

Fieldwork is the data collection stage of a research project, the stage in which you go into the world, employing specific research methods to collect primary data from research participants. Conducting fieldwork is a way to explore lines of enquiry related to the research question(s) you are investigating. In this section we delve into what it means to be in the **field** and problematise the colonial heritage of the concept. We also suggest practical ways to decolonise academic knowledge production based on fieldwork.

(i) What is 'the field'?

Traditionally, practitioners of qualitative and ethnographic research define the field as 'a physical setting, the boundaries of which are defined by the researcher in terms of institutions and people of interest, as well as their associated activities in geographic space'

DOI: 10.4324/9781032668727-17

(Schensul and LeCompte, 2012, p. 23). The field, then, is the place(s) in which your sphere of enquiry is located, the place(s) where you will find the data to analyse and answer your research question(s). The boundaries of the field may be easily defined – they may be predetermined by existing administrative boundaries, for example – or they may be amorphous and shift over the period of the research. Field settings for GenUrb include the municipality of Sacaba in Cochabamba, Bolivia; a resettlement colony tucked away in the North Eastern part of New Delhi, India; Sophia, a neighbourhood based on the site of a previous sugar estate in Greater Georgetown, Guyana; Challenge, a neighbourhood in Ibadan, Nigeria; Ein Qiniya, a village on the north-western edge of the municipal boundary of Ramallah, Palestine; and Community X, a Workers' New Village located on the west side of Shanghai (Community X is a pseudonym).

The nature and location of the field will depend on your research question(s), theoretical framework, methods, and analysis. For example, when conducting ethnographic research in feminist urban studies the field may well be a specific building, street, or neighbourhood within a city. You may have more than one field site if you are doing comparative research in two or more cities. In the case of transnational research, the cities may be in different parts of the world. Or you might study a neighbourhood in a city and a locality in a connected rural area. In other cases, when conducting ethnographic research with an online community, the field might be located in digital spaces. In the case of a study of urban activism online, the field might encompass physical localities in a city as well as online communities. If you are taking an historical approach using archival research methods, the field might include archives as well as spaces within a city.

Feminist geographers have long questioned the boundaries – both spatial and temporal – of fieldwork (see Staeheli and Lawson, 1994, and the related special issue on the politics of feminist fieldwork in *The Professional Geographer*). They argue that the field is not merely a bounded physical setting: the field is dynamic, and fieldwork cannot be disentangled from the everyday lives of the researcher and participants.

Feminist geographer Cindi Katz (1994) has challenged the normative boundaries between doing fieldwork and everyday life. She discusses the distance between her positionality as a white researcher from the global North with that of participants in her research living in the villages of Howa in Sudan and Harlem (a predominantly Black and Latinx neighbourhood) in New York City. While her approach to research has changed with different projects over the years, she says that 'I am always, everywhere, in "the field"' (Katz, 1994, p. 72). Katz's subject position as a researcher and those of participants in her research are constituted in 'spaces of betweenness', and she argues that people exist in 'multiply determined fields', enabling us to find commonalities (p. 72). Since then, feminist geographers and ethnographers have written about the many axes of difference that bleed across the field site and everyday life, such as disability (Jokinen and Caretta, 2016), embodied vulnerabilities and privileges (Caretta and Jokinen, 2017), race (Kobayashi, 1994; Faria and Mollett, 2016), masculinity (Vanderbeck, 2005), desire and sexuality (Cupples, 2002; Schneider, Lord and Wilczak, 2020), pregnancy (Madge, Noxolo and Raghuram, 2004), menstruation (Nash, 2023), and the power dynamics of researchers from the global North doing work in the global South (Wolf, 1996; Nagar and Geiger, 2007; Sultana, 2007; Peake and de Souza, 2010).

These feminist interventions build on a history of anti-colonial interventions by anthropologists and geographers in the history of fieldwork and conceptualisations of the field. The next section discusses the colonialism in the study of 'other cultures' within the disciplines of anthropology and geography and suggests how you might disrupt such practices in your own field research.

(ii) Disrupting colonial practices in fieldwork

Data collected during ethnographic fieldwork and the conclusions drawn from its analysis have long been used to legitimate colonial endeavours. The European period of Enlightenment sought to subjugate Indigenous peoples and knowledges and establish Western knowledge as superior. Research into the roles of anthropology and geography in colonialism and the construction of modernity illuminates the economic, political, and social conditions under which the two disciplines developed and the role played by anthropological and geographic knowledge production in the dispossession of lands from Indigenous people (Willis, 1972; Lewis, 1973; Hudson, 1977; Godlewska and Smith, 1994; Pels and Salemink, 1994; Pels, 2008).

Colonialism is not in the past. Scholars across the Americas have argued that colonialism and settler colonialism are ongoing processes in relation to knowledge production as well as in other respects. Decolonisation requires doing the daily work of engaging anti-capitalist, anti-racist, and anti-patriarchal practices seeking to counter interlocked systems of oppression. Decolonising academic knowledge production requires more than one person (or, indeed, a large transnational feminist project with multiple members) can do through fieldwork alone. As we outline in Chapter 2 on decolonising feminist knowledge production, it means ending extractive research practices, exploitative labour conditions in universities, joining in struggles for reparations (including the return of stolen lands), as well as acknowledging the knowledges and work of Indigenous, racialised, and trans scholars and listening to their calls for how to address ongoing structural inequalities within academia (Smith, [1999] 2012; Harney and Moten, 2013; Todd, 2016; Davis and Todd, 2017; Curley and Smith, 2020; McKittrick, 2021; Clarke, 2022; Kinkaid, 2022).

Decolonial theorists seek to disentangle colonial relations that remain embedded in academic knowledge production (Smith, [1999] 2012; Tuck and Yang, 2012; de Leeuw and Hunt, 2018; Naylor *et al.*, 2018). While the persistence of the very concept of the field as being somewhere else and of ethnographic fieldwork as comprising the collection and categorisation of data for analysis about people other than oneself illustrates the necessity of decolonising the larger system of Western knowledge production from which it stems, we focus here on decolonising fieldwork. Building upon what we write in Chapter 2, Section 1 Decolonising knowledge production in feminist urban research, we turn here to practical suggestions about how scholars can disrupt colonialism in fieldwork. They include:

- maintaining collaborative research partnerships with communities (Wilson, 2008; Peake and de Souza, 2010; Smith, [1999] 2012; Nagar, 2014; Weir, Woelfle-Erskine and Fuller, 2019; Mei-Singh, 2021);
- creating less exploitative and more respectful relations between researchers and those who are the subjects of research and between white and racialised researchers, including those in hierarchical research teams (see Chapter 2, Section 1 Decolonising knowledge production in feminist urban research);
- finding new ways to create knowledge in ways that are inclusive of people who have been historically excluded from knowledge production processes, expanding the

parameters of what counts as 'knowledge' and of who can produce theory (see Chapter 2, Section 1 Decolonising knowledge production in feminist urban research);
- investigating the place-specific ways that the colonial past exists in contemporary fieldwork (Smyth, 2023), including, as we state in Chapter 1, 'forms of epistemic violence that result from racist and ethnocentric forms of thinking and representations of colonised peoples and places';
- emphasising accountability to lands and peoples (Daigle, 2018; Smiles, 2020);
- drawing attention to new and emerging forms of imperialism and recognising new possibilities for solidarities and multidimensional acts of refusal (Simpson, 2017; Sylvestre *et al.*, 2017);
- connecting knowledge production with practical projects for addressing reparations, inequality, overcoming disadvantage, and enabling self-determination, such as mapping for political change (The Counter-Cartographies Collective, n.d.; Wainwright and Bryan, 2009; Sletto *et al.*, 2020; Pearce, 2021);
- being sensitive to the more-than-human, including animals and emotions (Wright *et al.*, 2012);
- and using methods like *cuerpo-territorio* (body-territory) – an Indigenous feminist concept translated into a research method by Latin American geographers – that centre the embodied experiences of research participants (Dos Ventos Lopes Heimer, 2021; Zaragocin and Caretta, 2021).

With these suggestions in mind, Reflection exercise 12.1 asks you to investigate the colonial histories of fieldwork where you plan to do research and brainstorm how to do fieldwork differently.

Reflection exercise 12.1: Colonial histories in fieldwork

Wherever you do fieldwork, there may be a colonial history. If so, it is also likely that some of the previous academic research done at the same field site(s) served colonial, imperial, or (neo)colonial interests. Consider where you will conduct fieldwork, and see what you can learn about who else has done research there before you.

- Who has conducted research from your discipline at the same field location?
- What has been done in past field research projects that you do not want to replicate?
- How might your research project relate to the place and the people there differently?

The following section turns to the 'doing' of fieldwork, taking you through its various stages, helping you prepare and determine the methods you will employ. It emphasises the importance of developing a feminist ethics of care.

Section 2. Doing fieldwork

This section first discusses research site selection and preparing for fieldwork. It briefly describes the main ethnographic methods for fieldwork, including observation, **participant**

observation, and the use of fieldnotes. Collecting data in the field requires organisation, and this section provides strategies for maintaining a routine during fieldwork. It concludes with a discussion of the necessity of developing a feminist ethics of care in the field.

(i) Preparing for fieldwork

Preparation involves being aware of your goals for the research, choosing a field site, making preliminary visits, managing data, and being self-reflexive during these activities.

Before embarking on fieldwork, it is important to have a clear idea of the goals of your research so that you use time and resources efficiently. Be aware that your research questions may change once you start fieldwork and you begin to encounter a range of barriers – political, cultural, or personal.

Once the goals of your study and the research questions that guide it are clear, you should select an appropriate site(s) for conducting fieldwork. Consider the following questions in relation to the site (summarised from Buch and Staller [2014] and Ravitch and Carl [2021]).

- Is a site available in which you can successfully investigate your research interests?
- What are the benefits and challenges of a particular site (e.g. as regards languages, gatekeepers, distance, access, financial costs, obtaining research equipment, mental and physical health needs, safety concerns, visa applications, immunisations, and insurance)?
- How will your presence affect the local population?
- If there is more than one such site, what criteria can you use to choose between them? How would locating the research in a different site affect the research?

Depending on your personal history and research topic, you may or may not be familiar with the research site you select. Some researchers leave their homes to collect data, and others remain in their hometowns and work with research participants they know intimately (Cuomo and Massaro, 2016). Either way, it is important that you become familiar with the location in which your research will be conducted. If you need to gain formal or informal consent from authorities or communities (e.g. relevant community members, state authorities, or Indigenous groups) for your research, set aside enough time to do so, and determine whether consent needs to be acquired formally or informally and from whom.

It may be useful to make a preliminary visit to the field or various field sites before committing to the location. Advance site visits – when and where possible – can help to shape or revise research questions and tools and to anticipate any potential issues. Consider the following questions in these visits.

- Are any accommodations for sight and hearing, health (mental and physical), mobility, and learning differences required to ensure accessibility for you and research participants at the field site, such as transportation, sign language interpreters, or printed material?
- Consider the costs of fieldwork: what will you personally need at the site to do the fieldwork you are proposing? For example, think about long-term housing or temporary shelter, transportation, appropriate clothing, and food purchasing.
- What is required for the research to be conducted? Consider, for example, data collection tools such as appropriate recording devices and storage options and the reliability of electricity supplies.

- What will you do in the field? Identify tasks such as meeting potential research participants, identifying participant observation locations, conducting interviews, following up with participants, and more.
- Consider safety: what kinds of risks might you be introducing into the lives of any research participants? What personal risks might you face?

Preliminary conversations with potential research participants as well as other researchers who are familiar with the area will help provide the background information needed to make informed decisions about conducting fieldwork, including the viability of the methods you plan to use. They may also help you determine whether or not the location and community is best suited for your project goals.

As you select the field site and evaluate methods, you also need to consider how you will collect, secure, and manage all the data that you collect (see Chapter 10, Section 1 Mainstream and feminist conceptions of data). Consider the following issues to guide data collection in the field.

- *Data type.* Will you be collecting images, text, audio recordings, geospatial information, or another type of data?
- *Data collection.* Will you record interviews with an audio recorder, camera, or smartphone? Will you record video or take photos? Will you collect archival material such as articles, brochures, posters, or photos? Will you take fieldnotes with a pen and paper or on a mobile device?
- *Data organisation.* What file formats will you use to store data (JPG, PDF, TXT, CSV, MP3, MP4, etc.)?
- *Data security.*
    - When and how do you remove identifying information (i.e. metadata, which can include names, age, gender, place of residence, and workplace) from the data that you collect?
    - If you are assigning pseudonyms to research participants, when and how do you do so?
    - How will you create data backups? Will you use Dropbox, Google, an external hard drive, a second device, or hard copies?
    - What precautions do you have in place to keep data and backups secure? This can include using passwords, encryption, locked boxes, or cabinets.
    - Where will you store data? Do you need to store data both physically and digitally?
    - What is your plan if data security measures are compromised?

Sticking to a routine while you are in the field is helpful for adhering to thorough and rigorous data-collection practices such as those outlined. Helpful preparation for data collection in the field requires you to consider the following issues:

- packing the necessary equipment (for example, a camera, audio recorder, mobile device, interview guide, paper, and digital copies of consent forms, pens, etc.) and backup equipment in case your original plan fails (including cables and chargers, batteries, a secondary device, pens, pencils, notebooks, etc.);
- packing other items you and research assistants might need while you are collecting data (including sun protection, extra clothing, water and snacks, and cash in the

appropriate denomination – small bills if change is hard to come by – or, if you are paying research participants, set aside the exact amount for them);
- and having a post data collection routine that includes typing up fieldnotes and backing up data at the end of each day.

Your data collection methods, fieldwork routines, and notetaking practices will depend on your research methods and methodology, the context of the field, your access to research tools, and, of course, your own preferences.

As you prepare it is important that you exercise reflexivity. How do you locate or position yourself in relationship to research participants? Self-reflexivity is central to meeting ethical obligations (see Chapter 7, Section 3 Feminist ethics and research and Chapter 11, Section 4 Feminist methodologies). You might usefully ask yourself the following questions before arriving in the field.

- How did I arrive at this topic of study?
- How might I be viewed, or positioned, by others in the field?
- What influences my perceptions of research participants?
- What privileges am I assuming when I think I can gain access to the field (Buch and Staller, 2014)?
- What biases do I bring to the field, and how can I challenge them while conducting fieldwork?

Writing about your positionality before and throughout fieldwork is a helpful activity for generating insights about and your relationship with research participants, as well as your research topic. For this you need to keep a field notebook or a dedicated file on your laptop. (For examples of positionality memos written prior to fieldwork, see Ravitch and Carl [2021], and for reflective narratives written during fieldwork, see Harcourt *et al.* [2022]).

Reflection exercise 12.2 walks you through the practicalities of preparing for fieldwork.

Reflection exercise 12.2: Preparing for fieldwork

If you have not already done so, think about what field site may be appropriate for meeting the goals of your study and answering your research questions. Use the prompts in the earlier bulleted lists to help you assess whether there is one potential site or several, the benefits and challenges of a particular site, and how your presence might affect the people there. Then think through the practicalities of doing research in that site.

- What accommodations will you need?
- What costs do you associate with doing research there?
- What will you need in the field?
- What kinds of data will you be collecting, and how will you keep data organised?
- What sort of daily routine will you keep while conducting fieldwork?

As you think through these logistical issues, critically situate yourself in the site.

- Why do you want to study this topic and be in this place?
- What might you practically put into place, as a part of your fieldwork routine, that would disrupt the colonial power geometries shaping fieldwork?

(ii) Methods in the field

A broad range of methods are used in fieldwork. Many researchers employ more than one method. In a mixed-methods approach these often include ethnographic methods of observation such as observation, participant observation, and taking fieldnotes. This list is not meant to be exhaustive – you may also want to conduct interviews, focus groups, or surveys (see Chapter 14, Section 1 (i) Types of interviews) or use geospatial techniques or visual methods (see Clifford, Cope and Gillespie, 2023).

Observation can range from unstructured to structured. In the latter the researcher knows in advance what it is they wish to observe while in the former the researcher is open to observing everything in their environment (within ethical bounds). Observation can also range in intensity from situations in which the researcher is simply an observer to that of the researcher as full participant. In the former, the researcher does not engage in any activities in the field – their presence is not intended to have any effect on the social world they are observing – while in the latter the researcher may become a functioning member of the community they are researching, a practice known as **participant observation.** This involves the researcher being immersed in the everyday life of a social and geographical setting, recording their experiences and observations by taking regular fieldnotes about direct observations and understandings of everyday interactions, encounters and events, as well as self-reflexive memos on their own participation, and constantly engaging in reflexivity. The goal is to gain an in-depth and intimate familiarity with participants. This cannot be achieved in a few weeks but is built up over time through repeated visits (as in geo-ethnography) or a single period of months in the field (as in ethnography).

Fieldnotes must be taken to record your observations and thoughts while you are in the field. Fieldnotes are the foundation for helping you build detailed observations from the ground up. You need to take fieldnotes every day, and you should plan to type them up daily as well. Record both what you can see and make sense of and what puzzles and upsets you: both will inform your eventual analysis. The feminist anthropologist Lisa Cligget (2005, p. 46) describes daily reflection on ethnographic fieldwork:

> While rehashing the day's conversations and the week's activities, I would feel my mind sliding 'up', away from the precise conversations I had had, to a bird's-eye view of what I had learned, what I 'knew'. Along with the high-altitude vision, ideas . . . would float by like clouds, making the vision fuzzy, and then focused again. The next morning . . . I would feel my mind slide 'back down' and focus on what questions needed to be asked that day. The distant vision from the night before had clarified my immediate tasks, and both viewpoints shaped my days in the field.

You will use fieldnotes primarily to help you remember details of a situation that you wish to describe, reflect on, or analyse. There is more than one kind of notetaking in the field, and we list several here:

- *Jottings* include phrases, quotes, key words, mnemonic devices, etc. that can serve as a guide for more substantive notes that you will make later.
- *Observational fieldnotes* are based on observations of the field site and occurrences there such as community events, markets, political activities, children's activities, etc.

- *Fieldnotes based on interviews* are in addition to the recording (if you make one), and provide an overview of the context in which the interview was conducted (including time and location), salient points covered in the interview, and your thoughts on whether the interview went well or not and why.
- *Full fieldnotes* contain detailed notes, both descriptive and analytical, that are expansions of jottings and the observations of the day. They will still need to be tidied up, with details to be filled in and expanded upon later.
- A *diary*, which is a place for personal stories, reflections, and insights about yourself and your positionality on topics including emotions, relationships with others, and more while you are in the field.

You may want to organise your fieldnotes into categories as you collect data in the field: for example, dividing fieldnotes between interviews, personal diary entries, reflections on methods, daily observations, etc.

While not a method, workshops can be employed while in the field to facilitate various methods. GenUrb example 12.1 documents how workshops were held by the Shanghai City Research Team (CRT) to employ the visual method of Photovoice.

GenUrb example 12.1: Photovoice workshops in Shanghai

In spring 2019, the Shanghai CRT collaborated with Vox Photo Project, a non-profit organisation founded by Dolce Wang, the former creative director at Walt Disney Imagineering. Together, we organised a Photovoice workshop for eleven of the women participants. During the workshop, the women learned basic photography techniques and were guided to create a visual journal utilising three themes: everyday objects, everyday feelings, and the 'good old days'. The women used the themes to capture their observations of their homes and their social and physical environments and the emotions they felt in their everyday lives, living in run-down Workers' New Villages.

These photos formed the basis of the book *Lens in Bloom: A Photography Journal*, which illustrates women's memories of living in Shanghai and how they currently experience their lives in the city. It has been used to share with the broader public how disadvantaged lower-working-class women in Shanghai see and experience their lives and live through daily hardships (a PDF of *Lens in Bloom* is available on the book's website, Document 12.1 www.routledge.com/9781032668680).

With the support of two artists, Li Xiaofei and Guo Qingling, we also organised an art exhibition in the Yuwatai Gallery, close to the women's neighbourhood, to exhibit the photographs taken by the women. The art exhibition attracted the attention of local officials, media, and students. Through the Photovoice workshop and art exhibition we aimed to increase the capacity of lower-working-class women to call public attention to the social problems they endure.

By Penn Tsz Ting Ip

(iii) Cultivating a feminist ethics of care in the field

Building and maintaining accountable relationships with people in the field requires trust and respect. Exercising a feminist ethics of care for the places where you are doing fieldwork and for the people living there is central to the praxis of feminist research. This entails a commitment to the well-being and concerns of research participants and researchers. Before, during, and after fieldwork, you should account for the emotional impact of your research on participants and yourself (see Chapter 7, Section 2 Towards a feminist ethics).

Maintaining accountable relationships with people in the field is part of the feminist ethical obligations you owe to individual participants who are directly involved in your research but also to the broader community or communities in which you are conducting your research (see GenUrb example 12.2). Feminist scholars have written about the complexity of ethical issues that arise between researcher and field assistants (Caretta, 2015), co-workers, family, and friends (Cuomo and Massaro, 2016) and the need for greater self-reflexivity about sexuality and desire in cross-cultural fieldwork (Cupples, 2002), to cite just a few examples.

Accountability also stretches beyond your time in the field. It includes making sure that the results of your research are made known to the participants in ways that are relevant and accessible to them. Engagement in such practices can include writing up the results of the research in plain-language summaries and sending these to participants. It may also include returning to the field and organising meetings at which the results can be discussed and any further actions (for example, grant writing, making contacts with donors) can be planned. Though sharing academic outputs with participants may be meaningful, it is possible they may have very little resonance with the everyday lives of participants. Indeed, you cannot assume that participants will always be interested in your research results (see Bain and Payne, 2016). While you may not know in advance of being in the field the best way to share research results, you do need to include the time and cost of doing so in your planning; otherwise, this practice runs the danger of falling by the wayside.

Reflecting about being a scholar-activist, feminist geographer Laura Pulido (2008, p. 351) writes,

> you are embedded in a web of relationships, some of which demand a high level of accountability to a community or other group of individuals. It is accountability that will hopefully ensure the relevancy of your work in the effort to create social change.

In her theorisations of Indigenous relations of care and decolonial possibilities, feminist urban geographer Michelle Daigle (2018, p. 201) argues that scholars should 'not only think about but embody the relational accountability that is tied to living and working on one's own Indigenous territory, or as a(n) (uninvited) visitor on stolen and occupied Indigenous lands'. While you might not identify as a scholar-activist or be working on Indigenous lands in a settler-colonial context, embodying relational accountability to lands and peoples where you do research is always important for your commitment to the well-being of yourself and others in the research process and is part of cultivating a feminist ethics of care in the field.

GenUrb example 12.2 draws on the work of the Cochabamba and Shanghai CRTs to highlight how the ethical obligations that guide relationships between researchers and people in the field are specific to cultural contexts.

GenUrb example 12.2: Approaches to a feminist ethics of care in the field in the Cochabamba and Shanghai City Research Teams

The **Cochabamba CRT** comprises one researcher based in Toronto and the women's grassroots organisation Centro de Estudios y Trabajo de la Mujer (CETM) in Cochabamba. CETM has longstanding ties in Indigenous communities, and over the years they have fostered relationships of trust with the Quechua women with whom they engage.

Working with CETM provided open lines of communication with participants and ensured continuity throughout the research process and beyond. Given the commitment in the GenUrb project to reflexivity and accountability, we prioritised the well-being and concerns of participants, being aware of power dynamics, and exercising care in the research process. This included considering the time and travel commitments of the participants and providing adequate honorariums. We also took into account holidays, such as the *Virgen de Urkupiña* festivities in August, and participants' other responsibilities, such as their social reproduction commitments, in scheduling interviews and workshops. For example, we ensured there were spaces, activities, and snacks for children during workshops.

While the COVID-19 pandemic interrupted the research process, we were able to maintain contact with participants over the phone and through WhatsApp to carry out regular check-ins and provide financial support to the women and their communities during this difficult time (see Chapter 5, GenUrb example 5.2 Adapting research methods during the COVID-19 pandemic). Making use of GenUrb emergency funds, the Cochabamba CRT organised the distribution of food and personal protection equipment when women were facing a strict lockdown and their mobility and ability to earn an income were severely restricted. These gestures of care were possible given the relationships of trust among the research participants and the researchers and especially the trust women have placed in CETM.

Since 2017, the **Shanghai CRT** has been working with women in the Workers' New Villages located in Shanghai (Ip, Huang and Wang, 2023). The Shanghai researchers, together with students and volunteers, have engaged in a variety of activities for the women participating in the research.

Given that the lead researchers of the Shanghai CRT, Tsung-Yi Michelle Huang and Penn Tsz Ting Ip, were not local researchers, we found the most challenging and time-consuming part of the research to be establishing trust with participants and local government officials. To build trust and a positive social relationship between the researchers and participants, we organised a number of annual arts and cultural events. These events were also a means of collecting qualitative data and including feminist non-profit organisations and volunteers in the research, thus bringing triple benefits to the research.

In 2018, we co-organised a mental mapping event with the local NGO We and Equality (see Chapter 19, GenUrb example 19.3 The GenUrb opening ceremony in Shanghai). In 2019, we invited the international NGO Vox Photo Project to teach participants how to take photographs and then employed photo-diaries to collect data (see GenUrb example 12.1 Photovoice workshops in Shanghai). We also organised a COVID-19 diary-writing workshop in

2020, inviting another feminist scholar, Zhang Yu, and one of her students, Janet Xu, recruited as our research assistant, to help guide the participants in writing diaries related to their experiences in the face of the global pandemic (see Chapter 5, GenUrb example 5.2 Adapting research methods during the COVID-19 pandemic).

In addition to these arts and cultural events, the researchers visited participants on a regular basis to maintain a close social relationship with them. We employed an 'eating-together' method to chat with women casually, with lunch gatherings at their homes and local restaurants. Since these meetings took place in the settings of their everyday lives, the eating-together method allowed participants to share their thoughts and feelings more freely and enabled the researchers to observe the women's social lives. Another benefit of employing such a method was simply to get to know each other better, further establishing trust and positive researcher–participant relationships.

By Nasya S. Razavi and Penn Tsz Ting Ip

In the following section we turn to the ways in which everyday life intrudes into even the best-laid plans for fieldwork.

Section 3. Fieldwork as a 'messy' process

Fieldwork is a messy process (Mountz *et al.*, 2003; Sharp, 2005; Billo and Hiemstra, 2013). We have already considered how the boundaries between 'the field' and 'home' are not distinct and argued that fieldwork requires practising a feminist ethics of care and relational accountability, all of which demands 'radical vulnerability from those who inhabit different communities of meaning' (Nagar, 2014 p. 174). In addition, fieldwork is not a linear process that proceeds smoothly from one phase to the next (for example, from planning to recruitment to data collection to analysis and writing). In practice, the beginning and end of the fieldwork period can be blurry, so although you may set aside a discrete period of time for fieldwork, you should be prepared to be flexible with timetabling. In particular, do not underestimate the amount of time it can take to secure ethics approval for your research from your institution and also from community partners. Several stages of the research process may have to occur at the same time or may need to be repeated. For example, recruitment of research participants, conducting interviews in the field, data analysis, and writing may significantly overlap (Browne, 2005, p. 53). Or your research plan may be disrupted or taken in new directions. Examples of disruptions during fieldwork include but are not limited to the following.

- *External events.* These can include geopolitical conflicts or events such as elections or global pandemics and weather and climate change related disasters and displacements.
- *Problems in the field.* These can include, for example, running out of funding, failure of recording equipment, or loss of data.
- *Issues with research personnel.* This can include the breakdown of key relationships resulting from lack of support from your supervisors, university, or any institutions or organisations with whom you are working or challenges working with co-researchers.

- *Issues with participants*. These include, for example, challenges recruiting or working with research participants, encountering people engaging in political, social, or cultural practices that you disagree with, or witnessing or learning about the trauma of other people.
- *Issues you can face as the researcher*. This wide-ranging list can include harassment or assault, illness, pregnancy, caretaking obligations, or personal loss.

Many of the issues listed here were encountered in GenUrb (see GenUrb example 12.3). As you can see, fieldwork can include serious challenges, many of which take an emotional toll in addition to shaking the foundations of your research design and requiring you to reconceptualise aspects or all of your project (Pandey, 2009; Baxi, 2014; Davis and Craven, 2022). It is important to remember that although you may not experience all these issues, facing unexpected challenges in the field is part of the process of conducting fieldwork. Fieldwork disruptions will happen, and it is helpful to treat them as potentially generative moments rather than disruptions or failure. It is not wrong if your research changes course or moves more slowly than you anticipated. It helps to be flexible and reflexive about the research process. Be open to revising your approach, research questions, and strategies in the field.

GenUrb example 12.3: Issues faced by City Research Teams while conducting fieldwork

GenUrb CRTs confronted many obstacles while conducting fieldwork, which took place over a five-year period, from 2018 to 2023 (see Chapter 13, GenUrb example 13.2 Security issues).

- It was global crises that were to cause the biggest disruptions to fieldwork and the longest delays in completing fieldwork. COVID-19 added more than eighteen months to our completion schedule. Climate change also played a large part, from the heavier-than-usual monsoon season in Delhi that delayed fieldwork for weeks to torrential rains during the rainy season in Georgetown that prevented fieldwork from taking place on certain days.
- National events, such as elections in Guyana, Bolivia, and Nigeria, all made it unsafe to conduct fieldwork for several weeks, either as a direct result of the threat of violence or because being in the field during election time, for example in Georgetown, would result in people associating the researchers with the ruling party.
- Fieldwork took longer than expected in some cities because it proved difficult to recruit younger women to join the research – for example in Georgetown. Building trust among participants was a problem in Shanghai, where a number of women dropped out of the project. As noted earlier, this led the Shanghai CRT researchers to change their approach to participants, spending more social time with them, and engaging them in a range of activities in what, fortunately, were largely successful attempts to build up trust. Some researchers also had to take time out of the field to deal with the emotional trauma of working with women living in poverty who were contemplating suicide.

- Everyday mundane issues also affected fieldwork schedules, including running out of petrol when travelling, electricity blackouts, taking care of sick children, and becoming pregnant.
- Fieldwork in Ein Qiniya, Ramallah was repeatedly interrupted and burdened by Israeli settler-colonial violence and the aftermath of such violence e.g. incursions or/and assassinations leading to curfews, strikes, and cancellation of meetings and activities due to arising urgencies. In the first phase of the project, in August 2018, the home of one Ramallah CRT member was raided and turned up-side-down by the Israeli army in the middle of the night, and her husband was arrested for weeks before being released without charges. And although the Takhayali Ein Qiniya (Imagine Ein Qiniya) project has been exhibited in Ein Qiniya, Berlin and Paris, the war that Israel launched on Gaza on October 7, 2023, has led to the indefinite postponement of the 'Takhayal/i alWadi' advocacy film and campaign (that were scheduled for release that month) (but see takhayali.net, as well as Aruri, Al-Battat, Qawasmi with Adami (forthcoming)). Document 12.2 Takhayali Ein Qiniya (Imagine Ein Qiniya) presents a few of the images of how women residents in Ein Qiniya on the outskirts of Ramallah visually imagine its future development (with text in Arabic). www.routledge.com/9781032668680.

By Linda Peake and Natasha Aruri

Consider that the experience of being made uncomfortable by encounters and events in the field can be an important way in which you learn more about your research topic and about your own position as a researcher. Feminist geographer Richa Nagar (2014, p. 167), for example, writes about engaging 'difference, disagreements, mistakes, and dissonance' as a multifaceted process that can build trust through shared vulnerability. Ensure that you have adequate support while conducting fieldwork (from peers – friends and other researchers – for example) and that you build local support mechanisms in and around the field site if applicable. The networks of care and responsibility you strive to build may very well come through for you!

There are times, however, when you may have to abandon your research. Feminist scholars have brought attention to issues of safety that emerge in fieldwork, illustrating how physical danger and psychological distress can affect researchers, research collaborators, and, of course, participants and their communities (see Sharp and Kremer, 2006; Brigden and Hallett, 2021; Bundhoo and Lynch, 2021; Kocsis, 2024). Conducting feminist activist research has to prioritise researchers' safety, well-being, and respect for sociocultural norms above obtaining data. This may mean that the visibility of the feminist scholar-activist can vary and have different implications across different geopolitical and cultural contexts, affecting praxis, performativity, and even the language(s) in which you choose to talk and write about research. Even the threat of violence, to you or anyone you work with, should make you seriously consider leaving the field.

Research is a holistic, ongoing, and iterative process in which research questions, meanings, and analytical insights are being developed, tested, and revised at every stage of the research process. This process of questioning and revising your ideas is central to developing a complex and well-rounded analysis. Fieldwork is the stage of the research where your ideas and assumptions are often tested, opened up, and extended in response to what you encounter in the field. The potential for surprises, serendipity, challenges, crises, and doubts in fieldwork can also be the most rewarding aspect of fieldwork and the reason it is so central to the research process.

Reflection exercise 12.3 asks you to prepare a timeline anticipating possible disruptions to fieldwork. Be prepared for the mess of fieldwork; embrace it and learn from it.

Reflection exercise 12.3: Disruptions during fieldwork

Sketch your timeline for completing fieldwork. Where will you be, and when will you be there? Include lengths of time in a place and where you expect to live, down to the address and neighbourhood if you know it.

Using the timeline you have just drafted, zoom out and brainstorm the possible disruptions you may have to manage as you are conducting research. Some of these may be planned disruptions such as attending a conference. It may be possible to predict or prepare for some disruptions: for example, think about the ecological and geopolitical seasons of the field site. Will you be doing research during the rainy season or when a local or national election takes place? Other disruptions may be more difficult to predict, but you should still think about your responsibilities to other people. For instance, would you need to suspend fieldwork to help someone move house or care for a sick relative? Finally, consider yourself. When might you need a break, and what might that look like? What allowances are you making for yourself and your well-being in the fieldwork plans? What if fieldwork just takes more time than planned? How can you adjust your research plans following fieldwork if you encounter delays?

While you can't plan for everything, it will help to imagine what could disrupt fieldwork and how you might respond to it. That way, when something inevitably does come up, you may be better prepared to pause the research and attend to yourself and the well-being of the people around you.

Section 4. Summary

This chapter has discussed fieldwork as the stage of the research project in which the researcher collects primary data from the field site. It is one of many aspects of academic research that has roots in European colonisation and the military pursuits of imperial powers. Working with the understanding that colonisation is an ongoing process, scholars have sought to provide practical suggestions for how researchers might disrupt ongoing colonial practices in fieldwork. Feminist geographers have troubled the notion that the field is separate from the places in which researchers live their everyday lives. It is important to consider the many axes of difference that exist across the field site and everyday life as you proceed with selecting an appropriate field site for your research goals and fieldwork. Remember, as much as you plan for fieldwork, considering accessibility, accommodation, costs, methods for data collection and storage, etc., fieldwork rarely goes according to plan. Be flexible and reflexive about the research process and revise your approach as needed. Exercise reflexivity before, during, and after fieldwork so that you might meet your feminist ethical obligations of care to research participants and the broader communities where your fieldwork is located. Fieldwork is not a linear process and will undoubtedly be interrupted. Disruptions do not equal failure. They are to be expected, and they are more manageable with a support network that includes your research advisors, friends, and peers, as well as people at the research site.

13 Feminist geo-ethnography

Araby Smyth and Linda Peake

Learning objectives

After completing this chapter, readers should understand

- the distinctions between ethnography and geo-ethnography;
- the defining features of geo-ethnographies;
- and the key methods of geo-ethnographic inquiry.

Further, readers should be able to

- determine whether or not a geo-ethnographic approach is appropriate for their project;
- and implement geo-ethnographic methods in their research.

This chapter introduces ethnography and geo-ethnography as field-based approaches to research. Geo-ethnography emerged in the early 2000s and is taking root in feminist geographic scholarship. The chapter first provides a critical discussion of the colonial legacies of ethnography and situates geo-ethnography in the paradigm shift of feminist ethnography and feminist geography. It then turns to helping the reader apply a feminist geo-ethnographic approach in their own research project by outlining the defining aspects of this approach: namely, that it is place based, addresses the everyday, requires thinking across scale and tracing power, is relational and embodied, and engages with feminist activism.

Section 1. Ethnography

Ethnography is a research approach that has predominantly been associated with anthropology, although researchers in other disciplines have also adopted it. Geographer Steve Herbert (2000, p. 551) describes ethnography as an exploration of the 'tissue of everyday life . . . in all its richness and complexity'. This exploration has typically involved a researcher from one culture or community travelling to another to conduct an in-depth study of people, paying attention to lived experience, social relations, and cultural practices in that community. Ethnographers gain insights into the **everyday** reality, or lifeworld, of a community's inhabitants and seek to do so from the perspectives of those being studied. As such the ethnographer observes and participates in daily life with the aim of developing an 'understanding of lived realities . . . of the social norms, rules and practices that shape diverse forms of human sociality' (Buch and Staller, 2014,

DOI: 10.4324/9781032668727-18

p. 108). Analysis of the power dynamics in social interaction is thus tightly bound to the norms and perspectives of the ethnographer, which themselves are subject to continual interrogation.

In keeping with its focus on the richness of everyday life, the methods employed in ethnography are predominantly qualitative, with a heavy emphasis on participant observation. Ethnographers may also employ other forms of observation as well as methods, such as interviews, to collect oral or life histories (see Chapter 12, Section 2 (ii) Methods in the field). Ethnographies allow for both 'thin' description (describing what has happened) and 'thick' description, which 'seeks to provide a detailed account of an aspect of human behaviour through reference to the context in which it takes place' (Hammond and Wellington, 2013, p. 45). The term 'thick description' is commonly associated with the ethnographer Clifford Geertz (1973) but has earlier origins with the philosopher Gilbert Ryle (1968), who argued that 'actions took place in a world of shared meanings' (Hammond and Wellington, 2013, p. 45).

The origins of ethnography are problematic. It is deeply implicated in colonialism, 'both as a direct instrument and as an implicit reinforcer of colonial categories and imperial worldviews more broadly' (Anderson, 2012, p. 278; see also Pels and Salemink, 1994; Chapter 12, Section 1 (ii) Disrupting colonial practices in fieldwork). Ethnographers have therefore sought to problematise the heritage of applying a dominant cultural view of Western interests and analytical frameworks (Clifford and Marcus, 1986). Ethnographic research nonetheless involves entering into social worlds, with long periods of immersion in communities, moves that are riddled with power differentials and ethical implications (Schensul and LeCompte, 2012). Not the least of these is that such moves have invariably engaged white researchers from the global North engaging with racialised communities, either in their own countries or in the global South.

Feminist ethnography began to emerge amidst white second-wave feminist movements of the 1970s predominantly situated in the USA, with foundational texts considered to be those by Michelle Rosaldo and Louise Lamphere (1974) and by Rayna Rapp (1975). At that time there was a paradigm shift across the humanities and social sciences in which feminist scholar-activists, alongside a host of others from diverse radical movements, were calling for an end to legal discrimination, critiquing patriarchal and racist institutions and practices throughout society, and demanding greater equality for people marginalised by overlapping systems of oppression (see Stacey [1988] and Lather [2001] for contributions on the possibility of feminist ethnography). Although Black female ethnographers were working at the same time it took longer for them and other racialised women to become recognised in the developing canon. Hence, early calls for feminist ethnography critiqued male-centred research and sought to add women to the field, both as research participants and as researchers, and introduced methods and writing styles informed by feminist theories and ethics that address the relational nature of knowledge production (Buch and Staller, 2014). For instance, the feminist anthropologist Lila Abu-Lughod (1990) mobilised feminist critiques of objectivity from this critical turn, advocating for a feminist ethnography that places at its centre the lives and experiences of women written about by women, arguing that this has the potential to shift ethnography away from being a colonial methodology based on an East/West division and study of 'the other' to recognition of differences and building solidarity along lines of political affinities.

While feminist ethnographies started by addressing the gendered power dynamics of social interaction they have evolved not only to detail the experiences of women but also to address other groups, such as people experiencing poverty and racialised, queer, and disabled people (see Sandoval, 2000; McClaurin, 2001; Naples, 2003; Smith, 2005). Subsequent waves of feminist activism have continued to influence ethnographic scholars (see Davis and Craven [2022] for a history of feminist ethnographies). Feminist ethnographies are now diverse but are generally shaped by the themes of opposition to oppression, empowerment, and adopting a multi-voice approach to create a space from which people can be recognised (Gobo, 2008). Moreover, feminist ethnography often takes an advocacy approach to produce scholarship that 'may contribute to movement building and/or be in the service of organizations, people, communities, and issues we study' (Davis and Craven, 2022, p. 13). From the 1990s to the present, feminist ethnographic work has expanded to include a global perspective with the proliferation of transnational and multi-sited research projects, as well as more analyses of masculinity and LGBTQIA2S+ experiences. These moves open the potential for locating the oppressive power dynamics of social interaction and the sites of power from which they emanate, an important tactic for decolonising knowledge production (Kapoor, 2017).

In the following section we move from ethnography to a discussion of **geo-ethnography** and their similarities and differences.

Section 2. Geo-ethnography

It is likely that you have heard of ethnography, but what of geo-ethnography? How are ethnography and geo-ethnography distinguished from each other? A geo-ethnography can be thought of as an ethnography imbued with a geographic sensitivity. Feminist geographer Karen Till (2005) defines geo-ethnography as a blend of feminist ethnography and critical geography. It demands a 'geographic "thick description" . . . specifically in terms of the *spatial frameworks* in which [human actions] take place as well as the larger symbolic meanings those frameworks have for a particular society' (Till, 1996, p. 11). Geo-ethnographers thus bring a spatial approach to ethnographic methods, considering not only place-as-locale but situating that **place** in the broader spatialities of power relations that flow through it, connecting it to other places and scalar circulations of power, expanding its spatialities far beyond its physical boundaries.

In his research on gentrifying neighbourhoods in New York City, the critical geographer Christian Anderson (2020) also takes a geo-ethnographic approach. He immersed himself in the life of residents on one block in the New York City neighbourhood of Hell's Kitchen in order to study urban structural inequalities. His approach illustrates how spatialities, subjectivities, and agency beyond the local are incorporated into ethnographies. He turns away from the concerns of early ethnographers with culture and community and space as a grid or container and from those of more contemporary ethnographers who produce 'stories about shifting connections between different categories of people in relatively static and binary spaces' and that gloss over processes involved in the production of spatialities (Anderson, 2012, p. 281). Instead, Anderson (2012, p. 282) argues for a geo-ethnography that focuses on processes, that brings to light 'obscured and sticky agencies', and that identifies 'social–spatial forms while also recognizing the contingencies and inequalities of those forms and even suggesting means by which they might be undone where necessary'. His focus is on the relationality of uneven socio-spatial processes and our increasing inability to box these into discrete hierarchical scales like local and global. He remarks, for example, how the feminist scholars Geraldine Pratt and Victoria Rosner (2006) chose not to separate analytically but to pair ' "the global and the intimate" to highlight the extent to which the two are intricately interwoven, rather than mutually exclusive or hierarchically ordered' (quoted in Anderson, 2012,

p. 277). This conceptualisation also points to the fact that many processes do not have distinct scales at which they are realised and experienced. Relationality thus brings into focus questions of fluidity, difference, identity, and mobility and the situated nature of the complex interrelations and spatialities of people's social worlds, with the potential to connect with efforts to promote social change and transformation.

A further feature distinguishing between ethnographies and geo-ethnographies relates to temporal differences. Ethnographers spend long periods of time in the field, often one year or more. Participant observation, the principal method used by ethnographers, generally requires a significant time commitment and extensive engagement with participants, allowing researchers to immerse themselves in the cultural setting (Buch and Staller, 2007). Geo-ethnographies are characterised by shorter but repeated visits to the field, such as daily visits that do not require staying overnight or visits that require staying overnight every week or month but may include annual fieldwork sessions that may last weeks or months and that take place over several years. Anderson's (2020) research in New York involved accompanying people during their daily routines and collecting audio recordings with a recording device strapped to his neck over a three-year period, while Till's work in Berlin used interviews and informal conversations, the collection of printed documents, and visual analysis over a ten-year period. Shorter-term engagements over protracted periods still allow for the building of close relationships with research participants, and researchers may well spend as much time overall in the field as ethnographers. Such engagements can be equally productive in developing trust, enabling intimate insights into everyday practices to be obtained, revealing the embodied power underlying language and performance, and making it possible to access data vital to understanding place-embedded social, economic, and political contexts, subjectivities, and conjunctural shifts. Rather than 'parachuting in' to extract data, geo-ethnography requires slow and sustained relationship building.

GenUrb example 13.1 documents the maintenance of relationships over time with participants by the GenUrb City Research Teams (CRTs). It speaks not only to the differing periods of time spent in the field during data collection but also to how these periods were influenced by, in some cases, pre-existing relationships and how the completion of the data collection did not necessarily signal an end to those relationships.

GenUrb example 13.1: Lengths of time spent in the field by the City Research Teams maintaining relationships

The **Cochabamba CRT** collaborated with the local feminist organisation Centro de Estudios y Trabajo de la Mujer (CETM) to recruit participants to the GenUrb project. CETM has longstanding ties to the communities in Sacaba where the participants reside, given their decades of activism and presence in the region. CETM had already established years-long relationships of trust with the women who joined the GenUrb project as a result of their previous participation in CETM activities and workshops. Over the lifespan of the GenUrb project, the Cochabamba CRT has been in contact with the participants through GenUrb interviews and storytelling workshops and regular check-ins that took place both in person and over WhatsApp. Coordinating with CETM, the Cochabamba CRT was able to provide additional support to the participants during COVID-19 by distributing food hampers to their communities. Beyond the interviews, CETM has continued to maintain these relationships and provide other types of support to the participants.

The **Delhi CRT** has remained in touch with some of the participants through the three field assistants who are members of the same community and who are still engaged in other projects there. The CRTs association with the field assistants is based on Swagata Basu's acquaintance and feminist friendship with them, which pre- and post-date the GenUrb project. Swagata Basu continues to maintain links with the field assistants in a personal capacity, conveying Anindita's greetings to them, exchanging festival greetings and family news, and offering solidarity. In these conversations, participants sometimes recount the dignity kits comprising cooking oil, spices, pulses, and other small items they received from the Delhi CRT during the time of COVID-19.

Red Thread in the **Georgetown CRT** has kept in touch with the participants in Sophia in a variety of ways. We phone the women periodically to check on how they have been managing since COVID-19, with Red Thread members offering advice or resource support, often about family situations. We also stay in touch by inviting the women to events we host, such as workshops, protests, picket demonstrations, and other research projects. In addition, when Red Thread has the resources, approximately three to four times a year, we provide women with food hampers. We are mostly still in touch with individual women who need help. Red Thread, for example, was able to help one woman pay off her debt to her landlord (to help her avoid imprisonment) and provide an advance payment for a house plot, and others we have connected with an organisation that is able to supply monthly hampers. We have also managed to stay in touch with women who have left Sophia (such as one woman who has moved to Barbados with whom we stay in touch by WhatsApp or by visiting her when one of us travels there). A few of the women have not been reachable, however. Red Thread plans to keep on continuing these relationships as part of its regular activities.

In the **Ibadan CRT** all the participants have our phone numbers, and greetings are often exchanged over the holidays via a phone call, text, or WhatsApp message, and most of the women keep us updated about life events (e.g. births, weddings, deaths in their family, tertiary school admission, training completion, and so on). The participants also sometimes ask for monetary support. We started hosting lunch meetings twice a year with all the participants in August 2019, although COVID-19 restrictions prevented these from taking place in 2020 and 2021, when we still called the women periodically to check in. During our meetings, the participants are given an opportunity to discuss their concerns about life in Ibadan, specifically, and Nigeria more generally. They discuss topics such as food insecurity, transportation, education, gender-based violence, and electoral politics. What is most important about these meetings is that they are also a form of community building. Many of the participants who did not know each other beforehand have since exchanged numbers and stay connected with each other, and they also support each other in times of need and celebration. The last meeting, to wrap up the project, was held with the women in June 2023. During this meeting, based on the women's previous collective decision to produce a play as one of our knowledge-mobilisation activities (delayed because of COVID-19), a representative of the University of Ibadan drama group provided an overview of the drama they are working on with us, and the women provided feedback and suggestions regarding how they want the play to be scripted. The hope is that the play will go ahead and continue the development of community building.

The **Ramallah CRT** has remained in contact with the interviewees in a variety of ways, which developed organically, during and after the interviews, workshops, and other research activities. Participants and CRT members have been included on social media sites (including Facebook, WhatsApp, and Instagram), giving everyone a chance to check on each other and to learn what is happening in their lives. Researchers have been invited to social events such as weddings and lunches. Mai Al-Battat continues to visit the village, and since the population is relatively small, this has provided the opportunity to meet up with many of the people she interviewed. She also continues to meet some of the people she interviewed at public events in Ramallah.

The **Shanghai CRT** has kept in touch with some women through sporadic contacts, including WeChat and telephone communications, neighbourhood visits, and home visits. When the CRT lead researcher, Penn Ip, was in Shanghai, she invited the women to meet for a walk or tea in the neighbourhood and sometimes had lunch or dinner with them in restaurants or at their homes. She also organised a community service group at the School of Media and Communication at Shanghai Jiao Tong University in order to have students visit the women with her. When Penn was in the USA or Hong Kong, the Shanghai CRT's research assistants were then able to help contact and visit the women on her behalf. The assistants would bring fruits and other food to give to the women during the home visits. Contact finished in late 2023 when Penn moved from Shanghai to Hong Kong.

By Nasya S. Razavi, Anindita Datta, Joy Marcus, Grace Adeniyi-Ogunyankin, Mai Al-Battat, and Penn Tsz Ting Ip

The temporal difference between ethnographies and geo-ethnographies lends itself to different methods. Elana Buch and Karen Staller (2007, p. 203) claim that ethnographies' main methods are participant observation, talking with participants and/or people in the field, formal or informal interviews, and analysis of social artifacts such as 'documents, photographs, shopping lists, and diaries'. In geo-ethnographies a wide range of methods is also employed, including different types of interview, visual documentation, surveys, and the use of geographic information science technologies (GIS). While some geo-ethnographers employ participant observation as a primary method, for others this plays a less important role.

We should note that there are competing definitions of geo-ethnography, with some researchers referring to it as a form of multi-method research specifically tied to GIS. Here the prefix 'geo' is tied to place as mapped through GIS. Examples include the work of Stephen Matthews, James Detwiler, and Linda Burton (2005), who combine ethnography and GIS in their research on the spatial and temporal restraints experienced by low-income families. Nazgol Bagheri (2014) combines mapping and ethnographic methods to study women's spatialities in Tehran. These scholars argue that by deploying GIS and ethnography they can use mapping to show patterns in how people use neighbourhoods and cities and use interviews to capture the complexities of people's feelings about these processes. While their definition of geo-ethnography differs from ours in the central emphasis it places on GIS, the use of geo-ethnographic methods resonates with ours in that it focuses on research participants' sense of place and embodied experiences.

In the following section we explore how feminist scholars have engaged with geo-ethnographies.

Section 3. Feminist geo-ethnography

In this section we ask: what is feminist about geo-ethnography? Caroline Faria, Sarah Klosterkamp, Rebecca Torres, and Jayme Walenta (2020) write about feminist geographic ethnographies in their research on courtrooms. For them **feminist geo-ethnography** is an approach to research that traces spatialities of power, aims to enter the social world of those being studied, and is grounded in feminist epistemology, 'connected, political, relational, [and] performed' (Faria *et al.*, 2020, p. 1096). It is guided by a feminist ethics and the production of antithetical knowledge. In what follows we review the central concepts of feminist geo-ethnographies – it is place based; it addresses the everyday; it requires thinking across scale and tracing power; it is embodied; and it engages feminist activism – building on the work of Anderson (2012, 2020), Till (1996, 2005), Faria *et al.* (2020), and GenUrb's own contributions to feminist geo-ethnography as an emerging approach to urban research.

(i) It is place based

First and foremost, geo-ethnography is grounded in place. The discipline of geography has a long history of developing analytical approaches to place from the perspectives of spatial science, the critical social sciences, and the humanities (see Johnston and Sidaway, 2016, for a comprehensive overview). These approaches can be parsed into understandings of place as location, locale, and experience: 'While location equates place with position in space', locale consists of 'the settings for everyday routine social interaction provided in a place', and sense of place is a form of 'identification with a place engendered by living in it' (Agnew and Duncan, 1989, p. 2)' (quoted in Adams, 2017, p. 3). Each approach illuminates different aspects of place, and researchers have mostly explored these separately in various paradigms and across time periods, sacrificing the multidimensionality of place for ontological precision. As Paul Adams (2017) asserts, however, places result from synthesis, from the coming together of dynamic and heterogeneous elements and processes. Places are more than geographic locations with distinct attributes; processes happen with places and not merely in them. It is through places that geo-ethnographers can explore the spatialities of social relationships and uncover systems of meaning, which as Herbert asserts 'are, at least in part, locally specific and frequently intrinsic to a particular place; they are both place-bound and place-making' (Herbert, 2000, p. 557). Geo-ethnography also encourages you to focus on constitutive and embodied practices in place, which can connect to systems of meaning in ways that you may not expect.

With globalisation, geographers have increasingly adopted relational views of place, with perhaps the best-known among these being that of the feminist Marxist geographer Doreen Massey's (1994) understanding of a global sense of place. Places, for Massey, are 'moments' in networks of social relations; situated in and within any place people are positioned differently in relation to flows of power, what she refers to as 'power-geometry', a conceptualisation of how flows of power, capital, and culture contribute to different groups' relationships to and constructions of place. She defines places not only through particularity, as distinct locales, but also through their dynamic connections to other parts of the world: 'what gives a place its specificity is not some long internalized history but the fact that it is constructed out of a particular constellation of social relations, articulated together at a particular locus' (1993, p. 67).

The concept of **place identity** builds on these conceptualisations of place and the importance of place to identity formation, both individual and collective. This concept, which comes from environmental psychology, proposes that a person's knowledge and feelings, as well as identity, develop through experience of their environment: 'A sense of place identity derives from the multiple ways in which place functions to provide a sense of belonging, construct meaning, foster attachments, and mediate change. [It] can inform their experiences, behaviors, and attitudes about other places' (Gieseking *et al.*, 2014, p. 73). Understanding one's own place identity can explain a particular attachment to one location versus another, as well as why one might feel at home in a new place or uncomfortable when returning somewhere familiar after a long absence. Feminist scholars have shown how such attachments and identities are felt, sensed, and thought through the body (Longhurst, 2005). As Till (2005, pp. 16–17) explains:

> Individuals perform their identities in particular places and through their bodies; by acting, speaking, dressing, and interacting in certain ways at different locales, they cite who it is they are supposed to be. In so doing they create bodies/places through which they experience, remember, and imagine the world, and through which they fashion an identity. While each person is physically and socially embodied in distinct ways, through routine and repetitive actions a person situates himself or herself in social spaces. Through those repetitive acts, each may experience a reassuring (or distressing) fiction: it is the fiction of the self, that there is some coherent person underneath all of their confusing actions (past and present), that there is someone that remains at least in some respects more or less the same. It is a fiction, of course, because a person is always a different self with each return and with every performance, a self styled according to specific needs and the particular contexts of the present moment.

Your analysis of place as a relational spatiality and your understanding of how research participants identify with the place in which they live are integral elements of engaging in geo-ethnography (see also Chapter 12, Section 2 (iii) Cultivating a feminist ethics of care in the field for a discussion of the need for an embodied, decolonial, relational accountability when working on stolen or occupied Indigenous lands).

Reflection exercise 13.1 asks you to consider the role of place in your research and how place will inform your research findings.

Reflection exercise 13.1: Questions to ask about place in your research

In your research, you will need to address a number of questions in relation to place, including the role of place in your analysis, empirical questions about your field site, and the relationship you and the research participants have to the place of your research. Consider the following questions.

The analytical role of place

- What economic, political, social, and cultural factors do you have to take into account in developing a relational understanding of the place of your research?
- How does place feature in the research? Beyond a relational understanding of place are you also taking into account place identity, land ownership, and the participants' sense of place?

The empirical role of place

- In what ways are place identities being constituted in and through particularities at the field site(s) you are working in?
- What is the significance of the chosen location (field site) to the research? i.e. how would the research differ if it were situated elsewhere? For instance, how would the research questions, research methods and methodology, and access to participants change if the field site were different?
- What are the relationships between the location(s) in which your research takes place and the activities that occur there?

Relationships to the place of research

- How are you defining the location and locale of the research?
- How does your own place identity inform the research? Consider your positionality and issues of reflexivity. (For example, to what extent can researchers with children or other care responsibilities spend extended periods in the field?)
- How does the research participants' place identities inform the research?
- What can you learn from participants' embodiment?

(ii) It addresses the everyday

Geo-ethnography addresses everyday life through the spaces, practices, and relationships that shape it. Indeed, the routine activities of people's lives, particularly women's lives, have long been of interest to feminist scholars (Mackenzie, 1989; Katz, 2004; Hall, 2019; Razavi *et al.*, 2023). At first glance, everyday life consists of mundane activities. However, 'everyday life is not a static phenomenon, but is rather a dynamic process which is continually unfolding and emergent' (Eyles, 1989, p. 102). Seemingly ordinary routines are full of details that convey meaning about the relationships between individuals and broader societal structures. The health geographer Isabel Dyck (2005) analyses care work that takes place in people's homes in various cities in Canada. By examining the daily rhythms of a nanny from the Philippines in Vancouver or the professional who lives in Ontario and travels to Scotland to care for an aging parent, Dyck is able to identify 'gaps in services left by neoliberal policy' and better understand how families are responding to global economic restructuring (Dyck, 2005, p. 238). Her work shows how everyday routines construct and reconstruct structures in the societies that we are a part of and how those broader structures shape the context of everyday activities.

A geo-ethnographic analysis of the everyday requires us to observe the spatial and temporal dimensions of routine activities and aspects of life, which may not always be obvious. The meanings of everyday objects and events are not always directly observable; they emerge indirectly through actions as well as words; they are often revealed through practices, reactions, cursory comments, and facial expressions. As Herbert (2000, p. 553) explains: 'These meanings are discerned through inferences, as daily practice is interpreted against the ethnographer's developing understanding of the larger cultural system of which it is a part'. In your research pay close attention to the habitual, spontaneous, and adaptive, asking how these practices align (or do not align) with dominant narratives. Consider asking people how they feel when they are going about their daily activities, and pay close attention to their body language and facial expressions. In addition to these nuances, pay close attention to changes to and obstacles within everyday life. For example, the feminist geographers Poppy Budworth and Sarah Hall (forthcoming) write about how disabled people have had to reshape their social lives to avoid highly populated places to mitigate risk of COVID-19 infection. Their focus on disability in everyday life offers insights into urban inequalities.

(iii) It requires thinking across scale and tracing power

Geo-ethnography is attuned to tracing such inequalities and relations of power given its sensitivity to the need to offer an analysis of meanings of place from various geographic scales (Till, 1996). Consider how the spatialities of the topic you are researching connect to the 'wider sociocultural, political, and economic circuits of power that stretch across scales of the local, regional, national, and international' (Faria *et al.*, 2020, p. 1104). Thinking across scale demands an understanding of how your research topic fits into and interacts with wider economic and geopolitical processes. It is feminist work to connect 'events, bodies, feelings, objects, and processes' across scales, including the 'privatized, trivialized, or otherwise quotidian' (Faria *et al.*, 2020, p. 1104).

Feminist geographers have advanced thinking about scale as relational rather than hierarchical. They have pushed us to think about how life's everyday intimate moments are interrelated, for example, with the geopolitics of race (Vasudevan, 2021), territory (Smith, 2020), religion (Gökarıksel, 2012), and nationalism (Militz, 2023). Feminist economic geographers theorising about space, place, and scale have reconceptualised our thinking about the local and the global not as nested distinct sites but rather as interconnected in a web of shifting power relations (Massey, 1994, 2005; Nagar *et al.*, 2002; Roberts, 2004; Werner *et al.*, 2017). And feminist urban geographers have long been interested in how the urban is an important spatial scale through which gender, race, sexuality, class, and ability are experienced and constituted (Peake, 1993; Bondi and Rose, 2003; Parker, 2011; Gieseking, 2020). Thinking across scale, not hierarchically but relationally, is useful for conducting research that is attuned to what we may or may not see, hear, and feel, to interdependencies and interrelationships, and to global intimacies of power. It is useful because it places embodied spaces, practices, and knowledges – such as those of patriarchy, fundamentalism, whiteness, and heteronormativity – at the centre of attention. Such spaces, practices, and knowledges are not always visible and not always accounted for within the Western academic canon but are nonetheless deeply interwoven into dominant systems.

Working across scale in geo-ethnographies also helps trace power within the topic your research project is exploring. Tracing power means asking how the research topic is understood within 'larger systems of gendered, racialized, classed, and sexualized power' (Faria *et al.*, 2020, p. 1105). For example, in her book *Golden Gulag: Prisons, Surplus,*

Crisis, and Opposition in Globalizing California, abolitionist geographer Ruth Wilson Gilmore (2007) seeks to explain the political and economic forces behind the extreme increase in the number of people incarcerated in US prisons since the 1980s. She carefully traces patterns of capital investment alongside attacks on radical struggles and labour movements to explain how surpluses of finance capital, labour, land, and state capacity were behind prison expansion. She has described this as tracing alternative geometries of power, and her findings illustrate how the place-based textures of everyday life in California have shifted over time and are produced in relation to broader conjunctural shifts (Gilmore, 2002, 2007). The ways in which her research traces power reflect a feminist political commitment to understand, respond to, and undermine or disrupt oppressive power relations within the research project and within knowledge production itself (see Chapter 7, Section 2 Towards a feminist ethics and Chapter 4, Section 3 Defending space for feminist activism in the neoliberal academy).

(iv) It is embodied

Embodied presence is vital in geo-ethnographic research, important for the creation of field notes, sketches, maps, etc. that contain sensorial details about the sights, sounds, smells, and what it *feels* like at the research site. In addition, your consistent presence and actions have an impact on the research and the lives of research participants. Beyond corporeal presence, embodiment has an affective and emotional aspect that bears upon geo-ethnography. Feminist urban geographer Jess Linz, for example, writes about sensorial cues that 'stick out' in major life events and re-appear in ordinary daily life, often hidden in plain sight (Linz, 2021, p. 286). She considers how the sounds and smells in contemporary Mexico City bring to the fore memories of the massive earthquakes that devastated the city and its residents on the same day 32 years apart: September 19, 1985, and September 19, 2017. Through their physical senses people tap into emotions linked to specific places and times in the city. It is through such symbolic markers and activities that involve the senses that attachments to place are created.

Faria *et al.* (2020) argue that having a consistent physical presence at the research site that is humble and open is key to cultivating relationships and building trust. Research participants may be more likely to tell you how the research could benefit their lives if you are available to them and if you remain respectful and open to hearing what they have to say. Intellectual humility and deep listening are feminist practices of 'critical reflexivity about our subject positions as researchers' (Koch, 2020, p. 52). A good listening practice is helpful for building a deeper understanding of human subjectivities, animal and plant life, weather, climate, and more at the field site (Gallagher *et al.*, 2017; Koch, 2020). In addition to taking fieldnotes about the sounds, sights, and smells that are present, think about what is *not* there – the absences and silences. Through consistent presence and employment of

Audio clip 13.1: City soundscapes

Audio clips of street sounds recorded by the GenUrb CRTs in Delhi, Ibadan, and Shanghai are available on the book's website www.routledge.com/9781032668680.

By Anindita Datta and Swagata Basu, Bukola Omolara Odunola and Sylvester Nmormah, and Penn Tsz Ting Ip and Judy Zhu your senses, you can build a rich repository of field notes about how it feels to be at the field site.

Audio clip 13.1 gives you a sense of the richness of navigating understandings of place by just listening to the sounds where you do research. As you design your own geo-ethnographic methods consider how place is experienced through the body.

(v) ***It engages feminist activism***

Feminist scholars adopt a progressive vision of social justice in their research, engaging with critical analyses that aim to contribute to changing unjust and inequitable power relations (see Chapter 4, Section 1 The academy and feminist scholar-activism). They are committed not only to making their work meaningful and relevant to the people they work with but also to having a political intent to promote social betterment.

Geo-ethnographies allow feminist scholars to engage in various ways with such a vision. In GenUrb, for example, the levels of engagement, publicness, advocacy, and activism varied a great deal due to a range of factors (see Chapter 4, Section 3 The challenges of feminist scholar-activist research in transnational contexts). Some of these considerations made engagement and public activities prohibitively risky or dangerous (see also Low and Merry, 2010). This is not to argue that it is not possible to show up consistently or remain committed to producing what Faria *et al.* (2020, p. 1106) call 'antithetical knowledge – a commitment to disruptive work that helps us better understand and respond to the machinations of power'. In some cases, it may be safer or more politically astute to engage not as an activist but as an interlocutor, as someone who can 'offer social critique' (Davis and Craven, 2022, p. 197). The bottom line in knowing when and when not to engage openly in activism is when it involves danger to you or anyone else in your project, invoking concerns about security. In GenUrb example 13.2, we outline some of the security concerns we have faced in GenUrb.

GenUrb example 13.2: Security issues

Security concerns have had to be considered in a number of the GenUrb teams. As we outline in the Introduction of the book, GenUrb example 0.1 How the GenUrb partnership evolved, we had to make the difficult decision to cancel plans for establishing a CRT in Tehran after the imprisonment of the team lead.

The increasing clampdown on feminist and queer research over the life of the project has also affected the ways in which the Shanghai CRT has been able to operate. The CRT had put together a network of Chinese feminist urban scholars working in China, the UK, the Netherlands, Canada, and the USA. It became too dangerous to conduct a workshop with this group in Shanghai, so it was agreed to hold a virtual workshop of the network members and other members of GenUrb in August 2021. We could not advertise this meeting nor let anybody attend whose politics towards the government of China were unknown.

In more than one context, individual GenUrb researchers have felt subject to surveillance or targeted by political authorities, by members of the public, or by their colleagues because of being female, feminist, or lesbian or because of their nationality. The concerns of one GenUrb researcher about her personal safety meant that our project manager had to work with lawyers at York University and Canadian immigration officials in case we had to help the researcher flee the country in which she worked.

We have also had to consider the material that we put on our website. Throughout the life of the project, in order to keep members safe, we have had to remove material that identifies specific individuals.

By Linda Peake and Leeann Bennett

Activism does not, of course, guarantee justice. It may not have the leverage needed. But what engaged geo-ethnographies can do is reveal how power can be negotiated and produce potentially generative tactics. Geo-ethnography is a research approach grounded in praxis, an actively engaging process that is embodied and relational, that involves not only working together in the hope of creating meaningful research but also contributing to efforts to bring about positive social change, sometimes practically and immediately and at other times strategically, by brokering knowledge and helping to create the conditions for such change.

Reflection exercise 13.2 asks you to consider how your research is already and might become more geo-ethnographic.

Reflection exercise 13.2: In what ways is your research geo-ethnographic?

Your research may not attend to all these questions, but by thinking through each question carefully, you may become more attuned to place identities, sensorial cues, spatiality, and power in the areas where you are conducting research.

- How is your research geo-ethnographic?
- How does it consider place as relational?
- How does it address the everyday?
- How does it think across scale?
- How does it trace power?
- How is it embodied?
- How does it engage activism?

Section 4. Summary

Feminist ethnographies emerged in the 1970s, building upon and critiquing the colonial origins of ethnography and addressing the dynamics of social relations to detail the experiences of groups marginalised by unequal power relations. Drawing from feminist ethnography and critical geography, geo-ethnography is an approach to research that is taking root in feminist geographic scholarship. It is grounded in the understanding that place identities are developed from people's experiences in their lived environments, including the structured inequalities and oppressions they might face. Geo-ethnography requires thinking relationally across embodied experiences of the 'global intimate' to learn about how everyday life, rooted in place, is interconnected with broader economic and geopolitical processes. The physical and affective embodied presence of the researcher and research participants and the relationships between them are central to conducting geo-ethnography. Doing geo-ethnography thus requires paying attention to place and the tracing of scalar spatialities of power, as well as to embodiment and relationships, in the production of knowledge. It also requires a feminist commitment by the researcher to engage in activist research while taking security concerns seriously in order to undermine and disrupt oppressive relations within the research project and knowledge production itself.

14 Feminist interviews

Araby Smyth, Elsa Koleth, and Linda Peake

Learning objectives

After completing this chapter, readers should understand

- the elements of a research interview;
- the function of research interviews in qualitative research;
- the different types and modes of research interviews;
- and the power-laden nature of interviews as research practice.

Further, readers should be able to

- select a type and mode of interview for their research project;
- and plan and conduct a research interview.

This chapter provides an in-depth discussion of types and modes of research interview. The research interview is a qualitative method that can generate rich data from conversations between researchers and participants. The chapter prepares the reader to conduct their own research interviews by outlining a range of relevant considerations: in particular sample size, participant recruitment, consent, the interview guide, compensation, the significance of location when interviewing, and the need to maintain contact with participants. Our discussions of these considerations are informed by feminist concepts like reflexivity to assist the reader with navigating the fraught power dynamics of research interviews.

Section 1. Interviews in feminist research

Interviews are conversations framed by questions in which people discuss particular topics. Research interviews are conversations between researchers and research participants, for which informed consent has been given, in order to gather information that serves as data. Research interviews are also a power-laden practice wrapped up in ethical challenges, requiring consent and preparation and a high degree of trust between the researcher and the participant. Trust is the cornerstone of the relational process that is the interview. Although feminist researchers are very inventive in devising appropriate methods, interviews are still the most common form of qualitative research method used by them. Interviews were the bedrock of the research undertaken in the GenUrb project.

Following the feminist sociologist Ann Oakley (1981), feminist researchers understand the interview as a relationship between researcher and participant, who jointly

DOI: 10.4324/9781032668727-19

produce knowledge. Interviewing is not an 'objective' exercise of extracting information from a passive participant – the participant is an active agent who can add more to the interview rather than merely respond to questions. Not only have research interviews proved a reliable and well-tested method of procuring data, but the ease with which they can be set up, as well as their relative cheapness and versatility – from fact collecting to storytelling – makes them an indispensable feature on the feminist research landscape. Interviews are particularly useful as a method for feminist research because they can serve various functions: they provide insight into the lives of women and other gendered subjects and into participants' unique subjectivities, and they enable participants to narrate their experiences in their own terms.

We now delve further into understanding different types and modes of interview.

(i) Types of interview

The types of interview you conduct should be derived from your research questions, which, in turn, speak to the aims of the research and its underlying philosophical approach (Edwards and Holland, 2013). You can think of different types of interview as existing on a continuum from the most structured (such as those that use **questionnaires**) to semi-structured (which use an interview guide) to unstructured (for example, oral and life history interviews).

Structured interviews employ closed questions to collect data. The researcher asks a set of pre-determined questions and takes care to phrase them in the same way with each participant. As part of a survey these questions are formalised in a questionnaire. Structured interviews are commonly used to elicit information on topics that can be addressed with quantitative data or by a short reply. Structured interviews are helpful in supporting a comparative analysis of responses, but they are inflexible: they do not allow the researcher to probe or pursue lines of inquiry that are not considered in the original questions. In addition, they produce a hierarchy of power that places the participant in a subordinate position to the researcher (Oakley, 1981, 2016). Despite their limitations, structured interviews as part of survey-based research have been successfully used by feminist researchers to obtain information on a range of topics. See, for example, the research undertaken by Red Thread on the extent of domestic violence in Georgetown, Guyana, which is based on a questionnaire-based survey of 360 women (Peake, 2000, 2009).

Semi-structured interviews are conversations between researcher and participant that unfold in response to the researcher's questions. They enable the exploration of emergent themes and can provide insight into the lives of participants and how they understand their experiences. The researcher uses an interview guide or schedule of questions. Semi-structured interviews are unlike structured interviews in that the researcher can ask follow-up questions to clarify the participant's response, prompting them for additional information and context. Although the semi-structured interview is less formal than its structured counterpart, the power relation between researcher and participant is still one in which the researcher is in charge. Recent examples of feminist urban research based upon semi-structured interviews include Stephanie Butcher's research (2021) on the everyday politics of the urban poor in Kathmandu and Yui Hashimoto's (2021) investigation of racialised discourses of work and colour-blind redevelopment in Milwaukee.

Unstructured interviews are an open-ended way to explore meaning(s) about personal and emotional subjects and to develop a detailed biography of the participant, told from their perspective. Also referred to as non-directive or open-ended interviews, unstructured interviews can be time-consuming and may require more than one session. Although unstructured interviews have no 'prespecified protocol' (Ravitch and Carl, 2021, p. 252), they may involve one opening question, usually a prompt related to the research purpose that sets the scene, such as, 'Can you tell me what your childhood was like?' It is largely the participant who determines the direction of the interview, although the researcher can probe for further detail, circumscribe, or 'rein in' the interview. Unstructured interviews may yield unexpected findings that can be generative for the research project. Sapana Doshi and Malini Ranganathan's (2017) analysis of land dispossession and corruption narratives in Bangalore and Mumbai and Dorota Golańska's (2022) research on shifting temporalities of political violence in the West Bank, Palestine, are both based on unstructured interviews.

In what follows we outline two types of interviews – **oral histories** and **life histories** – that are part of a repertoire of biographical methods undergoing a revival in the social sciences due to an increased interest in the role of agency in social life. Such narrative interviews focus on research participants' sense of identity, 'draw[ing] not only on their own experiences and understanding, but on culturally circulating stories that help them interpret and make sense of the world and themselves in it' (Edwards and Holland, 2013, p. 35).

Life histories are a long-standing method in history, anthropology, sociology, and psychology and involve a type of in-depth interview in which participants are asked to discuss specific periods, key moments, or certain aspects of how their lives have unfolded, with an emphasis on their own interpretations of their experience. Life histories thus give a voice to the lives of participants and, given their commitment to storytelling, can often have a wide appeal beyond academic publications. Typically, life histories make use of thick description, pay attention to the social, historical, and geographic contexts in which they are narrated, and have a methodological concern for the ways in which the life history has been constructed. They also pay attention to criteria such as the trustworthiness of the narrative (Hammond and Wellington, 2013).

Life histories may be conducted with a small number of participants whose narratives may speak to events in which the researcher has an interest. They can be demanding on the researcher and participant because they work in depth over prolonged periods and may therefore uncover strong emotions or because information may be revealed that is disadvantageous to the participant. They also require a great deal of time, such as several 60- to 90-minute sessions. They can extend even longer, over days as opposed to hours, if participants are willing and if the researcher can maintain active listening.

Oral histories are based on listening to individuals talking about their lives that can produce richly detailed and intimate data. They enable an 'understanding [of] the past, the relation of past to present and the lives of others through time' (Gardner, 2006, p. 206). Developed in the 1960s, oral histories have been used to uncover the voices of groups marginalised by unequal power relations and to recover collective memories, hence the preponderance in many research projects of elderly participants. As with life histories, the major pitfall of this method is its reliance on memory, which can be unreliable, but this problem can be mitigated by the employment of factual secondary sources.

For some researchers, life and oral histories are synonymous. Indeed, both utilise loosely structured interview formats in which the researcher and participant engage in conversation. Their primary difference is their focus: life histories primarily address the life of the individual, while oral histories often centre around historical questions. Despite the challenges to employing these methods, they have an unparalleled capacity to generate rich qualitative data.

GenUrb example 14.1 introduces the types of interview used in the GenUrb project. Given that interviews as a research method are not inherently feminist, we discuss how these interviews are integral to our feminist project.

GenUrb example 14.1: Types of interview used in GenUrb

The GenUrb project engaged in different types of interview with two distinct constituencies.

- *In-depth semi-structured interviews and oral histories with women living in low-income neighbourhoods.* The choice of women participants in the neighbourhoods in each city was determined by the local City Research Teams (CRTs). The criteria for the neighbourhood's inclusion was that it was politically, economically, and socially marginalised and its residents subject to risks and vulnerabilities.
- *Structured interviews with 'policy-shapers' in each city.* The policy-shapers included people who had a direct and indirect impact on influencing, formulating and enacting policies in relation to women and cities, with a particular focus on those who were familiar with Sustainable Development Goals (SDGs) 5 or 11. They included public officials at national and local levels, members of local and international organisations, including NGOs, and activists or members of civic society based in the city where the research was taking place.

The interviews with women living in low-income neighbourhoods served a feminist purpose in that they were the conduit through which a commitment could be made to these women. For some of our CRTs, their engagement with the women participants went beyond the time of the interviews to include interventions that could address facets of their everyday lives. For example, participants were involved in the dissemination of their interview data, through radio programmes (in Georgetown), through the production of books (in Shanghai), through postcards (in Ramallah), and through exhibits for the general public (in Ramallah and Shanghai). These activities brought material in the interviews (for example, on housing costs, domestic violence, pollution, environmental destruction, and urban planning) to broader audiences, including urban planners.

The interviews, including the further interviews conducted during the COVID-19 closures and lockdowns, also served to provide some temporary monetary relief to individual women, who were given financial support for their time spent in the interviews. Some CRTs were also able to spend more time with women who they had identified as needing help to secure government monies to which they were entitled.

The interviews with policy-shapers were feminist as they enabled the collection of data enabling a critical interrogation of the SDGs. Our initial plans to use the interview findings for public education with municipal governments in each city were disrupted by COVID-19. For example, while engaged in the first round of policy-shaper interviews, I gave a talk in the Guyana parliament to parliamentarians and policy aides on the collective right to the city and housing issues in relation to SDG 11 (on cities) and SDG 5 (on women). Despite the interest this generated, COVID-19 then prevented all parties from following up on this initiative.

By Linda Peake

(ii) Modes of interview

As with the types of interview you conduct, the modes of interview you choose should be derived from your research questions and approach to your research topic. We outline here the modes of interview that are commonly used in feminist research, including one-on-one interviews, group interviews, **focus groups**, mobile interviews, and virtual interviews.

A one-on-one interview is typically what people think of when they imagine a research interview. One-on-one interviews involve the researcher and one research participant. They can be structured, semi-structured, or unstructured, depending on what you want to achieve with the data collection. While group interviews may provide diverse perspectives on a topic, one-on-one interviews focus on an individual and provide more privacy for participant responses.

A group interview is a group of people 'engaging in a collective discussion of a topic previously selected by the researcher' (Edwards and Holland, 2013, p. 36). The researcher guides the discussion with a series of questions or another kind of prompt to elicit data. Group participants might have important commonalities such as age or gender, occupation, or class, or they might be different, depending on the research design. It is important to consider these participant attributes as they will affect the power dynamics of the group interview, sometimes in ways that cannot be predicted.

Focus groups were initially utilised when researchers were interested in the nature of interactions between participants in a group, but now the terms 'focus groups' and 'group interviews' are commonly interchanged with each other. Focus groups are typically conducted with a small-sized group (approximately six to eight members), and the researcher guides a collective conversation around a predetermined topic or a prompt to elicit qualitative data that reveal group interactions concerning what participants think about a topic and why they think the way they do (see, for example, Gieseking, 2020). GenUrb example 14.2 outlines why GenUrb reversed an initial decision to conduct focus groups.

GenUrb example 14.2: Deciding (not) to use focus groups

In GenUrb, we initially intended to include focus groups in the methods we would employ but then mostly made decisions not to engage with them. Interestingly, each CRT came to the decision not to continue to use focus groups separately, with only one CRT deciding to stick to the original plan to conduct focus groups. There was more than one reason behind these decisions.

- It was proving much more difficult to secure women participants to engage in focus groups than in one-on-one interviews. For the most part, women did not have the time to meet. Whereas interviews can be arranged to suit the schedule of the individual interviewee, focus groups are not within the control of any one participant. So even the prospect of being paid for their time was not enough of an incentive to secure women's participation in a meeting that was potentially inconveniently timed for them.
- The venue for the focus group needed to be a quiet space, large enough to accommodate all the participants. In some cases, this meant the women travelling beyond their neighbourhood to an agreed place, which was beyond what prospective group members could manage.
- In a few cases the recording equipment available was not of sufficiently high quality to record a group meeting in which members could be talking over each other or where there could be a lot of background noise.
- Focus groups took a lot of time and effort to set up, and it took a lot of time to transcribe the results. Given the time and effort already being put into other methods, therefore, the 'value-added' of focus groups was questioned.

By Linda Peake

Conventional interviews are conducted sitting down with researchers conversing with participants in a single location. Mobile interviews are an increasingly common alternative, which produce 'spatially grounded and place-specific data' (Finlay and Bowman, 2017, p. 263). The term 'mobile' (as opposed to 'walking') is explicitly employed so as not to discriminate against people with disabilities and to recognise the multiple mobilities that people employ in their everyday lives. During a mobile interview, the researcher and participant move through different spaces. The route and setting provide context for the interview and can be the source of prompts and probes. The mobile interview can not only produce insightful spatial data that can generate theories grounded in participants' sense of place but also serve to put participants at ease, leading to more forthright conversations (Finlay and Bowman, 2017).

Virtual interviews allow researchers to connect with research participants across time and space. Tools such as Facebook Messenger, FaceTime, Signal, Skype, Telegram, WhatsApp, and Zoom allow users to connect with video, audio, phone, and chat functions that operate on desktop and mobile device applications (see Chapter 20, Section 2 A guide to social media tools and their uses). Virtual interviews can be conducted asynchronically, via email or text/voice messaging, or synchronically via audio or video (see O'Connor and Madge, 2017). The GenUrb team turned to virtual interviews when COVID-19 made it impossible to conduct in-person interviews in most cities (see Chapter 5, GenUrb example 5.2 Adapting research methods during the COVID-19 pandemic). Virtual methods may increase accessibility and provide different modes for participants to express themselves (for instance, by sending photos). However, while virtual interviews are flexible, they can (often) be less intimate. For example, you may not always be able to make eye contact, read body language, or know as much about what is happening in the participant's surroundings during a virtual interview. There are other potential downsides. Not all participants may have access to platforms or the cellular networks or internet connections they run on. Even if they do, privacy and security on virtual platforms is always a concern (see Chapter 20, Section 4 Challenges faced using social media in feminist scholar-activism). Virtual methods also require that researchers and participants are familiar with the relevant technology.

Based on what you have learned about types and modes of interview in this section, Reflection exercise 14.1 asks you to think about your own research project and determine how you might use interviews to collect data.

Reflection exercise 14.1: Types and modes of interview

Reflect on the aims of your research project. To help determine the type(s) and mode(s) of interview that will yield the data that are best suited to your research, explore the following issues.

The epistemological concerns embedded in the project (see Chapter 11, Section 3 Feminist epistemologies)

- How important is it to your research that you explore who is the knower in your research?
- How will this affect the type of interview you choose to employ?

The data analysis process

- Are you comfortable with a more open-ended and exploratory approach to analysis, or do you have a set process in mind?
- What type or mode of interview can support your analysis process?

In the following section we turn to the issues you need to take into consideration when planning research interviews.

Section 2. Planning research interviews

When planning interviews, it is important to consider the following: sample size, the means used to reach out to potential participants, gatekeepers, consent, the interview guide, compensation for participants, the interview location, and the need to maintain contact with participants. These planning aspects may seem simple, but they have wide-reaching implications for participants and researchers. In particular, they all have a role to play in establishing and developing trust between the researcher and participant. Document 14.1, on the book's website www.routledge.com/9781032668680 can help you with the specifics of how to prepare for conducting an interview.

(i) Sample size

The simple question of how many people you will interview is not always easy to answer. Determining sample size is governed by a number of factors. The primary determinant of sample size for qualitative research often relates to a subjective understanding of how many people you need to interview before you can recognise patterns in the data. In this sense, you may not know a desired sample size before you start interviewing. Questions relating to the resources at your disposal will also play a part. How much time do you have to interview people? If you are paying participants, how much money do you have? It is important to the success of your research to have a clear sampling strategy and to understand how it relates to your research questions, your resources, and the lives of participants.

GenUrb example 14.3 details how sample sizes were determined for the research interviews in the GenUrb project.

GenUrb example 14.3: Determining sample size for different groups of participants

The GenUrb interviews were conducted with women living in marginalised neighbourhoods and with policy-shapers in each city.

Data were collected via two types of interview with women participants: in-depth and life history interviews. The data collected encompass women's understandings of their everyday lives and their entanglement in social and economic relations that stretch within and beyond the city. We planned to interview both young women (18–25 years old) and women in their 40s and older, based on the assumption that there may be differences in their attitudes to work and their perspectives on the future and in their use of technology. After discussion between the CRTs we arrived at a mutual decision as to how many women we needed to interview in each city neighbourhood in order:

- to secure a large enough sample size in each city, with roughly equal numbers of younger and middle-aged or older women;
- to be able to recognise stories that were emerging from the data;
- and to enable comparison to take place across the six cities.

Based on their previous experiences of conducting research, a couple of CRT members felt that a sample size of twenty women would be enough. However, the majority felt that 30 would be a more appropriate figure. In some of the cities, there were high levels of instability and mobility, and we knew that a number of women who may have been available at the start of the project would probably not still be there four to five years later when the second round of interviews would take place. Having 30 participants allowed for five to ten women to have left the neighbourhood during the research process but still leave a sufficiently large sample of interviewees to enable meaningful analysis.

For the interviews with policy-shapers, there was mutual agreement that a sample of twelve to fifteen would be sufficient for each city in order to secure (mostly factual) material related primarily to SDGs 5 and 11 as well as to urban development plans and policies in each city.

In total, we conducted interviews with 152 women participants – 152 in-depth interviews and 97 life histories – and 143 interviews with policy-shapers. Of the women, 56 also participated in further interviews during the period of COIVD-19.

By Linda Peake

(ii) Finding participants

Finding participants (also known as recruitment or outreach) involves negotiating access to potential participants and discussing with them the importance of the study (England, 2002). The recruitment stage is vital for building relationships with participants, but the work involved in this stage of the research, in developing the connections needed, is often unseen and underestimated (Jackson, 2021). Choosing who to interview is 'often a theoretically motivated decision', meaning that the choice of participants is determined by a range of factors, including your research questions, research methodology, the resources available to you, and the context in which you are conducting your research (Valentine, 2005, p. 112).

GenUrb example 14.4 shares how the Georgetown CRT i.e., the women's organisation Red Thread, recruited participants.

GenUrb example 14.4: Participant recruitment by the Georgetown City Research Team

Finding women participants in the low-income neighbourhood of Sophia

We had long-standing connections with women in the neighbourhood of Sophia, where the research was conducted. This made it easy for us to introduce the research to women who were already known to us and then ask them to invite other women in Sophia to join in. Once we knew the parameters of the sample, such as size and age range, we started by contacting the ten or so women in Sophia we had worked with previously on various projects. This process was interrupted, however, when the Ministry of Housing started demolishing squatters' houses in Sophia as part of an Inter-American Development Bank project on community improvement, and we started to engage with the residents caught up in this process. We had to delay the start of the research as we helped the squatters to picket the Ministry of Housing, which then stopped the demolitions, allowing the squatters to apply for expedited allocation of house plots in a resettlement site located far from Sophia.

Once the squatters' issues were resolved, we started recruiting other women in Sophia. The sample was divided into younger and older women, with the intention of interviewing equal numbers of women in both groups, but recruitment was not easy. It took numerous visits to secure the total number of 30 women to participate in the research, although it proved easier to secure participants in the older age group. Finding enough younger women to interview through snowball sampling was difficult, but we made up the number of younger women by recruiting relatives – three daughters and one granddaughter – of the older women participants. Younger women were either too busy to engage because they were working or they were less interested in the research or less settled in Sophia, having only recently moved into the neighbourhood. Recruitment was hampered by a sense that people kept to themselves in Sophia: a common response was 'I don't know anyone, you would have to check with them yourself'. However, as a technique, snowball sampling worked well enough to secure the participation of the agreed-upon number of women.

We took the decision to pay the women participants for their engagement in the research, although we did not communicate this decision to them until after the first round of interviews was complete. We did not offer money up front as we wanted a commitment to the research from the women based on their interest in the objectives of the research: to engage women in talking about the conditions of their everyday lives in Sophia and Georgetown; to talk about these conditions publicly (through a series of radio programmes); and then to engage further in public education work (which is ongoing).

The main challenge in organising the interviews, apart from initial recruitment, was the danger of being in Sophia, a neighbourhood with a high level of street crime, gun crime, and sexual assaults, not helped by the lack of street lighting and the desolate nature of parts of the neighbourhood, which led the interviewers to fear being robbed or raped. As Joy put it: 'You had to put on this brave appearance when all the time you would be trembling with fear inside'. A further challenge for some of the interviewers was dealing with the emotional toll of listening to and then living with the women's stories, especially the women who were squatters, living 'hand to mout' with their children, often with no family support. We undertook to address the precarious living conditions of these individual women not only by helping the squatters but also through other activities we conducted in the neighbourhood. These other activities included helping participants leave violent domestic situations and providing financial assistance to one participant whose house was burnt to the ground by a violent partner, and to another woman who was squatting and needed help to pay court costs, as she was facing imprisonment for non-payment.

Finding policy-shapers

Because Georgetown is such a small city and Red Thread and its co-ordinator, Karen de Souza, are well known, there was no problem in securing the first round of interviews, conducted in 2018 and 2019. Many of the policy-shapers were personally known to Karen. However, by the time of the second round of interviews in 2022, there had been a change of government. The colossal oil and gas discoveries off the coast of Guyana had come to dominate the economy, and access to government ministries and officials was much harder to achieve. Although we were still able to secure the participation of many policy-shapers, the mode of engagement with those who were employed by the government was markedly different in the second round of interviews. In the first round there had been some conversations and exchanges that dipped into the 'reality' about the status of the SDGs as opposed to the official government line. By the time of the second round, this exchange of sentiments was absent. A further challenge in conducting these interviews was that many of the participants had very little to say about the SDGs, given that they have not been a government priority in Guyana.

By Karen de Souza and Joy Marcus

(iii) Gatekeepers

You may need to work through gatekeepers to access participants, particularly if you are dealing with vulnerable populations. A **gatekeeper** could be an individual, a group, or an organisation who controls or can influence access to potential research participants. Gatekeepers can be useful, as when assisting with access to interviewing elites and inaccessible social networks. They may also have a role in protecting vulnerable people, or they may exercise some power over the circumstances of potential participants, which may raise other ethical issues, for example, about whether participation is voluntary and consensual. GenUrb example 14.5 discusses CRT experiences with gatekeepers.

GenUrb example 14.5: Encountering gatekeepers

One of the **Ibadan CRT** members approached the *Iyaloja* (female leader) at Adelabu market in Challenge to help recruit participants. However, after attempting to interview someone the *Iyaloja* recommended (the first interview), the Ibadan team lead made the decision that using an *Iyaloja* was not an appropriate option. Why? Because the participant we attempted to interview 'consented' but was not at all forthcoming. She barely answered the questions and was practically silent throughout. It got to a point (about one hour in) where the interview was cut short. It was deemed a waste of time, and we did not use it. The participant had had no choice but to agree to 'participate' because she was recruited by the *Iyaloja*, the most powerful woman in the market, who people could probably not say 'no' to. As a result, we avoided using gatekeepers to recruit other research participants.

The **Shanghai CRT** contacted a professor from Fudan University who has a strong connection with policy-shapers and local officials in Shanghai. With his help, we were referred to various 'street offices' and neighbourhood committees. One of the government officials working at a street office, Director Z, accepted our request and offered to help us to recruit women who were officially identified by the government as 'destitute and disadvantaged' (*pinkun funv*). We translated the interview guide into Chinese to share with Director Z in order to obtain her approval and to adjust questions that might be considered inappropriate for the social and cultural context in Shanghai. The street office oversees sixteen neighbourhood committees located in Community X. Director Z shared our GenUrb project description and the Chinese version of the interview guide with the leaders of these sixteen neighbourhood committees, allowing them to choose whether or not to join in and to support our project. Ten neighbourhood committees chose to do so and helped recruit women for the research.

The **Georgetown CRT** did not encounter gatekeepers, although in the second round of policy-shaper interviews, we experienced how others were constrained by gatekeepers. During these interviews, public-service employees were afraid to speak without permission of their supervisors or political bosses, and they were less than forthcoming in volunteering information and opinions.

By Grace Adeniyi-Ogunyankin, Penn Tsz Ting Ip, and Joy Marcus

(iv) Obtaining consent

It is important to ensure that you acquire voluntary and fully informed consent from participants prior to interviewing them (see Chapter 8, Section 1 Professional standards in the academy). This includes making sure that participants are fully briefed on the nature of the research, the institutions and funding sources supporting the research, what is expected of them, and how the findings of the research will be disseminated. It is also important to have the consent of people with whom you engage in participant observation and of the leadership of community groups, local governing bodies, or other organisations where applicable.

Your university's research ethics protocols may require you to obtain a signed consent form from participants (see Chapter 8, Section 2 Ethics policies). Where this is not possible, researchers should develop a script to guide discussions with research participants and acquire verbal consent prior to conducting the interview. The consent form or script must detail how you will maintain anonymity of participant identities in your research project if that is what they wish. It must stipulate that participants in the project always have the right at any time to end their participation or withdraw their data (i.e. interview recordings and transcripts). You must also reach out to and obtain the permission of participants for any changes in your research plans. The GenUrb CRTs used a mixture of written consent forms and verbal consent scripts, and these are shared in GenUrb example 14.6.

GenUrb example 14.6: Consent forms

A number of informed consent forms used for policy-shapers and women participants in GenUrb are available for download on the book's website www.routledge.com/9781032668680; see Documents 14.2 to 14.8. They include the original informed consent forms written in English, as well as those adapted for use by the Delhi and Ibadan CRTs. There is also a copy of the script used to secure verbal consent and a copy of the confidentiality agreement form used for research team members. Finally, there is a form for consent of use of images.

Obtaining informed consent requires more than explaining the research project with each participant. Rather, it is a continuous conversation about the research process. Be sure to ask participants regularly if they have any questions or comments about the research and to make yourself available to them before, during, and after the project. They may change their minds about participating in the project, want to know how data are being stored, or read publications that you write. It is vital that you update them about the project and that they can reach you, even after the data collection portion of the research has finished.

*(v) **Interview guides***

Different types of interviews exist along a continuum. No interview is fully scripted, and none is completely unscripted. Creating an interview guide or interview schedule is a helpful way to prepare (see GenUrb example 14.7). A typical interview guide contains the themes that you want to cover and key questions that you would like interview participants to speak to for each theme. Interviews usually begin with 'easy' questions to develop a rapport and build up to more difficult queries. The different types of questions include:

- simple questions with minimal setup that can efficiently elicit responses from participants;
- open-ended questions that can elicit rich responses (i.e. questions that cannot be answered with 'yes' or 'no');
- and probing questions (e.g. 'Can you tell me more about that?') to encourage participants to speak in more detail on a particular issue.

GenUrb example 14.7: The interview guide

The GenUrb interview guide for the in-depth interviews with the women participants and for the life histories was initially drawn up by the Georgetown CRT, including the principal investigator and other members of the organisation Red Thread. As a women's activist organisation with a long-term, close relationship with women living in marginalised neighbourhoods, Red Thread members were well equipped to understand the aspects of women's lives that the research needed to engage with. Their long history of engaging in research was a further advantage in terms of their high degree of awareness of what kinds of questions could be asked. It took a total of five days to complete the guide. In addressing women's everyday lives some themes were easy to define while others were less obvious, and much time was devoted to discussing the themes to be addressed as well as the individual questions within each theme. The questions were intended to illustrate the possible dimensions of each theme, with the expectation being that they serve as interviewer prompts rather than be asked verbatim, enabling the interview to be conducted as far as possible as a conversation.

Upon completion, each CRT then worked with the interview guide to adapt it to their local context, with a few teams adding new themes or questions (see Chapter 15, GenUrb example 15.2 Translating the interview guide).

The final guide has semi-structured questions that address eighteen themes, namely: personal background, household and family characteristics, social networks, the neighbourhood, the city, employment, sources of income, debt, social reproduction and care activities, leisure, housing, land, infrastructure, mobility and transportation, use of technology and communications, health, identities, and personal ideology.

Mai Al-Battat from the Ramallah CRT explains their extensive process of reviewing and testing the interview guide:

> The questions were translated to Fusha standard modern Arabic by the York team and then to spoken Palestinian Arabic by the CRT. We then conducted a pilot interview, making a new draft of the questions, discarding those that did not apply to the context, and finessing the guide before conducting a second pilot interview. We then held a consultation with an action researcher, also Head of the Excellence Centre at Bethlehem University, Rabab Tmaish, after which we made more changes to the questions. We conducted a final review of the questions with Amal Juma', an NGO programme manager at Tam Shu'oun alMar'a (Women's Affairs Technical Committee), who is also a feminist journalist, trainer, and researcher. Lastly, we did a third and final pilot interview. We then felt the guide was ready to be used.

By Mai Al-Battat and Linda Peake

Audio clip 14.1 offers further reflections from the Delhi CRT about modifying the interview guide for patriarchal contexts and conducting interviews after the COVID-19 lockdown in the city.

(vi) Compensation

Consider whether to provide monetary or other forms of compensation or reimbursement to research participants. If compensation is being given, you need to take care with its mode of distribution. In some cases, the offer of compensation may interfere with the validity of your findings or the ethical requirement for voluntary participation: for example, if compensation affects the responses that participants give to your questions or if participation is dependent on the compensation that you offer. When participants come from a low socio-economic background, however, it is an ethical obligation for researchers to offer compensation for participants' time. Reimbursements differ from compensation in that the former involve money given to participants for any expenses

Audio clip 14.1: Conducting interviews in Delhi

An audio clip of Delhi CRT team members Anindita Datta and Swagata Basu discussing how they conducted interviews in Delhi is available on the book's website www.routledge.com/9781032668680.

By Anindita Datta and Swagata Basu

incurred as a result of their engagement in the research (e.g. transportation, childcare costs), while the latter refers to money or in-kind payments (e.g. food) provided to the participants for the time and effort they have spent in participating in the research.

(vii) Location of interviews

The geographers Sarah Elwood and Deborah Martin (2000, p. 649) describe the significance of an interview's location, which 'produces "micro-geographies" of spatial relations and meaning, where multiple scales of social relations intersect'. Chi Hoong Sin (2003, p. 307) also notes that 'the space in which an interview takes place can yield important information regarding the way respondents construct their identities'. Examples of the spatial factors in question include whether the interview is conducted in person, in writing, or over the phone/videoconferencing; whether it is in a private (e.g. home, office) or public space (e.g. library, cafeteria); the ambient noise levels; and the presence of other people. In mobile interviews, the route and setting can give important context to the interview and can be the source of prompts and probes.

When you are selecting the location for an interview, you will want to consider safety, cost, accessibility, and comfort. Give careful thought to any biases arising from the location or to potential harm, threats, or conflicts you and/or the participants might face in that location or from or with other people in it. Think about the costs associated with how you and participants will access the location including travel to and from the location as well as any cost associated with using that location for the interview. Offering participants reimbursements for their time commuting and interviewing in the form of cash or gift cards for local businesses they frequent may be welcomed. Take accessibility into account: participants should be able to access the necessary transportation options and location without difficulty. Consider whether or not participants will feel comfortable for the duration of the interview at that location. Participants may not feel comfortable talking about sensitive issues in a public place. They might be more comfortable in their home, or, on the contrary, prefer somewhere away from family members or friends. Select a location that will keep discomfort and interruptions at a minimum. Visiting a prospective location ahead of time is a good idea so that you get a sense of safety considerations, the commute, noise levels, and other people present.

GenUrb example 14.8 describes how GenUrb determined the location of interviews with women in low-income neighbourhoods.

GenUrb example 14.8: Determining the location of interviews

Most of the in-depth and life-history interviews with women living in low-income neighbourhoods were conducted at the women's homes or, in some instances, at their place of work. For example, many of the research participants in Ibadan are vendors, and interviews were held in the locations where women sold products each day. Interviewing inside or directly outside women's homes contributed positively to the participants being comfortable during the interviews. In some instances, women were hesitant to talk candidly. For example, the women were not always alone in their homes or places of work, which might have made them hesitant to share information that they did not want the people around them to overhear. While the home is usually considered a place of safety, it can also be a site of fear and danger, making it difficult for some women to speak openly.

The COVID-19 pandemic affected how GenUrb researchers could conduct interviews (see Chapter 5, GenUrb example 5.2 Adapting research methods during the COVID-19 pandemic). In order to prioritise the well-being of participants and researchers and to comply with changing and uneven travel bans and restrictions on movement, the Cochabamba and Shanghai CRTs experimented with virtual methods. However, participants in Delhi did not have reliable access to mobile devices, cellular networks, or the internet, so the researchers conducted in-person interviews outside women's homes, in local parks, or at nearby restaurants.

By Araby Smyth

(viii) Maintaining relationships

Maintaining relationships with participants may facilitate the process of 'giving back' to the community where you are doing research or co-creating knowledge with that community (Davis and Craven, 2016). Depending on the needs of the community in question and their relationship with you as a researcher this could be done by holding workshops for the joint analysis of data; providing education, advocacy services, or training to communities; or co-authoring papers with research participants (Smith, [1999] 2012) (see Chapter 19, Section 1 Knowledge mobilisation). It is important, however, to respect the wishes of participants in cases where they refuse further contact or invitations to participate or collaborate (Jones and Jenkins, 2008; Smith, [1999] 2012). Reciprocal relationships that are attuned to and challenge unequal power dynamics between those involved have the potential to decolonise knowledge production.

Use Reflection exercise 14.2 to assist you in determining how you will plan your interviews.

Reflection exercise 14.2: Planning your interviews

Once you are certain about the purpose of the interviews and the data you would like to collect, consider the following.

- How many people do you need to interview?
- Who will you ask to participate in your research project?
- How will you ask participants to consider being interviewed? Will you offer incentives?
- How will you obtain participants' consent and, if necessary, ensure data anonymisation? (Note that data anonymisation may not always be necessary or desired by research participants.)

- Will participants take part in more than one interview? For example, if you are conducting life histories or longitudinal studies that follow participants over time, several interviews may be required.
- What kind of interview guide, if any, will you employ?
- Where will the interview take place?

Section 3. Summary

This chapter has discussed interviews as a qualitative research method that can generate rich data from conversations between researchers and participants. Different types of interview generate different kinds of data, and the type and mode of interview, as well as the interview guide, must be tailored to the data needs of the research. There are many parameters to consider for conducting research interviews, including sample size, participant recruitment, consent, gatekeepers, reimbursement and compensation, location, and the need to maintain contact with participants. Interviews are not simple conversations: rather, they are institutionalised research methods shaped by legal and ethical requirements, and as such, they are a power-laden practice in which participants share their experiences in their own words, which are then interpreted by researchers. Feminist researchers negotiate these fraught dynamics through reflexive consideration of their social position and positioning participants as active agents in the research process.

Part V

Feminist data analysis

15 Feminist practices of translation and interpreting

Carmen Ponce

Learning objectives
After completing this chapter, readers should understand

- the most common forms of translation and interpreting;
- the role of translation and interpreting in a decolonial feminist research project;
- the integration of translation and interpreting into various stages of the research;
- and the main challenges involved in translation and interpreting in research.

Further, readers should be able to

- evaluate the role of translation and interpreting in their research project;
- anticipate some of the challenges of practising translation and interpreting;
- and create strategies to address those practical challenges.

This chapter engages with the most common forms of translation and interpreting and discusses them in a feminist decolonial context. It introduces you to the practicalities of translation and interpreting, which can play a role in nearly every step of the research process, including the literature review, the translation of interview guides, work with interpreters in the field, translating transcripts, and knowledge-mobilisation activities. To this end, it discusses how translation and interpreting have been employed in the multilingual project of GenUrb and provides practical guidance about forms of translation and interpreting that you may use in your own research.

Section 1. Introducing translation and interpreting

Translation and **interpreting** refer to transferring a message from one language to another (for example, Urdu to German or Spanish to American Sign Language). While translation involves written texts, interpreting involves spoken messages. Although translation and interpreting are discrete activities, in practice they frequently overlap because translators and interpreters are increasingly required to work with multimodal texts that integrate verbal (written or spoken text) and non-verbal (images, music, gestures) components (see also Chapter 3, Section 1 'Translation', 'interpreting', and making meaning across difference). The interactions between verbal, written, and visual meanings are essential to the formats of these materials. They underscore the complexity of different contexts in which translation and interpreting now take place and the fact that translation and interpreting are much more than simply converting one set of words to another (Perez-Gonzalez, 2014; Boria *et al.*, 2020).

DOI: 10.4324/9781032668727-21

Translation studies scholars distinguish between three ways to translate a verbal sign ('translate' is used here as an umbrella term that encompasses all translation and interpreting activities).

- *Interlingual translation* is the translation of speech or text from one language to another, the most common form of translation and interpreting.
- *Intralingual translation* is the translation of speech or text from one cultural context to another within the same language.
- *Intersemiotic translation* is the translation of meaning between two sign systems: for example, translation and interpreting from English to American Sign Language (Hatim and Munday, 2004; Gottlieb, 2017), as well as multimodal translation (the translation of a text constituted by multiple sign systems, such as text, images, music, or gestures: see Boria *et al.*, 2020).

Some of the most common ways in which these forms of translation and interpreting are employed in the research process are illustrated in Figure 15.1. They include:

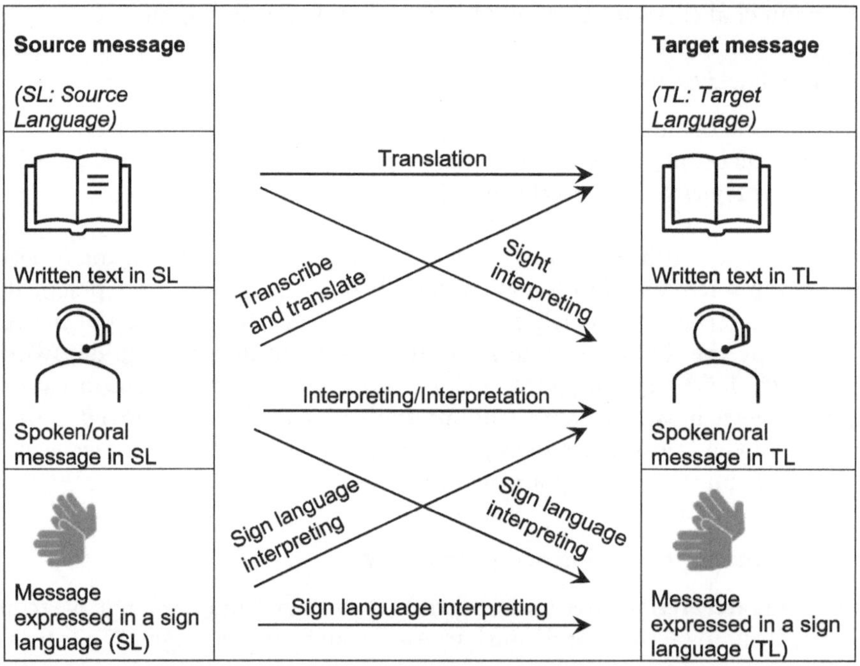

Figure 15.1 The most common translation and interpreting activities in research
Source: Carmen Ponce

- *general textual translation* (as opposed to forms of specialised translation required for literary, health, legal, or scientific texts);
- *translation of spoken messages* (for example, translation of recorded and transcribed interviews conducted as part of qualitative research);

- *simultaneous interpreting* i.e. the interpretation of speech almost immediately after it is spoken (for example, in conference interpreting, either in person or remotely);
- *consecutive interpreting* i.e. the interpretation of speech that occurs during intentional breaks in the conversation (for example, in practical interventions, such as those occurring at health services facilities, either in person or remotely);
- *sight interpreting* (sometimes called sight translation) i.e. silently reading a text in a language and speaking it in a different language (in person or remotely);
- *sign language interpreting* i.e. the interpretation of speech into a sign language or between two sign languages (in person or remotely).

Globalisation and the development of digital information and communication technologies have enhanced the development of other forms of translation and interpreting.

- *Multimodal translation* most commonly involves subtitling or dubbing, i.e. the substitution of words in foreign films voiced by actors with a textual or spoken translation (also called fansubbing and fandubbing when performed by non-professional fans) (Hatim and Munday, 2004; Munday, 2016). Multimodal translation is increasingly common, as translators and interpreters must now work with complex forms of media, such as live performances or audiovisual presentations, i.e. written material that is placed alongside an image on a screen such as film, television, or computer games.
- *Localisation* is commonly used in fields as diverse as activism, gaming, and marketing and refers to translating a message or campaign in a way that is linguistically and culturally appropriate (Hatim and Munday, 2004; Munday, 2016).
- *Automatic video transcription and translation* uses artificial intelligence algorithms and is widely used in applications such as YouTube and Zoom. Its accuracy varies greatly depending on audio quality and speech clarity, among other factors.

Most professional translators use computer-assisted translation software, which includes features that speed up the translation process without compromising quality (such as a translation editor, terminology management, translation memory that can be updated by the user, spell checker, and collaboration and messaging with other translators to receive feedback, among other features). Machine translation – using, for example, Google Translate, Microsoft Bing Translator, DeepL, or Systran Translate – differs in providing only an automated translation. It is increasingly being used by non-professional and professional translators as a first step to speed up the translation process, and it is especially useful when translating academic or literary sources. Recent developments in machine-learning technology, leveraging advances in neural-network computing, have enabled it to become more accurate (Stahlberg, 2020). Document 15.1 on the book's website www.routledge.com/9781032668680 includes examples of widely available open-source and commercial translation software and advice on their use. It is important to emphasise that machine translation cannot fully substitute for human translation, and translators need to conduct intensive reviewing and editing to have confidence in the accuracy of the final version.

The following section turns to the practical roles that translation and interpreting play in decolonising feminist research.

Section 2. The practicalities of working with translators and interpreters

Historically, translators and interpreters remained invisible to their target audience, not only in academic research but also in other work domains such as government, healthcare, education, and business. In recent decades, however, it has been increasingly recognised that they are not neutral parties in the communication process and that translation is a power-laden process fraught with affective tensions and political difficulties. Translation studies scholars have also developed substantial research on the politics of translation and on the consequences of misinterpreting and miscommunicating ideas (Fernández and Evans, 2018). From a decolonial feminist perspective, translation and interpreting are practices of meaning-making across difference involving power-laden processes that need to be taken into careful consideration. In Chapter 3, we discuss in detail decolonial and feminist approaches to translation and interpreting and the role these processes can play in dismantling the legacies of colonisation.

When engaging in translation and interpreting, you need to determine in advance whether you or other members of the research team are competent in the languages and local dialects the participants speak (see Martínez-Gómez, 2017). If this is not the case, you will need to determine the additional budget and time required for translation and interpreting activities: selecting and employing translators and interpreters, organising their work, training, and quality control. In the context of decolonial feminist research projects that use translation and interpreting, researchers need to care about both the process and the outputs of translation and interpreting. Researchers in these projects must prioritise building respectful, non-hierarchical relationships with local partners, including translators and interpreters. In the selection process, you will need to decide whether you will employ professionals or non-professionals. This decision will depend both on the local availability of translators and interpreters (for the participants' language and dialect) and on the personal characteristics of interpreters that will make the participants feel comfortable and safe during the interview (a non-professional interpreter may be a better option in certain circumstances). Beware that interpreters or translators may lack the appropriate training in the type of interviews or academic discussion for which you require their assistance (for example, because of the use of a particular lexicon, the challenging conditions in which interviews will take place, or difficult themes to be addressed). You may need to adapt training and briefing sessions accordingly.

Qualitative researchers are aware of the importance of carefully drafting interview guides and preparing interviewers with respect to participants, and the same level of care should be applied when working with translators and interpreters. To ensure the well-being of participants and interpreters, researchers need to plan for their training and briefing sessions (and for their compensation) before the interviews take place. The interpreters' questions and feedback during training sessions can contribute to improving the experience for the participants as well as the quality of the interview. Though translators will not interact with the participants, they are part of the research team and should also be trained and briefed about the project. Their questions and feedback may also contribute to the quality of the interview guides and data analysis. We cannot stress enough that high-quality translation and interpreting are key to feminist decolonial research, not only because they contribute to ensuring data reliability but also because of the ethical importance attached to representing women's voices accurately (see Chapter 7, Section 3 Feminist ethics and research).

In the remainder of the chapter we provide guidelines to the implementation of translation and interpreting in different stages of research and discuss how the GenUrb project adapted its schedules and budget to the COVID-19 pandemic and the specific local challenges faced by each City Research Team. These challenges affected translation and interpreting work both in the field and in data processing and coding.

Section 3. Planning for translation and interpreting throughout the research process

Translation and interpreting may be used throughout the research process. Figure 15.2 shows the most common forms of translation and interpreting involved in different stages of the research process. Bear in mind that this is a simplified representation and that the research process is rarely linear and may involve more stages than those shown.

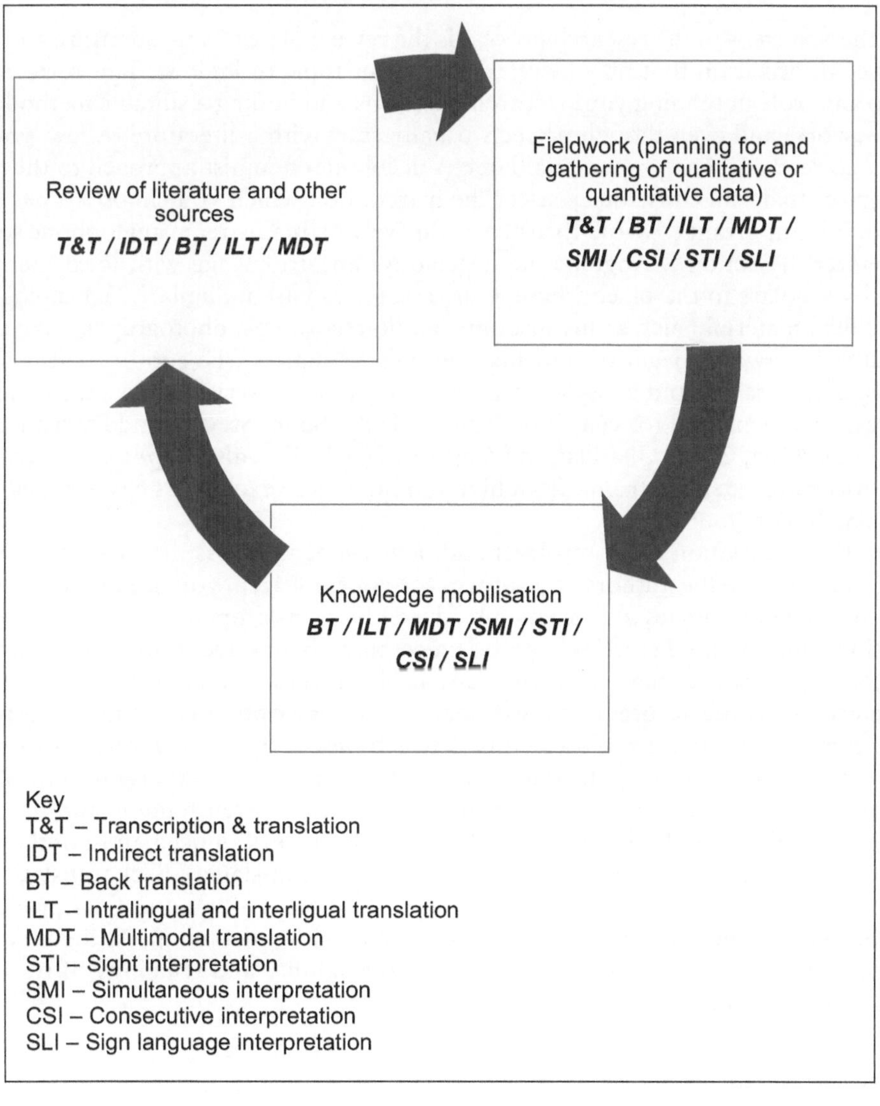

Figure 15.2 The most common forms of translation and interpreting in different stages of the research process

Source: Carmen Ponce

When developing your research plan, engaging in translation and interpreting will require that you be aware of your own language competencies, the languages of participants, the languages in which the literature you need to review is written, the availability of professional and non-professional translators or interpreters for fieldwork and for knowledge-mobilisation activities, and the time and funding available for translation and interpreting within your research project. You should not underestimate issues of time and funding: engaging with translation and interpreting in research can be a difficult, fraught, and time-consuming process.

In the remainder of this section, we discuss translation and interpreting in three stages of research, namely literature reviews, fieldwork, and knowledge mobilisation.

(i) Literature reviews

One of the first steps in the research process is the review of academic literature and other sources of information that may be relevant to your topic of interest. This review plays an important role in refining your research questions and finding a suitable methodological framework, and even though projects usually start with a literature review, you will continue to read and analyse material later. A decolonial feminist approach to the literature requires that you consciously select the material on which you choose to base your research. It is important to avoid focusing exclusively on mainstream anglophone sources (see Chapter 3, Section 2 Anglophone hegemony) and to engage with local literatures written by scholars in the place where your research is based. Similarly, when engaging with archival material (such as manuscripts, audio recordings, photographs, maps, film, architectural drawings, legal documents, journals or diaries), be sure to reflect on the structure of the archive and consider what materials were preserved, who authored them, who preserved them, and to what ends (Spivak, 1985; Sonn, Stevens and Duncan, 2013; Basu and De Jong, 2016; Ghaddar and Caswell, 2019). It is also important to consider whose stories are excluded from the archive and how these exclusions have shaped existing scholarship on your topic.

Given that translation is an imperfect and power-laden process, in the same way that you carefully choose the authors and sources for your research, you need to identify and understand potential issues and biases introduced by translators if your sources include translations. Biases and distortions can be exacerbated by indirect translations (material translated from a source that is itself a translation). If these are important sources, you will need to assess the accuracy of these documents. One option is conducting a partial or total retranslation (i.e. back translation), which means translating a text back to the source language to identify potential biases and distortions in the original translation that may compromise key arguments or ideas in the text. Given funding and time constraints, you will need to limit the extent of retranslation work (indeed, most of the translated research material in social science research is not retranslated). Back translation can be expensive and can lead to much confusion and a potentially worse final product. It would be better to find different sources (see Koopman, forthcoming). At the very least, you should discuss any concerns you have with the validity and reliability of translated sources in your methodology.

(ii) Fieldwork: Working with interpreters

While in the field, you will need to choose carefully how much you will engage in translation and interpreting and how you will fund it. The practice of translation and interpreting in the field is often constrained by the availability of local translators and interpreters and by research funding that typically limits translation and interpreting activities to those deemed strictly necessary. While the professional translation and interpreting industry has grown rapidly in the last few decades, services for minority languages across the globe or majority languages (and their local specificities) in the global South are usually expensive and not always available. Fortunately, non-professional translation and interpreting assistance is usually available in the field. The study of non-professional translation and interpreting is a growing field of research, and you may find this research helpful in relation to certain translation and interpreting challenges you may encounter in your fieldwork (Antonini *et al.*, 2017; Rodrigues, 2018).

If you are not fluent in the local language(s) and the research topic requires you to conduct interviews or focus groups in person, you may need to recruit an interpreter to assist you. Some points you will need to keep in mind when working with an interpreter are as follows.

- Interpreters should be considered members of the interview team and should be made aware of and follow the research ethics guidelines of the project (see Chapter 14, Section 2 (iv) Obtaining consent).
- Interpreters play a key role in the data-collection process and should be briefed about the project and the lexicon of the field before the interpreting practice or translation work begins, especially when the research topic is a sensitive one.
- Since interpreters will be in direct contact with the participant, you also need to consider their interpersonal skills and other personal characteristics that may affect the participant's trust or comfort during the interview. The profile of the interpreter you want will depend primarily on the research topic. For example, if you need to ask female participants questions about domestic violence or reproductive health, you may need to work with a female interpreter who is not related or known to the participant (excluding female relatives, friends, or neighbours from serving as interpreters).
- You need to determine the interpreter's proficiency in both the participants' and your own language(s). For example, fluency is critical for semi-structured and unstructured interviews and, to a lesser extent, for structured interviews too (see Chapter 14, Section 1 Interviews in feminist research).
- You will need to practise with the interpreter before conducting interviews with the participants. While ideally an interpreter's role is limited to bridging the language gap between you and the participant, in practice they mediate your communication with the participant.
- When practising with the interpreter, bear in mind that the interpreter is not the only one who needs practice and feedback: you need it too. The quality of the interpreting process depends not only on the interpreter's skills but also on your ability to pause and speak slowly. The practice requires a third person, who will play the role of the participant and who will allow you to practise how to communicate through (instead of with) the interpreter during the interview.
- Make sure that the participant consents to the interviews being recorded. This allows for critical review of potential interpreting bias during the data analysis stage of the research process.
- It is important that you record key information about the interpreters you engage with in data collection in the field. This information may include age, gender, race, ethnicity, nationality, formal education, and other factors that you believe could influence participants' responses. Including information about interpreters can

help you identify potential misinterpretation or bias in the interpreting process. Further, you should anonymise this information if your data will be archived for future investigation or deposited in a public repository (see Chapter 10, Section 3 (iii) Archiving data).

- Fieldwork involving participants of the d/Deaf community will require sign language interpreting. Investigate the sign language that participants use (e.g. American, Indo-Pakistani, Chinese, or Egyptian sign language, or Langue des signes Québécoise). When employing the sign language interpreter, allow time and resources for training and briefing (including their questions and feedback).

In the GenUrb project, the City Research Teams (CRTs) had bilingual or multilingual members, so the interviews were conducted in the participants' preferred language, without the mediation of interpreters (see Chapter 3 GenUrb example 3.1: Languages and dialects spoken by members and participants in the GenUrb project).

Taking into consideration the points raised in this section, turn to Reflection exercise 15.1, which asks you to reflect on the potential language differences you may encounter during fieldwork.

Reflection exercise 15.1: Language differences in fieldwork

Determine what languages are spoken in the city where your fieldwork site is located, the languages that research participants may speak, and the language(s) in which you are proficient.

- What issues may the differences in languages spoken give rise to (if any)? For example, participants may speak the same language, but the way in which you speak the language (e.g. in a specific dialect or with a particular accent) may reflect social differences, or you may find that older and younger participants in your research speak different languages to different degrees of proficiency.
- How might you negotiate language differences during your research? Take into consideration the time and money you have available as well as whether you need to employ both interpreters and translators.

(iii) Fieldwork: Working with translators on interview guides and transcripts

Translation can play a significant role in the field once you have started to collect, transcribe, and annotate your data. You will need first to translate any documentation you produce that relates to how you will collect data, such as interview guides (also referred to as schedules, protocols, or instruments) or questionnaires that need to be in the language(s) spoken by research participants. Qualitative data collected in the field may also need to be translated into another language for data analysis. GenUrb example 15.1 details the use of translation in the GenUrb project.

GenUrb example 15.1: The use of translation

Translation was critical to the GenUrb project: we translated interview guides and transcribed and translated data collected from the interviews and life histories with women living in marginalised neighbourhoods. We did not, however, translate field notes, which varied tremendously in structure and volume, and we did not engage in full translation of the interviews with policy-shapers. While the interviews with the policy-shapers were transcribed by each CRT, it was agreed that only sections of them would be translated into English. This decision was taken because of the extremely time-consuming process of translation already undertaken with the women's interviews and life histories. This partial translation process was considered a suitable compromise given that the data collected in these interviews was largely factual and that extensive reports detailing the results for both rounds of policy-shaper interviews were produced by each CRT.

The GenUrb project worked with researchers conducting interviews with women in more than a dozen languages and dialects across six cities (see Chapter 3, GenUrb example 3.1 Languages and dialects spoken by members and participants in the GenUrb Project). These interviews included 127 in-depth interviews and 100 life histories with women participants and 143 interviews with policy-shapers, yielding a large quantity of data. Beyond the agreement for each CRT to translate their interviews verbatim, the project employed a decentralised, context-sensitive approach to translation, empowering each CRT to make decisions that best suited the particularities of their locale and the needs of their participants. Honouring such a decolonial feminist ethic of translation proved to be challenging work. Due to a number of factors, the transcription and translation of interviews and their uploading into NVivo software took over two years, more than twice as long as planned. We had underestimated the length of time needed for translation and transcription (see Section 3 Planning for translation and interpreting throughout the research process). CRT schedules were also severely disrupted by COVID-19, leaving insufficient numbers of CRT members to ensure a timely co-ordination of activities between the CRTs, with the uploading of data into NVivo being particularly affected.

By Linda Peake

Before interviewing can start, the interview guide must be translated into the language(s) spoken by the research participants, and the language(s) used should address the local context and culture (i.e. both interlingual and intralingual translation processes may be involved). It is important that you or someone in your research team conduct pilot interviews and adjust accordingly the translated questions to make sure that the interview flows in a natural and comfortable way for participants. Once you are sure that the interview guide works well in the local context in which you are working, you can start to conduct the research interviews (see also Erkut, 2010). In GenUrb example 15.2, we describe the processes GenUrb followed in relation to translating the interview guide.

GenUrb example 15.2: Translating the interview guide

The GenUrb in-depth interview guide (See Chapter 14, GenUrb example 14.7 The interview guide) for conducting interviews with participants as well as recording their life histories was devised by the **Georgetown** CRT, who, at the time, were the first team ready to work on such a task and who already had extensive experience of conducting qualitative interview-based research. The Georgetown CRT researchers wrote the interview guide in standard English, but each CRT translated and adapted it to their unique contexts: in Cochabamba, translation was into Spanish; in Delhi, into Hindi; in Ibadan, into Yoruba; in Ramallah, into Arabic; and in Shanghai, into Simplified Chinese.

The **Shanghai CRT** used Simplified Chinese as opposed to Traditional Chinese, as it is the official written language in mainland China and has been promoted by the People's Republic of China government and widely taught in schools since the 1950s. This is compared to Traditional Chinese, which is officially used in the Hong Kong and Macau special administrative regions. Simplified Chinese is easier to write and learn as its characters are a simplified version of Traditional Chinese. For example, the term 'woman' in simplified Chinese can be written as 妇女 (funü/funv) with the character 妇 (fu) meaning 'married woman', and the character 女 (nv/nü) meaning 'woman'; the term 妇女 (funü/funv) is commonly used in government documents to refer to women in general. In Traditional Chinese 妇女 is written as 婦女 with additional strokes in the 'fu' character.

In the process of translating the interview guide CRTs took slightly different approaches to the extent to which they engaged in verbatim translation. The **Delhi** CRT, for example, took out or modified some questions relating to income, past relationships, and violence, that were considered intrusive or socially unacceptable and could not therefore be asked in their cultural context. While this reduced the capacity of the project to engage in comparative analysis, it was necessary to ensure that the interviews could take place and not be stopped prematurely.

The **Cochabamba** CRT kept all the questions but changed the wording to make questions more accessible. Nasya Razavi explains:

> In the section on 'personal ideology', we adapted the questions, 'How do you approach political and social issues?' and 'Do you have a particular political point of view?' We asked the women directly whether they agreed or disagreed with the policies and programmes of the governing national party MAS, 'Movement Towards Socialism', as a way to open the conversation about political ideologies.
>
> In the questions that asked the women to describe their religious or spiritual beliefs, we added prompts around Pachamama, or Mother Earth, a goddess in Andean cultures: 'Do you believe in Pachamama? When is Pachamama punishing or merciful? Do you make offerings to Pachamama?' If we asked only about their religion, the women might only have spoken to their Christian rituals and not to the practices and festivities in honour of Pachamama.

In the theme on 'performing identities' there were a series of questions around clothing and dress. Because the women we interviewed in Cochabamba are Quechua, we asked about their use of the *pollera*, a type of skirt worn by Indigenous populations in the Andes: 'Do you wear a pollera? Where and when? If not, did you wear polleras before and what made you change your habits?' The answers to these questions helped us understand how the women might move differently through different parts of the city, different notions of attractiveness, and experiences of racism.

The dynamics of social relations prevalent in Yoruba culture combined with the unique context of each interview also altered the specifics of how questions were asked by the **Ibadan** CRT. Team leader Grace Adeniyi-Ogunyankin explains how, before asking a question that the researcher perceived as sensitive:

> We might tell an elaborate story or first say something like, 'God forbid something may happen!' or say a prayer for them after some of their responses that shared something terrible that had happened or suggested that something bad *might* happen.

The Ibadan CRT did not take much creative license in changing interview questions, putting more emphasis on the comparative aspect of the research. As a result, the interviewers ended up asking questions that were sometimes addressed with laughter (such as whether the participants had access to currencies other than the Nigerian naira or whether they always had access to electricity). However, this did not result in the Ibadan interviewers being unable to complete the interviews, although they did experience difficulties asking culturally sensitive questions, as Grace explains:

> We were sometimes wary to ask, 'How is your mental health?' and sometimes skipped it because we didn't want them to misunderstand, since talking about mental health can sometimes be taboo. This is one of those cases where we may apologise first before we ask the question. Another question that some participants weren't always forthcoming about was 'Do you have a loan?' Perhaps because having a loan or owing money is seen as shameful by some. Most don't . . . want to be known for owing money. It would be later, perhaps via another question or conversations outside the interview questions that we would find out that they have a loan and the mental pain (*gbomu le lantern*) that can cause them. [*Gbomu le lantern* literally means to place one's breast on a hot lantern and is used as a term to describe the pain of taking out low-collateral, high-interest loans in Nigeria].

The **Ramallah** CRT rearranged the order of the themes and their associated questions in a way that they felt would give more freedom to the participant to provide her own interpretations and point in the direction of the main issues that she would like to discuss. They also removed some questions that did not fit the context or cultural norms or were perceived as inappropriate. They reformulated other questions such as 'What is your ethnic background?' and 'What language do you speak at home?' As Mai Al-Batatt said:

> All the participants spoke only Arabic, and they found it weird when we asked them about other languages, which made no sense to ask in our case. This was the case with other questions as well, such as the one asking about ethnic background. All the participants are the same ethnicity, that of the dominant Palestinian group, east Mediterranean, descendants of Arameans. And the question on religion: all were Muslims, and for us, this is not a question that needs asking.

Like the Delhi CRT, they removed some questions on violence.

> It is very sensitive to ask this question in our case, and it requires a special arrangement of first providing protection to the participant, which we are unable to provide. And the fact that it is a very small community and many of the participants knew each other, there is no way of ensuring that no one would mention that we asked this question, which would have put us and other participants and us in danger and highly likely could have led to blocking us from the site by the men who oppose such investigations.

By Penn Tsz Ting Ip, Anindita Datta, Nasya S. Razavi, Grace Adeniyi-Ogunyankin, and Mai Al-Battat

Once the interviewing stage has started, the processes of transcription and translation can also begin (see Chapter 16, GenUrb example 16.2 Strategies for transcribing interviews). The translation of interview transcripts can be conducted manually or with the assistance of machine-translation software. As mentioned in Section 1, translators are increasingly using machine translation as a first step in the process, but it may be less efficient and less accurate than manual translation (see GenUrb example 15.3 on the employment of machine-based translation in GenUrb). It is important that the translator and researchers share information and communicate any doubts during the translation process in order to fully understand the meaning of the participant's words.

GenUrb example 15.3: Employing machine-based translation

In GenUrb, the use of machine translation was hampered by a variety of factors. Machine translation was not available for some languages and dialects (e.g. for Guyanese Creole), or it may produce lower-quality translation for minority languages or languages with significant variety across dialects, like Quechua (spoken in Cochabamba), or tonal languages such as

Yoruba and Shanghainese. Furthermore, cultural references and idiomatic expressions are especially challenging for machine translation. In addition, sound interference reduces the accuracy of machine translation, and such interference was common among GenUrb's interview recordings, particularly for interviews conducted in densely packed neighbourhoods or in the street.

By Linda Peake

GenUrb example 15.4 illustrates how members of each CRT engaged in the process of the translation of interview texts.

GenUrb example 15.4: Strategies for translating interviews

Each CRT had to negotiate a number of challenges raised in translation, with a major practical challenge being the time-consuming and laborious process of translation. As researchers, we wanted to produce the highest-quality translations possible to capture the women's voices from our interview recordings accurately and to do justice to their stories. At the same time, we also had to adhere to deadlines, maintain our budgets, and deal with the real-life issues of the researchers, who were navigating other work commitments and familial responsibilities. These constraints required certain choices to be made about who engaged in the translation process. The translation process itself also required us to deal with a number of issues and problems that we encountered. The ideal we aspired to was not always possible.

Ideally, for example, we wanted to keep the translation process in-house, using GenUrb researchers in each CRT who were familiar with the research and participants. This was not always possible, however, primarily due to time constraints. The **Delhi** CRT lead researchers, Anindita Datta and Swagata Basu, translated their interview guide into Hindi. A team of twelve graduate students worked on the Hindi transcription of the interviews to generate the English translations of the transcripts. The graduate students possessed high proficiency in both English and Hindi, familiarity with the contexts of the study site, and the ability to catch the nuances of words that have both literal and contextual meanings. Finally, quality control of the translations was further ensured through a careful checking of all the translated transcripts by Anindita and Swagata.

The **Cochabamba** CRT recruited Victor Hugo Mamani Yapura, a Bolivian PhD student specialising in translation, to conduct the Quechua–Spanish translation required for the participant who opted to express herself in Quechua. All the other transcripts were translated from Spanish into English. A team of graduate students – Carmen Ponce, Carmen Ramirez, and Javier Garate Alfaro – and a development practitioner, Eleanor Douglas, all based in Canada, were contracted to undertake the Spanish–English translation, based on their previous experience of translation from Spanish. These team members had varying degrees of familiarity with the Bolivian context, requiring a final close review of all the translations by the CRT lead, Nasya S. Razavi, to ensure accuracy and to enable us to have a high degree of trust in their quality.

The **Ibadan** CRT decided to task a single scholar, Tunrayo Abimbola Adeyemi, a PhD student at Obafemi Awolowo University, with translating all eleven of the Yoruba transcripts in order to maintain quality control. Upon completion, the CRT lead, Grace Adeniyi-Ogunyankin, reviewed all the translations.

The **Shanghai** CRT decided that the team lead, Penn Tsz Ting Ip, would translate the interview transcripts into English, with the help of seven students: six based in Shanghai, two graduate and four undergraduate students, and one undergraduate at York University. The translation work was challenging for many reasons. For example, various words and phrases could not be interpreted without knowing about the women's lifeworlds and their daily practices. The research assistants, however, were all born after the 1990s and were therefore not always able to comprehend all the information in the transcripts, especially that relating to the socialist past (from the 1950s to the 1990s). The senior scholars in the CRT – Jing Wang, Zhang Yu, and Penn Tsz Ting Ip – provided advice to the translators on understanding the participants' life histories. To further improve the quality of the translation, the CRT recruited a non-Chinese-speaking graduate student, Tamires Lietti, to then proofread all the translations; she marked up words and sentences she could not understand for further checks. GenUrb's postdoctoral researcher, Araby Smyth, also joined in this quality-check process, reading the English translations for clarity. Penn then read over all the final translations, often going back to the Chinese transcriptions to ensure accuracy. Through this back-and-forth communications process between the CRT members, the graduate students, Penn, and Araby, the translations finally came to completion.

Given that the **Georgetown** CRT interviews were conducted in Creolese, no translation was undertaken as all CRT members read Creole.

The **Ramallah** CRT was unable to translate the interviews into English because of the cost. On one hand, Ramallah is a very expensive city, and the budgetary constraints made hiring professional translators who knew the Palestinian dialect impossible. Four interviews were translated by Egyptian student assistants at York University; however, their lack of knowledge of the dialect and cultural roots and meanings of certain expressions skewed the answers, and upon review of their work it was decided that correcting these translations required more time and labor than the CRT could afford. Neither are there any reliable software programs or applications for computer-generated translations from Arabic. Hence, this CRT's data have not been entered into NVivo and are used only sparingly in the comparative analysis of the project.

By Anindita Datta, Nasya S. Razavi, Grace Adeniyi-Ogunyankin, Penn Tsz Ting Ip, Joy Marcus, Natasha Aruri, and Leeann Bennett

Translating interviews often involves more than translating words and sentences. It also requires researching complementary sources of information that help the translator to understand or to confirm the meaning of certain words, such as jargon or acronyms. Annotations can provide clarification or complementary information to ensure full understanding of the participant's words. They also allow the translator to communicate with the researcher and share information the latter would not otherwise be able to access (given language barriers or lack of context-specific information). Further, using annotations can also help translators maintain, as far as possible, the participant's tone and way of telling their stories. GenUrb example 15.5 discusses these forms of communication in the translation of the Cochabamba CRT's interviews.

GenUrb example 15.5: Translation of the Cochabamba City Research Team's interviews

I was introduced to the GenUrb project by Nasya S. Razavi, the lead of the Cochabamba CRT. My first engagement took place towards the end of the translation process, and I translated seven interviews from Spanish to English. I am not a professional translator, but my mother tongue is Spanish, and I had previously conducted research with bilingual Quechua-and-Spanish speakers in the Peruvian Andes.

I started the translation process by reading the first interview to help me to understand the nature of the interview, the topics it covered, and the vocabulary used by the participants. Given the transnational nature of the GenUrb research, I made sure to include annotations to make local references easier to follow for non-local research partners. I made use of secondary resources to investigate the meaning of local words and the type of places that participants referred to (e.g. whether they were parks, marketplaces, schools, grassroots organisations, rural communities, or other specific places). I also posted queries for further discussion with other CRT members. Although a native Spanish speaker, I encountered issues with context-specific considerations. For example, the meaning of the word *bachiller* varies by country. While in Bolivia it refers to someone who has graduated from secondary school, in Peru it means that the person has graduated from a university undergraduate programme. To corroborate the meaning of the word, I searched the Bolivian Ministry of Education website to confirm that the participant meant that she had graduated from secondary school. In this case, I kept the word *bachiller*, with an annotation '[graduated from secondary school]' to ensure other readers would know the meaning of the word.

By Carmen Ponce

As with interpreters, in cases where you need to employ translators to engage in your research, it is important to keep in mind that they are members of the research team and that, as such, they must be briefed about the project and bound to research-ethics guidelines. You will also need to determine the translator's proficiency in both the participants' and your own language(s) and engage in training sessions until you are satisfied with the quality of their work. Regardless of whether or not the translated data will be made publicly available, you must also record information about the protocols used for translation, with a description of who translated each interview. As noted for interpreters, this information should include factors such as age, gender, race, ethnicity, nationality, formal education, and other information that you believe could have introduced bias into the work of the translators. If data are made public, this information can be anonymised, but the description will remain, allowing users to identify sources of any potential biases or systematic errors more easily.

(iv) Knowledge-mobilisation activities

Knowledge mobilisation in a decolonial feminist research project aims to foster the collaborative co-dissemination of knowledge and the development of non-hierarchical relationships in the sharing of research findings with academic and non-academic audiences (see Chapter 19, Section 2 Feminist approaches to knowledge mobilisation). Translation and interpreting play a key role in this context, especially for projects involving partners and audiences from diverse cultural and linguistic backgrounds.

Some of the most common knowledge-mobilisation activities involve publishing articles and giving presentations in conferences or workshops. It is important that you consider the audiences that you aim to reach with your work, the languages they speak, and the journals to which they have access. Most journals in the social sciences now publish abstracts of articles in two or three languages. Some journals, such as the *Journal of Latin American Geography* and *Antipode*, allocate resources to translating journal articles. Others have limited funds for translation to or from English, but they will publish articles in non-anglophone languages; for example, *ACME: an International Journal for Critical Geographies* and the *International Journal of Urban and Regional Research* (see Koopman, forthcoming). Whether a journal is open access is also an important consideration to keep in mind. In addition, consider whether the conferences that you organise or in which you participate will have oral or sign language interpreters available. Interpreting needs will vary depending upon your language competencies and those of your audience, and you may need interpreting services to ensure smooth communication and facilitate participation.

Similarly, when planning knowledge-mobilisation activities for non-academic audiences, you need to consider how language differences may affect communication. Even if everyone speaks the same language, there may be a need to translate academic concepts that do not travel easily between diverse cultural or geographic contexts and that can be difficult to understand.

GenUrb has had extensive experience with knowledge-mobilisation initiatives involving academic and non-academic audiences, implemented in different languages and dialects. GenUrb example 15.6 presents one experience that illustrates localised and multimedia translation implemented by the Cochabamba CRT.

GenUrb example 15.6: Relatos en voz alta – multimedia translation in Cochabamba

GenUrb researched the impacts of COVID-19 on the lives of women in Cochabamba. It revealed that there were few resources on COVID-19 available in Quechua, despite the 1.5-million-plus Quechua-speakers in the country. The Cochabamba CRT therefore produced a series of videos in both Quechua and Spanish relating to COVID-19 best practices and addressing myths and truths relating to COVID-19. The CRT teamed up with local Quechua graphic illustrator Phuyu (Facebook: @phuyudibujando; Instagram: @ phuyu.no) to produce the videos. These are available as both graphic stills and videos on the website, *Relatos en voz alta* (Sharing stories out loud) created by the Cochabamba CRT (see the book's website www.routledge.com/9781032668680). The videos were also uploaded to YouTube to be easily shared via WhatsApp, a messaging app that is popular among the research participants in Bolivia.

By Nasya S. Razavi

Reflection exercise 15.2 invites you to reflect on the potential role of translation and interpreting in your knowledge-mobilisation activities.

Reflection exercise 15.2: Translation and interpreting for knowledge-mobilisation activities

Reflect on your project's knowledge-mobilisation strategy.

- What audiences would you like to reach (e.g. academic audiences in the global South, activists, policymakers, civil society, local grassroots organisations)? What language or cultural barriers (including local knowledge-dissemination infrastructure) do you need to overcome to reach and communicate with these audiences?
- If your research project involves partners whose language preference or proficiency is not the same as yours, does the collaborative co-dissemination of knowledge or findings require translation or interpreting to be effective? How would the lack of funding and time for translation and interpreting affect your relationship with local research partners?
- Are you interested in joining academic or non-academic networks? Investigate other networks of interest that you cannot join due to language or cultural barriers. Are some of these networks critical for your knowledge-mobilisation strategy? What research outputs (academic papers, policy briefs, blog entries, podcasts, or videos) can you prepare to share your findings with and receive feedback from these networks?

Section 4. Summary

This chapter has addressed the practicalities of translation and interpreting. It has introduced you to the most common forms of translation and interpreting and the complexities encountered in processes of meaning-making across language and culture. It has investigated the issues researchers must consider in order to engage with interpreters and translators in a way that reflects the values of decolonisation. The chapter has also explored the ways in which translation and interpreting feature in different stages of the research process, especially in literature reviews, fieldwork, and knowledge-mobilisation activities. We emphasise how difficult, time-consuming, and potentially expensive translation and interpreting can be, while also discussing GenUrb's experience of translation and interpreting as integral to a decolonial feminist project that has aimed to work collaboratively across multilingual transnational partnerships.

16 Feminist approaches to qualitative data analysis

Linda Peake and Elsa Koleth

Learning objectives

After completing this chapter, readers should understand

- the purpose of data analysis and the relationship between data and findings;
- the most commonly used approaches to qualitative data analysis in the social sciences;
- the mechanics of qualitative data analysis;
- the practice of annotating and coding transcripts;
- the importance of reflexivity during data analysis.

Further, readers should be able to

- articulate an appropriate qualitative method of analysis for their research;
- prepare qualitative data for analysis;
- and address limitations and tensions of data through reflexive examination.

This chapter introduces two major approaches – deductive and inductive – to qualitative data analysis. It illustrates how these approaches are employed in the most commonly used forms of qualitative data analysis, such as content analysis, discourse analysis, narrative analysis, visual analysis, and grounded theory. The chapter then engages with the practice of qualitative data analysis: of preparing, annotating, coding, and exploring field data to yield meanings, insights, and conclusions. It ends with brief reflections on collaborative practices of analysis and the need to continue critically evaluating data.

Section 1. Data analysis

Data analysis involves the processing and interpretation of the data that you have collected in order to draw out key meanings, themes, and insights in response to your research question(s). It involves sifting through and organising data (i.e. selecting, coding, sorting into themes or clusters, and categorising) and providing for some form of organisation, usually through data reduction, enabling the exploration of relationships within the data as well as interpreting and giving meaning to the data.

Although there is general agreement on what analysis involves, there are differences in how it occurs, the major divide being between inductive and deductive approaches. The differences in these two approaches relate to the roles of social theory and hypotheses, as well as to how and when data are collected and empirical generalisations developed.

DOI: 10.4324/9781032668727-22

Deductive analysis is a top-down approach conducted in reference to existing theoretical understandings of social change and activity. It was first associated with the natural sciences and was introduced into the social sciences via the philosophy of logical positivism and in particular the hypothetico-deductive model. Malcolm Williams (2003) argues that positivism prioritises observation, is based upon verification of hypotheses, is value free, and only engages with things that can be measured. This is a far less exclusionary list of attributes than those of the scientific method, which is also concerned with explanation and prediction, is evidence based and 'truth seeking', and is considered objective and logical. In this model, knowledge production starts not with the accumulation of facts but with the generation of a hypothesis that relates to your research question. Once all the data are collected the hypothesis is tested to see if it can be held true or, if the testing challenges the hypothesis and the theory underlying it, the hypothesis can be rejected as false. Theories are constantly challenged and refined as more data are collected to test hypotheses. Social scientific knowledge is thus produced through a network of deductively related propositions and theories.

The strict prescriptive and normative process of deductive analysis is hardly applied in the social sciences (not least because it is recognised that no one model of scientific explanation fits all knowledge production). In a looser form, it is still evident in the orderly, staged nature of the research process, in which research questions are posed, a research design is chosen, and methods are administered to collect data, which are then sorted, organised, and interpreted, until finally conclusions are reached (Hammond and Wellington, 2013).

This deductive, 'top-down' research process is rooted in the epistemological view that the data collection and analysis processes yield objectively valid results against which to test the hypothesis. Indeed, deductive analysis is judged by its validity, which is considered independent of the relationship between the researcher and the reader of the research. Validity has no one specific meaning, but it typically refers to: 'the logic of the research; the clarity with which questions are formed; the fit of methodology and methods to the research questions being asked; the marshalling of evidence in support of propositions' (Hammond and Wellington, 2013, pp. 151–152). Feminist scholars have argued that analysis cannot be considered independent of the researcher and, that in deductive analysis, the supposed 'objectivity' of the scientific method is simply elevated in importance over the subjectivity of the researcher (Code, 1991). In practice, scientific knowledge production proceeds in contexts where non-scientific factors (e.g. racism, sexism, the making of errors) intervene.

Inductive analysis seeks to explore relationships in data without (explicit) reference to a theoretical framing. It is based on a form of reasoning 'from statements about observed cases to statements about other, unobserved, cases or – more usually – to a general claim about most or all cases of the same kind' (Hammersley, 2006, p. 146). Martyn Hammersley (2006) argues that while there is no form of logical induction from knowledge of particulars to knowledge about universals, there are other legitimate forms of induction. These are: enumerative induction, which involves generalisation from a sample to a population, as in a survey; hypothetical induction (abduction), where the most probable explanation in terms of a theoretical principle is inferred from observations; and probative induction, which involves drawing a conclusion by reasoning from the evidence collected to test a hypothesis.

In an inductive approach, data analysis need not wait until all data are collected. It can start while you are conducting your fieldwork and continues into the process of writing up your research. Analysis is fluid in nature as opposed to staged, with the researcher constantly amending their coding schemes as more data are collected. This approach allows the researcher flexibility, in that data are not forced into pre-existing categories and there is close attention to context. Inductive analysis allows for conclusions to be drawn in an open-ended and dialogic manner, reinterpreting existing theoretical conclusions based on what the data reveal about a particular research question. While it can be difficult for researchers not to bring to the research pre-existing assumptions about concepts, theories, and relationships, induction enlivens data with an agency in the research process.

Inductive analysis makes knowledge claims that are provisional and is judged by 'the strength of the claims to knowledge the researcher is making' (Hammond and Wellington, 2013, p. 146). This strength is commonly understood in terms of the research's trustworthiness and is interpersonal, i.e. it is dependent upon the relationship between the researcher and the reader of the research. The reader needs to be convinced that the research has been conducted in a rigorous manner and that it is 'confirmable, credible, transferable and dependable' (Hammond and Wellington, 2013, p. 147). While the research does not portray 'the truth', it is considered more credible than simple observation, and the quality of the research can be judged by others.

Although deductive analysis and inductive analysis are usually seen as diametrically opposed (a divide further cemented by the association of deductive analysis with quantitative techniques and inductive analysis with qualitative techniques), they can be used interchangeably as in an abductive analysis: 'Exploratory inductive analysis may lead to the articulation of propositions to be tested at a later stage in a deductive manner, while deductive propositions can be re-examined in the light of findings' (Hammond and Wellington, 2013, p. 10).

In the following section the focus turns to the analysis of qualitative research.

Section 2. Approaches to qualitative data analysis

Qualitative approaches are based on language (words as opposed to numbers), with an emphasis on participants' interpretations and understandings of their social worlds. They are underpinned by a range of (sometimes conflicting) philosophical and theoretical stances including phenomenology, hermeneutics, critical theory, feminism, antiracism, postmodernism, poststructuralism, and post-colonial theory. Determining which stance you take is not straightforward and is influenced by a number of factors, including your political beliefs; how, and by whom, you have been taught about research; and your range and depth of knowledge of these political and theoretical stances. It is important to take the time to think about your position as stances differ not only in how the role of the researcher is understood but also in what approaches you may take to the analysis of data.

Some common approaches to analysing **qualitative data** are discussed in what follows.

(i) Content analysis

This is a straightforward but somewhat limited method of analysis of the content and structure of texts that uses quantitative measures of the frequency of terms or categories of meaning (of words, phrases, or images) to generate a picture of the range of meanings in a text. It can be employed both inductively and deductively. When used inductively concepts are derived from the data themselves, whereas deductive content analysis is operationalised on the basis of previous knowledge and is most commonly used if the aim is to test a previous theory or to compare categories, for example, over different time periods.

Although content analysis employs descriptive statistics, it is commonly classified as a qualitative approach because the emphasis is to uncover the internal meaning of a text while understanding that the meaning of terms in a text depends on its context and that this internal meaning can differ from the message intended by the text's producer(s) and the messages received by the text's audiences. John Scott (2006, p. 40) states that the aim of content analysis is to 'identify clear and coherent categories that highlight salient aspects of the message conveyed and to use objective and reliable methods of calculating their relative significance in the overall message'.

(ii) Discourse analysis

Discourse analysis employs both induction and deduction and includes a range of techniques to analyse texts, spoken and written, and the contexts in which they exist (e.g. imperialism, underdevelopment) to reveal how knowledge is produced and reproduced in particular ways and through institutional practices. It is important, therefore, that the researcher is also familiar with these contexts. Analysis focuses on the purpose of texts and 'the connections between language, communication, knowledge, power and social practices' in order to reveal how 'institutions and individual subjects are formed, produced, given meaning, constructed and represented through particular configurations of knowledge' (Muncie, 2006, p. 74). Particular attention is paid to authorship, intention, and audience.

The theoretical underpinning of discourse analysis was devised by the philosopher Michel Foucault (1972) and is referred to as critical discourse analysis. Composed of multiple theoretical and methodological approaches to the study of language, it engages with how texts are 'organised by grammar, structure, vocabulary and so on, but also how texts are produced and "consumed" within a society . . . to show how power and dominance are exercised as discourse' (Hammond and Wellington, 2013, p. 53). Foucault employed the notion of discourse to oppose positivist claims to know the 'truth' by exposing 'the mechanisms whereby some versions of "truth" come to be accepted and internalized, whilst other readings are marginalized, discredited or discarded' (Muncie, 2006, 75). Hence discourses 'are practices which form the objects of which they speak' (Foucault 1972, p. 49).

(iii) Narrative analysis

This term refers to various **hermeneutic** approaches to texts that are based on stories. Narratives can be entire life stories (common in anthropology) or extended accounts of events or themes (as in sociology and psychology). Catherine Riessman (2006) recognises different types of narrative analysis.

- *Thematic analysis* emphasises context. Researchers collect stories and use induction to create conceptual groups from the data. This leads to a typology of narratives (e.g. narratives about experiencing violence) organised by theme, with case studies used to provide illustrations.
- *Structural analysis* focuses on the telling of the story, making language itself an object of investigation in terms of 'the communicative work it accomplishes' (Riessman, 2006, p. 187). Given this focus, structural approaches focus on detailed linguistic

analysis and thus concentrate on small numbers of stories. Strict structural approaches can risk ignoring the context of the story.

- *Interactional analysis* emphasises the interaction between the storyteller and the interviewer. Narratives may be studied in specific contexts (such as courts: See Faria *et al.*, 2020) in which the storyteller and the interviewer participate in conversation. While attention is paid to content and narrative structure the focus is on storytelling as a process of meaning-making.
- *Performative analysis* approaches storytelling as situated, embodied performance. It employs the visual through staged performances to 'narrative as praxis – a form of social action' (Riessman, 2006, p. 188). It is useful for identity-based studies – of how the storyteller wants to be known and how they involve the audience in performing their chosen identity (see also Chatterjea, Wilcox and Williams [2022] on *Dancing transnational feminisms*).

All these types of narrative analysis allow researchers to engage in meaning-making, reimagining lives through narratives, connecting individual lives to their social worlds.

(iv) Visual analysis

The use of visual images and technology is increasingly common in research (Cope, 2019; Rose, 2022). It allows the researcher to interpret visual texts, inductively and deductively, as 'a way to gain a subjective understanding of the cultural codes contained therein' (Craine and Gardner, 2016, p. 284). Images (e.g. photos, maps, videos, films, and art) tend to be used in research in three ways: to supplement or produce data (e.g. as prompts in interviews); to generate primary data (e.g. mental maps or the use of Photovoice); and in reporting data (e.g. films) (Hammond and Wellington, 2013). Existing images may also be analysed (as a form of secondary data analysis).

Although the ways images are used will vary, researchers often interpret the image with a focus on the processes used to create the image, the content of the image, and/or the context in which the image was taken (Pink, 2006). They focus on 'what is made visible, who sees what, and how seeing, knowing, and power are interrelated' (Craine and Gardner, 2016, p. 275).

James Craine and Colin Gardner (2016) identify various methods of visual interpretation.

- *Psychoanalysis and semiotics*. Feminist psychoanalytic visual analysis, for example, has focused on 'how the unconscious of patriarchal society has structured the visual form' (p. 277). Following Laura Mulvey (1975), some feminists have focused on the 'male gaze' as a way of looking. The ways in which visual codes create meanings has been at the centre of semiology: how images are constructed as signs that make meanings, where 'something' stands for something else, in particular times and places. These meanings are relational rather than fixed, in that signs derive their meanings from other signs and the wider sign system.
- *Discourse analysis*. This approach focuses on how and under what circumstances visual representations are created. While discourse produces subjects, it also produces the technologies (e.g. maps) that enable particular ways of viewing the world.
- *Geo-visualisation*. This is the application of visual analysis to virtual and digital material to provide insights into the social construction of space and place.

(v) Grounded theory

Grounded theory is an inductive approach to generating theory that applies to a particular context. It is commonly applied in ethnographic research, allowing social order to reveal itself through extended fieldwork, with the ethnographer's insights emerging from their progressive socialisation into the life of the group (Eyles and Smith, 1988; Lofland, 1995). The approach was developed by the sociologists Barney Glaser and Anselm Strauss (1967) as a response to the then-dominant orthodoxy of the hypothetic-deductive model in the social sciences; hence its concern with rigour and systematic attention to empirical data. It is 'grounded in' or generated from the data collected by the researcher, and it is theoretical in that it seeks to explore the relationships between the categories (codes) that it generates (Hammond and Wellington, 2013). In this sense, grounded theory is concerned with not only induction but also abduction, as it both tests and generates hypotheses.

The practice of a grounded theory approach involves a set of data-analysis procedures and techniques.

- *Coding* is the breaking down of data (from field notes) into component parts, usually by theme, into codes. While there are differences in the sequence in which phases of coding occurs in grounded theory, there is general agreement that the process involves a movement from generating:
  - descriptive codes that stay close to the data;
  - to axial codes i.e. selective and abstract ways of conceptualising the data, namely concepts, considered the building blocks of theory;
  - to categories, which involve more than one concept, thus making a category a higher level of abstraction than concepts.
- *Constant comparison* is the systematic and continuous exploration of categories in relation to each other. Sections of the data are constantly compared with each other to allow categories to emerge and for the relationships between these categories to become apparent. This process only finishes once saturation is achieved and no new categories can emerge. It is this process of constant comparison that allows grounded theory to achieve a high degree of rigour.
- *Theoretical saturation is* the point at which coding no longer produces any new information or insights.

Difficulties with grounded theory include putting aside your awareness of existing concepts i.e. not making 'theory-laden' observations from the start of a research project. Neither does grounded theory necessarily result in the development of theory, which may or may not be problematic depending on the purpose of the research. Practising a grounded theory approach is time consuming, and the practice of dividing data into chunks can result in the loss of narrative flow. These issues result in some researchers employing a partial version of grounded theory: for example, using it as a coding strategy that does not work towards constant comparison nor aim for theoretical saturation.

(vi) Secondary data analysis

Secondary data analysis may be considered more of a research practice than an approach to analysis, because the approaches to analysis reviewed so far can be applied to **secondary data** as well as primary data. **Secondary data** are those which have been previously collected, commonly by some other individual or entity, to which a researcher can gain access. They can be quantitative or qualitative, including data collected by other researchers or (usually) government departments and agencies. Common sources include large datasets such as censuses (from which smaller datasets can be derived). Data can also be obtained from library collections and archives, newspapers and magazines, websites, and other online sources, as well as from private sources, including personal items such as diaries.

Analysis of secondary qualitative data can help build up a picture of dimensions of your topic that may fall beyond your own capacity to investigate, such as its historical development or social structure. Alternatively, you may engage in secondary data analysis to save time and money, to engage in longitudinal analysis, or to analyse data relating to a specific group if a large dataset is available; or you may think re-analysis could produce different results. There are also disadvantages to consider, such as issues around access and ethics, your lack of familiarity with the data, your lack of control over data quality, and that there may be missing data that you would have wanted to analyse. Secondary analysis of qualitative data also has to contend with the disadvantages that the context in which qualitative data were collected and intangible factors, such as the relation between the researcher and researched, cannot be replicated, especially if a lengthy period has passed since the data's initial collection.

Secondary analysis of quantitative data is much more common than analysis of another researcher's qualitative data (Hammond and Wellington, 2013). While digitised quantitative datasets are becoming increasingly available, using them may not be straightforward. Many quantitative datasets are also large and complex and can be difficult to navigate (Clark *et al.*, 2021) while the context in which they were created may not be clear, **metadata** may not be available, and the question of how valid or reliable the data are may not be answerable.

Reflection exercise 16.1 asks you to consider some questions pertaining to analysis of secondary qualitative data.

Reflection exercise 16.1: Secondary data analysis

Although you may be a qualitative researcher interested in generating your own data, and may therefore consider that you will never have to engage in analysis of secondary datasets it is useful to think beyond your expectations.

- Are there circumstances in which you can imagine yourself analysing secondary qualitative data?
- What are the advantages of engaging in analysis of secondary qualitative data?
- What are the disadvantages of engaging in analysis of secondary qualitative data?
- Examine recent back copies of a prominent urban studies journal (for example, *Urban Studies*, *Urban Geography*, or the *International Journal of Urban and Regional Studies*). Can you find an article that is based on either analysis of secondary quantitative or qualitative data? What do you conclude?

Section 3. Stages of qualitative data analysis

It is not uncommon for researchers to dip into the specific approaches discussed here but use only specific parts of them, although these should be employed in ways that meet the epistemological and methodological frames within which you are operating. Moreover, they have in common the generic practice of classifying data, whether by identifying themes, typologies, or codes. In what follows, we discuss analysis of the most common form of qualitative data used in feminist urban studies – that derived from interviews. The key stages of classifying data are preparing the data from the field, including creating and annotating transcripts, and coding transcripts and other data to draw meanings and insights. These are all activities that can be undertaken collaboratively, and we briefly discuss the challenges of doing so in Section 3 (iii). It is important not to underestimate the time you will need to process the data that you collect in the field in preparation for analysis.

(i) Data preparation: Description, transcription, and annotation

Data preparation involves three separate activities: data description, **transcription,** and annotation.

Description involves processing data to portray it in a way that facilitates its interpretation. At the start of each transcription you should have a title page that includes metadata such as the number of the interview, the location, the time, the duration, the name of the interviewer and the transcriber (if different), and any other important descriptive information. Description should also mention the context: the place where the data were collected; the social context (was it a paid interview? did the interview go well or badly? were other people present?); and any economic or political considerations that might affect data analysis (for example, data were collected during the time of COVID-19). You should begin the practice of data description as soon as possible, while still collecting data.

Transcription is the process of translating spoken words into text. Although field notes can be used for analysis, interview transcripts provide a more examinable audit of data. As Judith Lapadat (2000, p. 204) puts it: 'Verbatim transcription serves the purpose of taking speech, which is fleeting, aural, performative, and heavily contextualized within its situational and social context of use, and freezing it into a static, permanent, and manipulable form'. The process of transcription serves to make the researcher very familiar with their data, prompting you to address methodological and theoretical issues. In other words, transcription is not a purely technical exercise but one that also engages the researcher in preliminary interpretation and analysis.

Transcription (of recorded interviews, notes, and observations) should take place as soon as possible after conducting interviews, while they are still fresh in your mind. If possible, you should start on these activities on the evening of the day you conducted your interview.

You can transcribe by hand or using software.

- If you are transcribing by hand, provide yourself plenty of time to transcribe recorded interviews: it can take anywhere between four to nine hours to transcribe one hour of recorded material. Digital recordings can help you manually transcribe interviews:

software can enable you to slow down playback and simplify the process of pausing, rewinding, and fast-forwarding the recording.

- Voice recognition software for transcription can be purchased, but there are also free online options. You should check if your institution has a licence for a particular program. Consider testing out a few programs before purchasing any, but if you do choose to use a free, online program, find out whether or not the program is collecting and storing the data you input (known as data logging). It is vital that you do not unknowingly 'give away' the data or compromise the privacy and anonymity of the research participants (see Chapter 10, Section 3 Feminist perspectives on data management).

Document 16.1 on the book's website www.routledge.com/9781032668680 provides a list of manual transcription software as well as machine-assisted transcription software that are available for free (although note that the availability of this software may well have changed by the time you read this). While manual transcription software helps a transcriber to convert an audio or video into text (the transcriber types within the software program and can edit, pause, revise, and make sure that the outcome is accurate), machine-assisted transcription software uses speech recognition technology and algorithms to convert audio and video into text, providing you with a transcript that you can revise. There is a plethora of software packages that can transcribe recordings – they work in approximately 120 languages, although some languages are more amenable to the software – and this functionality is often built into online videoconference tools (both Microsoft Teams and Zoom offer transcripts in some languages) and some qualitative-analysis software, like NVivo. These tools expedite the transcription process (if your recordings are not too full of background noise), and they are reliable, but they should not be used without also manually checking the transcriptions. If you use machine transcription, you should do so carefully and critically. The quality of transcription can vary dramatically depending upon the language and dialect of the speech, because the language models that power machine transcription software are trained on limited datasets that privilege hegemonic, anglophone dialects.

Strategies for transcription will vary, depending on 'the project, paradigmatic stance, analytical approach, and purpose' (Lapadat, 2000, p. 215). While there are situations in which only sections of an interview may be transcribed it is common practice to fully transcribe all interviews. In large projects, like GenUrb, a researcher may hire professional transcribers or train research assistants to engage in transcription. In such cases, it is essential that the researcher develop a transcription protocol and that people are given sufficient training to understand how to make interpretive decisions. Without such training, errors will take place (e.g. from deliberately tidying up sentence structures to interpreting only sections of an interview). Our experience taught us that training should not only focus on the technical aspects of transcription, but also address the role that assistants play in the research. They need to be familiar with the goals and organisation of the research and be provided with regular debriefings, while their performance also needs to be regularly evaluated.

GenUrb example 16.1 discusses the issues the GenUrb project faced when undertaking transcription, whether the process was done in-house, using research assistants, or by professional organisations.

GenUrb example 16.1: Strategies for transcribing interviews

All the City Research Teams (CRTs) had numerous interview- and life-history recordings with participants that were several hours long and sometimes recorded in more than one session. All were transcribed verbatim. The sheer volume of data in sound files to transcribe was daunting. The CRTs adopted a variety of approaches to organising transcription. Some kept the transcription process in-house, while others hired people outside the project. The further removed the transcription process was from the project, the more errors were discovered and the more time had to be spent redoing the transcriptions.

The **Georgetown** CRT started by tasking Red Thread members Joy Marcus, Susan Collymore, and Halima Khan, who had conducted the research interviews, to transcribe them. However, as Red Thread is first and foremost a grassroots women's organisation working, for example, on domestic violence or sexual assault cases, they could not devote all their time to transcription. Trying to keep to schedule, therefore, meant having to transcribe in the evening, but this became physically and mentally unsustainable. After approximately one-quarter of the interviews had been transcribed in this fashion, the project manager, Leeann Bennett, started to assist with the transcripts. Leeann had conducted graduate research in Guyana with Red Thread and was obviously familiar with the research. However, she faced the same problem of a lack of time, so she then hired a graduate assistant to work with the Georgetown CRT to finish the transcripts. This was Angela Stanley, a Guyanese PhD student in women's studies at York University, who spoke Guyanese Creole and was very familiar with the research neighbourhood. However, given that graduate assistants can work only 10 hours per week, transcription progressed slowly. The decision was then made to hire three other Guyanese transcribers – Karen Naidoo, Serene Paul, and Saaraa Esau, all graduate students at York University and each proficient in Creolese – to help complete the work. The Georgetown CRT had a high level of trust in their transcribers, as we outlined in detail, and frequently checked, the painstaking task of transcribing, which also involved, upon completion, listening to the sound file again to proofread the written transcript (all without the use of a foot pedal or software to help with this process). All the transcribers were trained to pay attention to the context of what each woman was saying and of not transcribing Creolese into 'standard' versions of English.

The **Delhi** CRT hired a commercial transcription company to transcribe their interviews. With transcribers fluent in Hindi, it charged a local per-minute rate from which the Delhi CRT could benefit as well as guaranteeing a two-week turnaround. The preference of the Delhi CRT's lead researchers would have been to undertake this work themselves, but this was not feasible, given their full-time jobs and family responsibilities. Upon completion of the transcription company's work, they hired twelve graduate students who were familiar with the study site and who had high proficiency in English, Hindi, and other languages and dialects spoken by the research participants. They read through all the transcriptions to pick up any errors due to literal rather than contextual transcription and to fill in gaps where participants had lapsed into native dialects from Hindi. This quality-control step allowed the Delhi CRT to have a high degree of trust in the transcriptions.

In **Cochabamba**, the mechanics of the transcription process were coordinated by the CRT lead researcher, Nasya S. Razavi, and GenUrb project manager, Leeann Bennett. Transcription was undertaken in-house. The transcription of all the interviews in Spanish was completed by the Director of the Centro de Estudios y Trabajo de la Mujer, Sonia Pardo Burgoa, and a Bolivian undergraduate student, Ingrid Baldivieso. The Cochabamba CRT also recruited a Bolivian PhD student in translation studies, Victor Hugo Mamani Yapura, to transcribe the interview of one participant who gave her interview in Quechua and to translate the transcription into Spanish.

In **Shanghai**, the CRT's lead researcher, Penn Tsz Ting Ip, hired seven bilingual research assistants – two graduate students and four undergraduate students based in Shanghai and one Chinese Canadian undergraduate student based at York University – to transcribe interviews into Simplified Chinese. They were asked to use the popular Chinese software Xunfei, developed by the information technology company, iFLYTEK, for the transcriptions. It became clear, however, that the use of this software was not producing the high-quality transcriptions that we required. The research assistants had to spend many intense days double-checking all the technology-generated transcriptions to ensure their quality. One issue was with the translation of specific words. For instance, as the gender pronouns – he, she, and it – in Chinese all sound the same – *ta* – it took time to check them. Additionally, some research participants came from rural areas and had strong accents that Xunfei could not accommodate, so the CRT members had to go back to the recordings to transcribe these entries. Given these problems and to ensure that the quality of transcription was acceptable, the CRT lead double-checked all the transcripts, comparing them with the audio files to ensure there was no missing material and deleting any personal information mentioned in the interviews. The process took an additional six to ten hours per transcript.

The **Ibadan** CRT also employed students to help transcribe the interviews. However, the students misunderstood what was required and summarised the interviews rather than transcribing them verbatim. As a result, the CRT had to start the process of transcription again. In an attempt to maintain quality, all the Yoruba transcripts were entrusted to one person, Tunrayo Abimbola Adeyemi, a Yoruba graduate student at Obafemi Awolowo University (nineteen of the interviews had been conducted in English, eleven in Yoruba). A member of the CRT, Bukola Omolara Odunola, then read the transcripts while listening to the audio tapes to verify their quality. Given the earlier problems experienced, the CRT then took the extra precaution of hiring Wumi Asubiaro-Dada (a Nigerian Canadian PhD student at the University of Toronto), to review all the transcriptions, those in Yoruba and in English. Finally, the CRT lead, Grace Adeniyi-Ogunyankin, also reviewed all the transcriptions. Needless to say, the Ibadan CRT then had a very high level of trust in the quality of the transcribed material.

The **Ramallah** CRT transcribed all interviews soon after conducting them. Some were done by CRT members and others by a hired professional transcriber.

By Joy Marcus, Anindita Datta, Nasya S. Razavi, Penn Tsz Ting Ip,
Grace Adeniyi-Ogunyankin, Natasha Aruri, and Leeann Bennett

The steps that each CRT followed to create a transcript are detailed in GenUrb example 16.2.

GenUrb example 16.2: Creating transcripts

The following steps can facilitate the creation of transcripts.

- On each page of text, leave a wide margin on the righthand side in which you can put notations and ideas. These will eventually lead to codes.
- You may also want to create a narrower margin to the left of the page in which you can jot down the pseudonym of the person who is talking.
- Have a running header on each page, with the project name and participant number.
- Number the pages.
- You may also want to number every line, as this makes it easier to source the location of potential quotations.
- Use standard transcribing conventions in the transcripts. We wanted verbatim documents that were contextual – recording false starts, overlaps, and interruptions – and we also wanted the practice to be systematic across the CRTs. While no transcription system is perfect, it is important to standardise the system you use, especially in cases where more than one researcher is engaged in transcribing. The most used transcription system for social (as opposed to linguistic) research has been Conversation Analysis, which treats talk as social interaction and thus has notations relating to pauses, intonation, pitch, elongation, and cut-offs. There are now multiple sets of convention codes available, and in GenUrb we adopted a very simple system. Because of the large number of researchers working on transcription, we decided to reduce the technicality of the transcription conventions as much as possible. We used the following standardised conventions:

(.)	=	a slight pause
(. . .)	=	a slightly longer pause
()	=	transcriber could not hear/understand what was said
(that?)	=	possible hearing
((laughs))	=	author description
italics	=	emphasis

- Upon completion make a copy of the transcription on both your hard drive and an external drive.
- Do not include any personal identifying information of the interviewee on the same file as the transcription. In other words, keep metadata on a separate file on which you link pseudonyms or a number you have used to identify each participant to their real name.

By Linda Peake

For those engaged in transnational research, data preparation may also involve the translation of the data, which will take place after transcription (see Chapter 15, Section 3 Planning for translation and interpreting throughout the research process).

Once the time-consuming activities of transcription (and translation) are complete, it is time to annotate the transcripts with notes, observations, and connections to other data. This can be done manually or within a software package (see Chapter 18, Section 4 Memos and annotations).

You can think of annotation as an informal or preliminary coding strategy. Use your field notes and any interview notes to refresh your memory about the interview and to help you with ideas about what to note. Consider the following strategies to help with your annotation.

- Ask interrogative questions, including who? what? where? how? and why?
- Consider a substantive checklist. How well do the data fit the themes of the research?
- Shift your focus. What details apply to the big picture of the data, and which ones apply to the little pictures?
- Use keywords, including those for emotions, attitudes, beliefs, and behaviours.

Once you have finished preparing the data, by transcribing, translating (if necessary), and connecting interviews with annotations from your fieldnotes, it is time to code.

(ii) Coding interview transcripts

Coding is a form of data construction and analysis that involves labelling and organising data to reduce them to a manageable size (Saldaña, 2015; Clark *et al.*, 2017). While coding allows you to conceptually organise the transcribed data, to make a large group of data meaningful by classifying them and by making connections between them, it does not provide an explanatory framework to interpret or draw conclusions from the data.

There are many approaches to coding. It can be used in an exploratory, inductive way, as in grounded theory (in which you generate theories from empirical data), proceeding by extracting meanings or themes; or it can be used to support a theory in a more deductive manner, having an *a priori* list of codes that you may expect to find based on your knowledge of the field and your reading of literature. You may well combine the approaches to your coding, using both descriptive and axial (sometimes referred to as analytic) codes, the latter for developing more abstract and explanatory categories. Open coding can also be used to explore associations between similar meanings or themes. The types of code you use will vary depending on the focus of your analysis.

Regardless of your approach, coding can be conducted using software (see Chapter 18, Section 3 Approaches to coding and codes) or by hand. The following points describe how to code data manually.

- The coding may flow from the questions asked but should also reflect what was said, which may have moved away from the focus you expected.
- Give simple descriptive labels to summarise words, phrases, or pieces of text that convey a similar meaning or theme and that you think are of importance to your research question.

- Start to group these labels into categories (codes) that relate to the data (some data may fall into more than one category).
- As you develop codes, list them in a codebook, i.e. a list of all the codes.
- After a first round of coding the data you will want to code again, developing both meta- and subcategories. For example, the code 'housing' could be divided into subcodes that address factors such as physical condition, size, ownership, labour conducted in the home, amenities (such as indoor toilet), and so on. Housing could also form part of a metacode that addresses, for example, 'physical aspects of a neighbourhood'.
- This process may be repeated several times until you feel that no new relevant codes or subcodes can be added or merged. This is described as reaching a saturation point in your coding.
- As a result of this process, you may end up reorganising codes, splitting, merging, or even discarding some codes.

Do not underestimate how long this coding process will take you. While some see manual coding as a laborious chore it has the benefit of allowing you to immerse yourself in and become very familiar with the data.

There is no guarantee that two researchers in the same project could code the same interview and arrive at the same codes, but processes can be put in place to increase the reliability of the process of coding (Morse, 2015) (see Chapter 18, GenUrb example 18.2 The coding process).

(iii) Working collaboratively

Research projects that are transnational, interdisciplinary, comparative, or participatory in nature may involve working collaboratively with others in producing transcripts and coding. Collaborative analysis practices can be very productive in terms of allowing for a division of labour between different people and broadening the range of perspectives on the data to produce a more rich and complex analysis. However, undertaking data analysis in a collaborative context also has its challenges, as Reflection exercise 16.2 indicates. Collaborative analysis requires ensuring that the frameworks for processing and analysing data are developed early in the project; that they are comprehensive, consistent, and communicated clearly across the research teams; and that they can be revised through consultation. If collaboration is occurring in the context of a participatory research project, where participants are involved in analysis as co-researchers, for example, strategies are needed to ensure that all data-related processes are conducted in a cooperative way that allows knowledge to be shared and different voices to be heard. For example, rather than circulating a completed codebook among researchers, a series of workshops could be held to work through the data and brainstorm a range of themes that could act as codes. Regardless of how careful researchers may be to promote equity in collaboration it can be difficult to achieve given the different aims, needs, desires, and timelines of those involved (see Chapter 9, Section 3 Transnational feminist praxis and teamwork).

Reflection exercise 16.2 asks you to consider a range of questions in relation to working collaboratively with data.

Reflection exercise 16.2: Practising collaboration

If you are working in a collaborative team, consider the following points.

- How will you agree upon the definitions of core concepts in cases in which people disagree about their meanings?
- How will each process (for example, data preparation, transcription, coding) be divided among the different people involved in the team?
- How often will you schedule team meetings, either in person or virtually, to evaluate how smoothly the process is moving, to maintain quality control, and to work through any challenges that present themselves?
- What is the best way to work with the skills, knowledge, and expertise that different people bring to the project?

The final section turns to what you need to do once your coding is complete; you still need to spend time with the data to continue your evaluation, ensuring that they are reliable and trustworthy.

Section 4. Critically evaluating the data

The completion of the coding process is not the end of your engagement with the data. A critical evaluation of the data requires that you undertake the following activities. You can start by fact-checking, an important step in assessing the quality of the data, to ensure that information such as dates, events, and descriptions is accurate. Next, you may want to contextualise the primary data you have gathered by comparing them with any available secondary sources that can help you understand the broader social, political, and economic context of the participants' experiences and thus strengthen your analysis.

Sitting with and listening to the data – that is, suspending your analysis to reflect further – are useful practices for gaining some distance from the data and your interpretations. Be careful not to take such a long break that you have to spend hours reviewing your analysis again, but don't be afraid to take some time to sit with the data and see what comes of it. Listen to the data to find out what they might tell you once you set aside your expectations as a researcher.

- Ask yourself whether participants are giving you answers they believe you, the researcher, want to hear. This is commonly referred to as response or participant bias. How will you determine if this has happened in your research?
- Pay attention to the silences in the data or to things that are *not* said to reveal new issues or connections not readily apparent in the first round of data analysis. Silences can include a reluctance to touch on certain issues due to shame or denial – for example,

violence or debt. Hence, recognising silences can be an important way to deepen your analysis by forcing you to ask new questions or to revise your existing questions or analytical and theoretical frameworks (Koch, 2020).

- Watch for data that do not fit with your analysis. As you begin to see common themes or ideas emerging and start to form hunches about arguments you can make, make note of cases that do not fit, that seem ambiguous, or that appear to contradict the patterns you've identified. Use this recalcitrant data to ask further questions about your analysis and to deepen your understanding of the data that you already have. These questions may include considering why and how these cases contradict others, what questions they raise, what these differences might be saying about your initial hunches, and how they challenge your current understanding of your research.

It is also important to recognise the limitations of the data. Ask yourself the following questions to ensure that your analysis adequately supports the arguments or conclusions you wish to draw. Do events after the fieldwork make some of the data irrelevant? If so, are the data still usable? What possible and actual limitations have been imposed on your analysis by the ways in which you have collected data or by the methods you have used? For example, can you extrapolate from the data (from the sample to the broader population)? (Note that unless you have utilised a random sample this is not possible.) Has your analysis raised concerns that the use of certain data could be harmful to participants? Has your analysis of the data made you aware of questions you did not ask that should have been asked? What can you do to resolve this situation? And do you have sufficient resources (including time) to complete analysis of all the data? If not, then how will you prioritise your time in relation to data analysis?

Finally, constantly reviewing and, where necessary, revising your analytical approach, conclusions, and even research questions is important to ensure that analysis is as thorough, rigorous, and well-founded as possible. Reviewing and revising during the data analysis process are also central to meeting ethical obligations by enabling you to exercise self-reflexivity about your positionality in relation to your research in terms both of the political implications of how you are interpreting the data and of how your analysis relates to the communities you are studying.

Section 5. Summary

This chapter has discussed different approaches to data analysis – in particular, deductive and inductive approaches – which involve the processing and interpretation of data to draw out key themes and insights for answering your research question(s). Your research questions, methodology, and methods, as well as the type of data that you collect, will be guided by your analytic approach, which, for qualitative data, can include: content analysis, discourse analysis, narrative analysis, visual analysis, and grounded theory. These approaches all include the generic practice of classifying data. The chapter then turned to the analysis of interview data, starting with data preparation: turning spoken words into written ones, creating and annotating transcripts, and coding, i.e. labelling and organising data, with an emphasis on manual as opposed to software-based analysis. The chapter concludes by engaging with the issues arising when analysis is undertaken collaboratively, as part of a team, and by reviewing the practices that underpin a reflexive approach to engaging with data.

17 Software-aided analysis for feminist research

Biftu Yousuf

Learning objectives

After completing this chapter, readers should understand

- debates for and against the use of computer software for feminist qualitative analysis;
- what NVivo is and how it can enable qualitative data exploration and analysis in feminist research;
- and the possibilities, challenges, and debates concerning the use of digital technology for qualitative data analysis in feminist research.

Further, readers should be able to

- contemplate the advantages and disadvantages of using NVivo for qualitative data exploration and analysis;
- and identify critical approaches to working with data analysis software.

This chapter starts with a discussion of the arguments for and against the use of computer software for qualitative research and then considers the various software packages that are available for qualitative data. It introduces NVivo as a research tool for feminist research and discusses why and when the transnational GenUrb project used this software to aid its engagement in feminist analysis. The chapter focuses on NVivo as a malleable software that can accommodate a wide range of disciplines and methodological practices. It provides an understanding of how you can use NVivo to facilitate the research process from start to finish. You will be able to contemplate the possibilities and limitations of NVivo and to identify critical approaches to working with software to aid analysis for feminist research.

Section 1. The use of computer software for qualitative research

The term 'computer-assisted qualitative data analysis software', **CAQDAS**, refers to computer-based programs that support the qualitative research process, including transcription, analysis, and dissemination. While the use of CAQDAS is widespread among qualitative researchers, misconceptions about using software to support qualitative research persist. This section examines feminist debates surrounding the use of CAQDAS and highlights why it can be remiss to overlook the strengths of software in aiding qualitative research. It then evaluates the various software packages for qualitative data that are available.

DOI: 10.4324/9781032668727-23

(i) Debates on the use of CAQDAS

Qualitative data analysis refers to the process of gathering, organising, and interpreting unstructured data. CAQDAS has long played a role in data management and analysis. While quantitative-data-analysis software, such as SPSS (Statistical Package for the Social Sciences), has long been in vogue among researchers in the social sciences, **qualitative analysis software** took longer to capture their methodological attention (see links to SPSS on the book's website www.routledge.com/9781032668680). Longstanding debates on the modes of knowledge production in academic research have shown quantitative and qualitative approaches to consist of different ways of viewing and addressing a phenomenon because they are embedded in distinct views of social reality that are informed by interrelated philosophical assumptions about the nature of existence/reality (ontology), the nature of thinking/knowing (epistemology), and the theoretical assumptions about the research process and suitable approaches to systematic inquiry (methodology) (Shaw, Dixon and Jones, 2010; see also Chapter 11, Section 1 Introduction to the philosophical fields underlying research).

Quantitative studies, supported by the positivist paradigm, have been based on an ontological position of a single reality, an epistemological position that believes in 'objective' knowledge, and a methodological commitment to the procurement of 'factual' scientific evidence. It was disagreements with these fundamental assumptions of positivist research and of its once-hegemonic position in social sciences that fuelled non-positivist approaches to research (Smith, [1999] 2012; Strega and Brown, 2015; Chilisa, 2019). Non-positivist paradigms, which lean toward qualitative research methods – but can also support quantitative studies (Mattingly and Falconer-Al-Hindi, 1995) – include interpretivist, realist, and feminist approaches to research (Cohen and Crabtree, 2006). These paradigms are grounded on the ontological assumption that social reality is variegated and contextually created, an epistemological position that believes in 'subjective' knowledge, and a methodological commitment to findings that emphasise embodiment and meaning (see Chapter 11, Section 2 Feminist ontologies).

This quantitative–qualitative paradigmatic divide characterises contrasting approaches to data analysis, underpinning wider debates on the politics of using CAQDAS. Early adopters of CAQDAS were engaged in research in the 20th century, at a time when prevailing attitudes towards computer usage were seen to rest within a positivist paradigm, as witnessed by the preponderance of software for quantitative data analysis (Tummons, 2014). As a result, the belief that CAQDAS is unsuitable for qualitative research lingers. Some qualitative researchers have been anxious about any use of computers due to fears that software replaces human endeavours and thereby enables the mechanisation of qualitative data analysis (Tummons, 2014). Others raised concerns that CAQDAS might guide qualitative inquiry in a quantifying direction by promoting the use of functions, e.g. word frequencies or string searches, which 'belong' to a positivist paradigm (Welsh, 2002; Tummons, 2014; Le Blanc, 2017). It has also been suggested that CAQDAS could potentially compromise the essence of ethnographic inquiry and other 'naturalistic', phenomenological methods of qualitative research, in that computers could generate 'too much closeness to the data' whereby researchers become entrenched in the coding process and do not have enough 'analytical distance' (Gilbert, 2002, p. 215). Others have seen the opposite as a problem, i.e. that technology creates distance between researchers and the data they have collected and that software-assisted data analysis thus results in a disembodied process.

A key set of assumptions about qualitative data analysis software has emerged through the implicitly gendered language used to describe its functions: that it 'systematises', 'disciplines', and 'manages' data analysis. By contrast, feminist methods (being about the everyday and the relational) are repeatedly referred to as being inherently 'messy'. This polarisation implies that 'systematising' data is somehow not a feminist practice, and by extension, software is inherently a tool for 'masculinist' research practices. Other qualitative researchers have sought to counter such misconceptions, rejecting the perceived disadvantages of the use of software for qualitative data analysis.

Feminist researchers have long contributed to these methodological debates and criticised the dualist thinking underlying the supposed divide between quantitative and qualitative methods (Peake, 2009). They have embraced alternative possibilities for advancing software-assisted data analysis in ways that are responsive to qualitative methodologies and that fit with feminist epistemologies and politics (Kwan, 2002a; see themed 'Viewpoints' section of *Gender, Place and Culture* [2002], vol. 9, no. 3). To this end, feminist methodological interventions challenge notions that the use of digital technology, such as computers and software, is an inherently quantitative and masculinist practice. Echoing feminist critiques of geographic information systems (GIS) (Kwan, 2002b; Kwan and Ding, 2008), the use of CAQDAS encourages researchers to transcend the binarised polemics characterising lingering scepticism about the role of technology in qualitative data analysis.

In GenUrb example 17.1, we outline the range of concerns voiced by members of the project about the use of software-based data analysis.

GenUrb example 17.1: Concerns about using NVivo

GenUrb's initial data management plan, as included in the grant application, anticipated the use of computer-assisted qualitative data analysis, with all data being translated into English and stored in a central repository at York University that would give all the teams access to each other's data. As the City Research Teams (CRTs) started working together, it became clear that this initial agreement was breaking down. Most of the CRTs were interested in using NVivo and were prepared to do the transcribing and translating of the data in-house. They felt that this was the only way in which each CRT would be able to explore the data from other teams in an in-depth way. A couple of the CRTs, however, preferred an approach in which members of each CRT would engage in discussions about their team's data with other CRTs and that these discussions would form the basis for analysis. This approach had the disadvantage that each CRT would only ever see their own data, with the data collected by the other CRTs remaining inaccessible to them. These CRTs were not in favour of the extra time that that would be needed to transcribe and translate their data into English to facilitate the collaborative use of NVivo, and they pointed out that this task would require extra funding for professional transcribers and translators. Differing views on the use of NVivo and engaging with CAQDAS for feminist analysis were also expressed by scholars from different disciplines: those from geography, for example, were more open to the use of NVivo than were those from anthropology.

The decision to use or not to use NVivo was finally determined by security concerns: two of the CRTs became reluctant to send data (digitally) across national boundaries to store in a central depository, making it impossible for them to participate in the original plan. Given the repressive and authoritarian nature of the state in both their countries, this was an understandable consideration and one which nobody wished to argue against. It came to trump all other considerations concerning the management and the analysis of our data. Faced with this impasse, GenUrb had to suspend its plans for the collaborative use of NVivo and adopt instead a plan for each CRT to organise its own data, either manually or using software.

This was not a smooth transition, and there were objections to this new approach. With not every CRT able to be present at every data analysis meeting, the comparative analysis could have been substantially weakened. The departure from the project of the two CRTs that preferred not to share their data led to a unanimous decision to return to the original plan to use NVivo and the much more rigorous and comprehensive approach to comparative analysis that this would allow.

Although there were several people in the project with firsthand knowledge of NVivo, it soon became clear that the project needed to employ an NVivo specialist who could oversee the systematic incorporation of our data into NVivo. The scale and complexity of the operation – in terms of the number of people involved, the quantity of data, the need to work across time zones, the number of NVivo licences needing to be purchased and the use of both PCs and Macs by GenUrb members – necessitated engagement with someone who had an in-depth knowledge of NVivo and who had the time to devote to ensuring that data were uploaded in the most efficient way. A search was made for an NVivo expert who was also knowledgeable about and preferably engaged in feminist research ethics and methodological issues. Biftu Yousuf, a certified NVivo expert, qualified instructor, and a geography PhD student at York University fulfilled all these requirements. She started work on the project in May 2021.

In early 2021, Biftu started to establish a plan for how to design an NVivo coding framework and how to facilitate collaborative coding in NVivo. This resulted in the collation of data from five CRTs (224 transcripts) into NVivo and the participation of sixteen people in a simultaneous coding process across twelve different time zones. The CRTs were overwhelmingly cooperative and responsive in learning how to use NVivo and how to engage in critical reflection on its use. The level of commitment and effort that this task required cannot be underestimated (see Chapter 18, GenUrb example 18.2 The coding process).

Despite differing attitudes towards the use of CAQDAS, use of NVivo goes some distance to unsettle assumptions about digital technology being an inherently masculinist tool or that engaging it results in 'disembodied' analysis. Our approach has been to underscore the mutual relationship between researchers and CAQDAS and to affirm agency in processes of knowledge creation (Kwan, 2002b; Jackson, 2017). In this sense, we endorse a '*methods-with* software' mantra when using NVivo, a view that acknowledges the mutual relationship between researchers and software (Jackson and Bazeley, 2019, p. 5).

By Biftu Yousuf and Linda Peake

Reflection exercise 17.1 encourages you to assess the advantages and disadvantages of using software-aided tools for research.

Reflection exercise 17.1: The benefits and drawbacks of using software-aided tools for research

Consider the following steps to make an informed decision about whether to use software that aligns with the goals and ethical principles of your research.

Step 1: Begin by contextualising your research: What are your research objectives and questions? What methods are you employing? What kind of data are being produced?

Step 2: List the potential benefits and drawbacks of using software-aided tools for your specific research design. Consider the following factors (among others discussed in this chapter):

- data management;
- methodology;
- approach to analysis;
- learning process;
- and collaboration.

Step 3: Reflect on your personal experiences and skills with software-aided tools for research. Have you previously used any? If so, what were the outcomes? If not, what challenges might you anticipate in learning how to use these tools effectively?

Step 4: Ponder the ethical issues related to using software-aided tools, such as data privacy, accessibility, and transparency. How might you navigate and address these concerns in your research?

Step 5: Reflect on the role of software-aided tools as supplements *to* rather than replacements *for* your research methodology.

Step 6: Consider whether you would like to explore further the option of using software-aided analysis in your research journey. If so, what steps will you take to acquire the necessary skills and resources?

Step 7: Develop an action plan based on your reflections, outlining how you will incorporate software-aided tools into your research.

Step 8: Write a well-reasoned explanation for your choice to use or to reject software-aided tools in your research. This justification will establish a connection between your research methodology and research goals.

(ii) Available software packages

Software offers several advantages in the process of analysis.

- It can help save time when organising and analysing large quantities of qualitative data.
- It provides tools for content searching and coding.

- It allows for the use of annotations to assist analysis and create links to other coded segments of text.
- It can engage with data visualisation, allowing for the identification of trends and patterns.
- It enables qualitative analysis by facilitating description, comparison, and hypothesis testing.
- It provides support for several languages if they use the same alphabet.
- Some software can support mixed-methods data analysis.
- Some software allows more than one user to work together in real time (see Chapter 18, GenUrb example 18.2 The coding process).

If you do use software for qualitative data analysis there is a wide variety to choose from, including: NVivo, ATLAS.ti, Provalis Research Text Analytics Software, Quirkos, MAXQDA, Dedoose, Raven's Eye, Qiqqa, webQDA, HyperRESEARCH, Transana, F4analyse, Annotations, and Datagrav. On the book's website www.routledge.com/9781032668680 in Document 17.1, we describe the pros and cons of three popular software packages used by university students for analysis of qualitative data, and we list open-source software packages for qualitative analysis that are in common use. Many researchers prefer open-source software because they believe that supporting freely available, collectively governed software is more ethical than using proprietary, commercial software.

Whether you will use software depends upon several factors.

- While you will often have a good memory of each interview you have conducted over time, this information will start to become hazy. When your memories of interviews and participants blend into each other, you may want to consider using software to manage the data you have collected. We recommend that once you have over 10 transcripts or multiple sources of data, managing the number of codes you may generate and the quantity of data you will have may not be manageable for manual analysis (see Document 17.1 on the book's website www.routledge.com/9781032668680).
- The cost of software may be prohibitive. Check whether your university has a licence, either for the software you want to use or for a similar package. You may be able to use this software at no or minimal cost. If you have only a small project and you feel your use of the software may not take longer than a couple of weeks, you may find introductory free offers to use software packages for a limited time. You may also consider using a freely available, open-source software package to avoid paying licensing fees.
- You may not have the time (or inclination) to learn to use a new software package. It is common to feel pressed for time and not to have set aside time to learn to use software. This problem may be exacerbated if nobody you know is using the same software. However, bear in mind that once learned, software packages become useful tools in your skill set and that once you become familiar with one type of software, it is not so difficult to pick up similar packages (your skills in using NVivo, for example, can quite easily be transferred to using MAXQDA or ATLAS.ti). Software developers usually provide resource guides, and there are also online groups of software users to whom you may turn with questions.
- You may not have the up-to-date computing equipment you need to run your preferred software packages. If this is the case you can always look for less sophisticated software packages that you may be able to run. For example, you may not be able to run NVivo, but you could replace this software with web-based applications like Dedoose that require less storage space on computers.

Reflection exercise 17.2 helps you to arrive at considered choices in relation to the software package that you use.

Reflection exercise 17.2: Choosing software for your research

The following list offers a framework outlining criteria to consider once you have made the decision to utilise a software-aided tool in your research. For this exercise, you can recreate the table and populate it with the details of each of the software packages you are evaluating. For example, you might create a table comparing NVivo, ATLAS.ti, Quirkos, and Dedoose based on the considerations mentioned.

Criteria	*Considerations*
System requirements	What are the hardware and software specifications e.g. operating system, memory, and storage capacity?
Data compatibility	What data files is the software compatible with? For example: textual data, audio/visual data, or demographic data?
Language packages	What interface and text content languages are available? What specific languages does the software support?
Cost and budget	What is the cost of the licence? What about the cost of renewing subscriptions for licences?
Availability	Does your university have a licence for the software? If not, does your university provide funds to support procurement of research software?
Ease of use	Does the software have a user-friendly interface?
Learning process	How much time will you need to invest to develop basic skills and expertise in the software?
Trial version	Is there a trial version or demo of the software that you can sample to evaluate the suitability for your research?
Ethical considerations	Does the software align with the ethical guidelines for research at your university, particularly if your research involves human subjects or sensitive topics?
Data securitisation	If your research involves sensitive or personal data, what security and privacy features does the software provide to protect your data and ensure compliance with ethical standards and regulations?
Resources	Is the software supported on an ongoing basis by the developer e.g. technical support, user-help guide, or support centre?

For the remainder of this chapter, we turn the spotlight onto the software package **NVivo.** The following section describes NVivo as a tool for qualitative data analysis and explains the reasons and conditions for its use.

Section 2. Introducing qualitative data analysis with NVivo

Digital research tools, such as NVivo, are neither inherently masculinist nor feminist. NVivo offers researchers a blank canvas and a set of digital tools to enable software-assisted data analysis, but the construction and implementation of a digital infrastructure for data analysis would not be possible without the driving force of human intellect and labour. In other words, NVivo can facilitate how researchers develop and use their analytical frameworks by virtue of the features and functions that it offers *and* those that it lacks. However, software alone cannot dictate how data analysis is conducted. Remember: NVivo is not a methodology; NVivo is a research tool that can facilitate a spectrum of methodological approaches. This view invites NVivo users to think about the interaction between epistemology, methodology, and technology as being reflective of an intentional, dynamic, and embodied process of analysis (see Chapter 18, Section 7 Reflections on using NVivo for feminist research).

NVivo is a leading digital technology for qualitative data analysis, produced by QSR International (now merged into Lumivero) and widely used in the academic, government, and commercial sectors (see links to NVivo and QSR on the book's website www.routledge.com/9781032668680). Although the previous section has shown that there is alternative software for qualitative data analysis, NVivo has become increasingly ubiquitous at academic institutions. It is used across a wide range of disciplines and can accommodate different methodological practices. This section describes NVivo as a tool for qualitative data analysis and explains the reasons and conditions for its use.

NVivo is an example of CAQDAS which is used by researchers to store, organise, manage, and analyse non-numeric or unstructured qualitative data. Researchers can use NVivo to code data, to create links to external files, to compare demographics, to search for complex patterns, to produce reports, and to export data. As Table 17.1 shows, these

Table 17.1 Data files compatible with NVivo

File type	*Examples*	*File Format*
Documents	• Interview transcripts • Life histories	Microsoft Word (.docx) Microsoft Word 97–2003 (.doc) OpenDocument Text (.odt) Rich text (.rtf) Plain text (.txt)
PDFs	• Journal articles • Archival documents • Web pages captured with NCapture	PDF (.pdf)
Datasets	• Online surveys • Social media data captured with NCapture, e.g. YouTube comments	Microsoft Excel (.xlsx) Microsoft Excel 97–2003 (.xls) OpenDocument Spreadsheet (.ods) Comma-separated text file (.txt)
Audio and video files	• Recorded interviews or focus group discussions • Observational videos e.g. YouTube videos	MPEG (.mpg,. mpeg,. mpe,. mp4) Windows Media (.avi,. wmv) QuickTime (.mov,. qt)
Images	• Pictures and drawings • Photo diaries • Archival photos	Joint Photographics Expert Group (.jpg,. jpeg) Portable Network Graphics (.png) Tagged Image File Format (.tif,. tiff) Windows bitmap (.bmp) Graphic Interchange Format (.gif)
Bibliographies	• Bibliographic references and related data files e.g. Mendeley, EndNote, Zotero RefWorks	Research Information Systems (.ris) Extensible Markup Language (.xml)

Source: NVivo user-help guide, e.g. https://help-nv.qsrinternational.com/20/win/Content/files/files.htm

modes of inquiry and exploration are applicable to a wide range of textual and non-textual data, including documents, PDFs, surveys, audio and video files, pictures, web pages, social media, and bibliographic references.

NVivo is a licence-based software that is available for use on both PCs and Macs (i.e. on Windows and macOS operating systems). The Windows and macOS versions are not equivalent in terms of functionality and interoperability: for example, they use different, proprietary file formats (.nvp for Windows and. nvpx for macOS). The Windows version of NVivo possesses the full range of features available in the software, including functions for creating links and annotations, conducting complex analytic queries, running advance queries, and producing a wide range of reports. Not all these features are offered in the macOS version of NVivo (although QSR has worked actively over the years to bring both operating systems into closer compatibility and parity). NVivo's website for Windows and Mac provides a comprehensive understanding of the differences between the two operating systems.

QSR discourages users from working across the Windows and macOS operating systems as part of an ongoing workflow because there are limitations to how these NVivo projects convert and merge together. This is a particular problem for research that involves multiple collaborators in NVivo where the likelihood of a large research team using both PCs and Macs is high. GenUrb example 17.2 highlights some of the issues we encountered with arranging licences and operating systems for multiple users.

GenUrb example 17.2: Issues with licences and operating systems

An enormous amount of work was conducted in the GenUrb project to conceptualise, design, set up, and launch the digital infrastructure for transnational comparison. There was no shortage of issues that needed to be contemplated and addressed, some of which were also new to NVivo.

The first issue we had to address was that there was a general unawareness that NVivo worked differently on different operating systems. Since some of the GenUrb team members use Windows and others use macOS, it was not feasible to have everyone use the same operating system. We wondered if the project would fail at the first hurdle. We contemplated buying PCs for Mac users given the greater analytical power of NVivo on PCs; however, the cost of doing so was prohibitive, and there was a reluctance among Mac users to learn to use PCs.

The next hurdle was obtaining licences for the team members engaged in the collaborative use of NVivo. It was possible to buy either a MacOS or Windows academic, non-academic, or student licence. All those GenUrb members involved in coding or using the project database for analysis needed their own NVivo licence. Biftu conducted an inventory to determine how many new NVivo licences needed to be purchased; how many (existing) licenses needed to be upgraded to the latest version of NVivo (everyone needed to use the same version of NVivo even if the operating systems were different); and how many of these licences would be connected to the different operating systems. Biftu worked with the project manager,

Leeann Bennett, to coordinate the purchasing of nineteen NVivo licences, including eleven macOS academic (perpetual), one Windows academic (perpetual), four Windows student (annual), and three MacOS student (annual). The project spent close to US$8,000 to acquire these licenses, not including the cost of renewing annual subscriptions for student licences as deemed necessary.

While it is possible to convert NVivo projects between the two operating systems, the limitations are such that project items created by Windows-only software features are not visible in projects converted to the macOS version of NVivo. In addition, the conversion time between the two different file formats is significant and varies with the quantity of data involved, which was a particular concern given the size of the GenUrb NVivo database. With these cross-platform issues in mind, Biftu made two central versions of the NVivo project: one for Windows and one for macOS.

Since additional tools for managing collaborative work in NVivo are indexed to different price points, it was important to consider the overall budget allocation for the software package. One option for team-based research was the NVivo Collaboration Server. Although it is the costliest option, this solution is ideal for projects on the scale of GenUrb because it allows for synchronous collaboration. However, the server is only compatible with Windows. Another (cheaper) option was the NVivo Collaboration Cloud. The cloud is a virtual space that allows team members to collaborate on NVivo projects. However, it is not a space that provides synchronous collaboration but a space to pass work between team members. The cloud is thus a hybrid solution in terms of collaborative capabilities. The third option was to oversee collaboration through a manual and decentralised management of the NVivo database, for example, by creating, distributing, and merging database copies independently of cloud assistance. While this solution would have been the most affordable, it would be quite taxing to streamline communication, version control, and integration.

The option we chose in GenUrb, the NVivo Collaboration Cloud, still required Biftu to update the main NVivo database with copies from team members every day. Notwithstanding such a laborious task, the NVivo Collaboration Cloud solution was the most efficacious to resolve questions about how to streamline a collaborative workflow across two operating systems.

The project purchased three Cloud subscriptions since each allows only five users to collaborate. The total cost was US$1,500, not including the cost of renewing annual subscriptions. It was necessary to set up two collaboration cloud workspaces, one each for Windows and MacOS users, and to troubleshoot various technical glitches in the cloud system, which was new not only to GenUrb members but also to NVivo. All the MacOS users worked in one cloud and Windows users in another.

In sum: efforts were made to address and resolve questions about operating system compatibility, licences, expenses, and cross-platform applications. Decisions were taken with the aim of accomplishing the objective of enhancing the collaborative workflow process.

By Biftu Yousuf and Linda Peake

The following section explains why we continued to persevere with a software package even though we sometimes felt that it created more problems than it solved.

Section 3. Why use NVivo?

The strength of NVivo lies in the software's malleability in relation to different research designs. In this regard, NVivo is a methodologically agnostic tool that can accommodate data from a wide range of qualitative approaches to research, including, for example, interviews, geo-ethnography, ethnography, and life and oral histories (see Chapter 11, Section 5 Qualitative methods in feminist research). Kristi Jackson and Pat Bazeley (2019) encourage researchers to become familiar with the methodological options available within one's own discipline before making the decision to use NVivo.

A primary appeal of NVivo comes from the software's ability to support systematic and rigorous data exploration and analysis. NVivo enables researchers to index segments of data to specific themes in a code structure and to link research/field notes to coding. Researchers also use NVivo to identify thematic trends and cross-examine data, which can lead to the development of rich analytical insights. And the visualisation features allow researchers to create diagrams, charts, or maps to present and clarify discoveries.

It is important to stress that NVivo functions aid the process of data analysis and do not analyse data on behalf of the researcher. NVivo is also not suitable for all research, however, it can be invaluable for managing large and diverse datasets for which collaboration between team members is important (see GenUrb example 17.3). For example, if a research project entails ten or more interviews that are an hour long, alongside focus group discussions and results from a survey, then a tool like NVivo is indispensable. Other important considerations that can help researchers determine whether NVivo is the right tool will depend on the research goals and specifically the fit between research questions, methods, and design.

GenUrb example 17.3: Coordinating NVivo across transnational research teams

To appreciate the scale of the GenUrb project's use of NVivo and to underscore the levels of commitment and effort that were required from team members, we outline some of the parameters of our engagement with NVivo as follows:

- members of the GenUrb team working across twelve different time zones, from morning in North America to midnight in Asia;
- taking into account the political implications of the movement of data across geographical locations and political landscapes – mutable landscapes shaped by colonial and neocolonial relations – and the economies of extraction and exploitation they engender;
- conducting a total of 249 interviews and life histories, of which 164 were transcribed and translated into English (from Yoruba, Spanish, Quechua, Hindi, Shanghainese, and Mandarin – those in Creole and Arabic were not translated) before being imported into NVivo (those in Arabic were not imported into NVivo);

- developing a code structure that produced 149 codes, 31 of which were 'parent' codes and 118 'child' codes;
- creating 127 NVivo cases (one each for the 127 women participants) to account for the data of the five CRTs, which included fourteen attributes and over 100 values (attributes provide additional information about the data you will analyse e.g. age, while values refer to the specific range limits that are assigned to the attribute values, e.g. 18–29, 30–39, etc.);
- co-ordinating contributions from a total of sixteen coders at various stages of the coding process;
- and co-ordinating between two operating systems (Windows and macOS).

By Biftu Yousuf

The chapter concludes with a discussion of the different stages in which NVivo can be employed in the research process.

Section 4. When to use NVivo

Though NVivo is used most for coding data, it is a comprehensive tool that can assist researchers at various stages of the research process. Coding data refers to the process of systematically arranging collected information or observations into meaningful and unified categories. Beyond this, researchers can draw on NVivo for all or part of the research stages as shown in Figure 17.1 below, from early conception to literature review to in-depth data analysis to report writing and to dissemination of findings.

During the early stages of research, NVivo can be instrumental in creating notes to record your thoughts including preliminary ideas, preconceived notions about the topic,

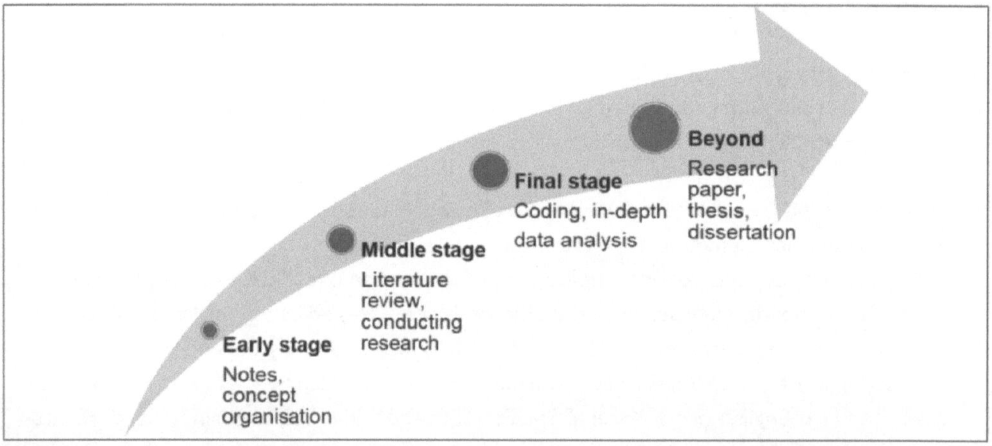

Figure 17.1 Using NVivo at different stages of qualitative research

Source: Biftu Yousuf

and key issues. These notes can help you recall important details and serve as a starting point for tracing your thoughts over the research life cycle. Notes can also be organised around specific concepts related to your research, allowing you to elaborate on their meanings, significance, or theoretical underpinnings.

In the middle stages of research, you can expand upon and integrate earlier notes as part of your literature review or analysis. NVivo is also compatible with the citation managers listed earlier (Table 17.1). By importing your literature into your NVivo project, you can perform an analysis of these materials by creating codes to represent key concepts or themes and coding relevant excerpts to them. This allows you to establish connections between the literature, concepts, and codes in your research. Building on this, you can use NVivo to gather, prepare, import, and organise data files. Once an NVivo project is set up in this way, researchers can access all their research material from their virtual database. This provides easy access to data and a means of carrying out complex searches and retrieval operations when needed.

Towards the final stage of the research, NVivo offers a wide range of features that facilitate coding and in-depth data analysis, including query-building, data visualisation, and the exploration of relationships between coded data (and literature). These analytical tools enable you to uncover patterns, associations, and themes within the data while supporting the interpretation and development of meaningful insights and an overarching storyline.

Beyond these three stages, you can leverage NVivo to facilitate the dissemination of your findings (e.g. research paper, thesis, and dissertation). Most of the project items in NVivo – for example, quotes and excerpts, query results, and visualisations – can be exported, such that you can then use them in presentations, posters, and publications. Researchers can export reports and extracts of information across selected criteria with the click of a few buttons. When used effectively, NVivo saves time during both the analysis and write-up phases of the research process.

In GenUrb example 17.4 we outline some of the concerns we had to address after having decided to use NVivo for transnational feminist research.

GenUrb example 17.4: Concerns that arose in using NVivo for transnational feminist research

Concerns	
Epistemological concerns	• Building the NVivo database and workflow process to facilitate the co-production of knowledge. • Specifying the relationship between the contextual specificities of the data in each CRT and the process of abstraction in developing a comparative framework. • Making decisions on whether there would be a deductive/concept driven approach to data analysis, an inductive/data driven approach, or hybrid.

Political concerns	• Encountering competing attitudes toward the use of digital technologies in feminist research (e.g. deep-seated disciplinary differences, differences in levels of trust of technology). • Negotiating and securing trust among CRT members in using NVivo. • Navigating the politics embedded in flows of data, ideas, and knowledge across national boundaries.
Practical concerns	• Developing and implementing procedures that were suitable for a long-term project of this scale. This was not straightforward because the teams were often at different stages, undertaking different tasks, and in different locations. • Questions around the amount of careful, attentive, and sustained work that would be required to keep the digital infrastructure both collaborative and iterative. • Overcoming the limitations of how NVivo currently functions as a digital technology.

Section 5. Summary

Prevailing attitudes support the use of software for quantitative studies but fears that it compromises the unstructured essence of qualitative research persist. Feminist scholars have challenged dualist notions regarding the role of technology in research producing critical scholarship at the juncture of software and qualitative methodologies. Echoing these feminist methodological interventions, this chapter has invited you to view CAQDAS as a tool that can embrace the messiness that often accompanies qualitative inquiry. We reclaim digital technologies as relevant to and appropriate for feminist research projects. The wide range of features offered, especially by NVivo, equips researchers with tools that support the research process from start to finish. Researchers can make standalone use of NVivo in Windows and macOS, or they can work across these platforms, as we have demonstrated through the GenUrb project. While the software is not suitable for all research, NVivo's malleability accommodates multiple disciplinary and methodological practices. After its initial setup, NVivo is timesaving, and in GenUrb, it played a crucial role in cultivating solidarity to advance our research goals.

18 Using NVivo in feminist research

Biftu Yousuf

Learning objectives

After completing this chapter, readers should understand

- how to conceptualise and create a design framework for project setup;
- how to work with a wide range of tools for qualitative data exploration in NVivo;
- and ways to operationalise NVivo for analysing feminist research.

Further, readers should be able to

- draw on feminist principles and practices to explore qualitative data with NVivo;
- and think creatively about how to use digital technologies for qualitative data analysis.

This chapter will help you better understand NVivo, from the nuts and bolts to advanced functionalities, providing a linear framework for how to use it to conduct qualitative data analysis. It starts by considering how to create a design framework before going on to the topics of research material preparation, coding, memos, and annotations and how to undertake data exploration. The chapter considers some of the complexities NVivo users often encounter and shows how GenUrb utilised an inclusive methodology that aligned with a feminist ethics of care. As you begin data analysis using NVivo, you may find it helpful to consult web resources that provide instructions (see links to NVivo on the book's website www.routledge.com/9781032668680). NVivo (Release 1) for Windows and macOS is the version referenced in this chapter.

Section 1. How to use NVivo

Using **NVivo** begins with creating a design framework: a useful tool for planning how to set up a project in NVivo that defines the data-organisation approach of your project. Given that each research project is unique, there is not one set of 'best practices' for how to approach the organisation and analysis of data. Different practices are required for handling each research project, but there are two useful strategies that can help you get started.

- You should develop a data management plan (see Chapter 10, Section 2 (ii) Devising your own data management plan) before you begin analysis in NVivo.
- You should sketch a design framework for how you will set up your database in NVivo.

DOI: 10.4324/9781032668727-24

To draw up a design framework, as the NVivo specialist on the GenUrb project, I needed to understand the rationale of the project: the project objectives, research questions, and methods. This started with a review of the project grant material and conversations with the principal investigator, the project manager, the project's postdoctoral fellows, the City Research Teams (CRTs), and the Comparative Team – all necessary starting points for securing this information. GenUrb example 18.1 shows the design framework for the GenUrb data from the interview and life histories with the women participants.

GenUrb example 18.1: Design framework for GenUrb interviews and life histories with women participants

Data type	*Units*	*Attributes*	*Coding framework*
In-depth interviews Life histories	Individuals	(A) Personal background • Place • Age • Ethnicity • Language/dialect • Number of languages/ dialects • Religion • Social class • Education (B) Household/Family • Relationship status • Household size • Family type • Household head • Number of children in the household	Hybrid approach combining deductive and inductive coding: • deductive (*a priori* template of codes) – common codebook for coding; • inductive (data-driven approach) – codes emerging directly from coding process.

By Biftu Yousuf

Reflection exercise 18.1 encourages you to think about key components of qualitative research design and how they are represented in NVivo by considering how to develop a design framework for your own research.

Reflection exercise 18.1: How to develop a design framework for your NVivo project

When starting a new NVivo project, it is important to ensure that the best possible use is made of your research data. One way to achieve this is by making sure the design of your research is reflected in how the project is set up in NVivo. This requires you to list the following aspects of the research.

- *Data types.* What types of data are included in your research (e.g. interviews, surveys, news articles)?
- *Units.* What 'units' of analysis, if any, will your research compare (e.g. individual participants, different communities)?
- *Attributes.* What are the attributes that correspond to the 'units' you are comparing (e.g. demographic characteristics for individual respondents, such as location, age, and gender)?
- *Temporality.* What is the time frame of your research (e.g. longitudinal, comparing data from different time periods)?
- *Coding framework.* What is your strategy for organising and categorising data (e.g. inductive, deductive, or hybrid)?

Review your research proposal to find the answers to these questions and sketch your design framework for NVivo. Look at the comparative analysis design framework developed by Biftu Yousuf in GenUrb example 18.1. For both the in-depth interviews and life histories it lists the attributes of the research participants in terms of personal background (place, age, ethnicity, language/dialect, number of languages/dialects, religion, social class, and education) and household/family characteristics (relationship status, household size, family type, household head, and number of children in the household). It also records how GenUrb employed a combination of inductive and deductive strategies for categorising data.

After developing the design framework, you can confidently set up your research project in NVivo. In Figure 18.1, you can see how the components of the GenUrb design framework are represented in NVivo.

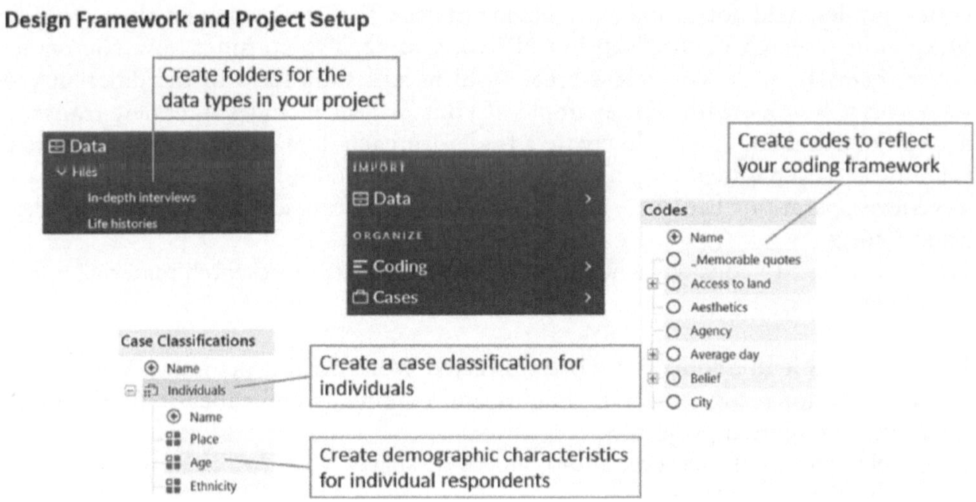

Figure 18.1 Components of the GenUrb design framework as represented in NVivo

Source: Biftu Yousuf

In the following section we turn to how you prepare data to input into NVivo.

Section 2. Putting data into NVivo

Gathering, preparing, importing, and organising research material in NVivo requires you to ask: what type of data do you want to analyse? While you can include a broad range of data in your NVivo database, the data need to be presented in formats that are compatible with the software, which includes textual data, audio/visual data, demographic data, and social media content. After you have collected the data and completed an inventory of those you want to analyse in NVivo, the next step is to prepare the data for import.

Data are never directly ready for import, as they arrive unmediated from the field. Data undergo a curation process led by the researcher to prepare them for importing into the software, and this preparation can include transcription (see Chapter 16, Section 3 Stages of qualitative data analysis) and translation (see Chapter 15, Section 3 Planning for translation and interpreting throughout the research process). In addition, the process may involve editing material, formatting content, verifying the quality and integrity of the data, and taking notes or recording memos. Once you have curated the data to your satisfaction, you can import the research material into NVivo. NVivo also provides several features that can support the curation process within the software, such as:

- creating synchronised transcripts, whereby transcripts are synced with the associated audio or video file, leading to the creation of new documents;
- creating a new demographic dataset;
- creating new memos;
- and editing select content.

Bringing data files into NVivo requires using the options on the Import tab. You can also consider importing all other research materials (e.g. research questions, consent forms, interview guides, field notes, and a project log of activities), to keep them all in one place.

Once your research materials are in NVivo, you need to consider how you want to organise them. To start, you might create folders and subfolders of the different types of data you will work with. For example, if your project includes interview transcripts and social media data, you could create a folder for each data type and then allocate the relevant files to each folder. Organising your data files into folders will streamline search-and-retrieve operations because you have the option to restrict your searches to files in selected folders.

Issues you need to consider when preparing and importing research materials include the following.

- It is not possible to import password-protected Word documents into NVivo.
- NVivo does not recognise the highlights and comments embedded in a Word document imported into a project.
- If you plan to modify picture, audio, video, or survey files, you need to do so before importing them into NVivo.

The following section turns to what happens once your data have been imported into NVivo and are ready to code.

Section 3. Approaches to coding and codes

Coding is a systematic way of condensing data into smaller, analysable units through the creation of categories, themes, and concepts, either derived from the data (inductive coding), applied to the data (deductive coding), or a combination of the two. In other words, codes can be developed inductively (derived from practical knowledge of what the data tell you) and deductively (what the literature tells us is important) or a combination of knowing what we want to ask about but also not knowing all the ways in which the data will unfold.

Coding brings into one place all references to specific topics, concepts, ideas, opinions, emotions, or experiences recorded in the data. It facilitates the organisation, retrieval, and interpretation of data, and it leads you to conclusions at the end of this process. The way you handle coding will depend on the amount and type of data you collected, your analytical framework, and the project time frame.

Coding must have a purpose, so as you reflect on a passage of text, consider the three common approaches to coding.

- *Topical, thematic, or 'broad-brush' coding.* What topics or themes are present in the data? For example, topics such as finances, family, housing, and violence.
- *Detailed or analytical coding.* What do the data discuss? Why is the content interesting or significant? Does it relate to my research question(s), and how? Do the data lead me to question other researchers' published findings? Consider what these questions mean in context and how they can lead to your identifying new analytical categories relating to the data, such as social reproduction, production, and place-making.
- *Descriptive or 'case' coding.* Who is the speaker? What place, organisation, place or other entity is under observation? For example, categories that describe the data, such as gender, age, household type, and geographical location.

These three approaches to coding relate to the different techniques used to assign codes to segments of data within NVivo and can be used jointly. These approaches map on to code structures, which refer to the organisation and arrangements of codes. There are three major types of code structure that can be used:

- *standalone codes* – codes that have no clear logical connection with other codes and that do not readily fit into a hierarchical structure;
- *tree codes* – codes that fit into a hierarchical structure from a general code at the top (the 'parent' code) to more specific codes at the bottom ('child' codes);
- *cases* – codes used to gather material about people, organisations, or places, with attributes such as gender, age, or education, which can be organised into hierarchies.

Once you have determined the technique you will use to apply codes and their structural arrangements, you can employ various strategies to code in NVivo.

- You can create a code structure first and code accordingly (deductive coding).
- Similarly, you can start with broad-brush coding to organise material into broad topic areas and then explore the codes for each topic to undertake more refined (parent-to-child) coding.

- You can create codes as you read through the data files line by line and code at the same time based on what you read (inductive coding). You can then combine and group codes into related categories for a more structured coding.
- You can code until you reach 'saturation', stopping when you cease to find new themes or ideas.
- You can code whole data files to codes or cases.
- You can consider coding to only three or fewer levels of a code structure (parent code, child code, grandchild code) to avoid decontextualisation.

The website links for NVivo for Windows and Mac (see book website) provide more tips on coding strategies.

Once you have decided how to code the data, using these strategies, you can start the coding process from within NVivo. Codes in NVivo can be understood as virtual containers, each with a specific label – buckets, containers, or bins – that captures the essence of the data you assign to them. These labelled containers (i.e. codes) act as repositories that house organised and categorised data segments.

If you have files in PDF format, there are some specific issues you need to consider when coding. While it is possible to select and code text or images in a PDF file, you must first convert it into a text-searchable PDF with optical character recognition (OCR) before importing it into NVivo. Scanned PDFs that contain images and text are not text-searchable. You can, however, circumvent this issue by selecting and coding parts of a page using the 'regions' function in NVivo. When you opt to code a 'region' of a scanned page, you are coding an image of the text rather than the actual text itself. You cannot, however, employ lexical queries, searching for matches of words or word variants, on this selection.

Reflection exercise 18.2 asks you to brainstorm how you will approach coding and begin to develop a code structure for your project.

Reflection exercise 18.2: Coding an interview

Even if you have not collected any data (such as interviews or field notes), you can begin to outline approaches to coding.

- Begin by making a list of the broad themes that you anticipate will be present in the data collected for your project. If you are unsure about where to start, read scholarly articles on the topic you are researching to see what themes are addressed in these publications. If you have written a project proposal, you will already have covered common themes in your literature review.
- Then, make a list of the concepts that you anticipate will be applicable to the data that you collect. The scholarly literature you have read will be very helpful here. For example, if your project is about investment properties, then gentrification and 'ghost cities' are concepts that your research might address. If your project is about eviction, then dispossession may be a relevant concept.
- Finally, make a list of attributes that you anticipate will be helpful for organising the data you collect into categories, such as demographics (gender, age, education, etc.), or the names of neighbourhoods, organisations, or businesses.

From these three lists, begin to outline your first draft of a coding structure.

- Are there codes that fit into a hierarchical structure (with a general parent code and more specific child codes)?
- Are there codes that apply to cases about people, organisations, or places? Can these codes also be organised hierarchically?
- Are there standalone codes that have no logical connection with other codes?

The brainstorming you do for this reflection exercise will no doubt change as you start data collection and begin data analysis with coding, but it is helpful to have a first draft before you begin. Sharing your anticipated approach to coding is important in project proposals and in grant applications.

In GenUrb example 18.2, we outline the coding process used in GenUrb, providing an overview of the decisions and steps that were taken to enable the coding of data from the in-depth interviews and life histories.

GenUrb example 18.2: The coding process

To devise an NVivo process that would be grounded in the GenUrb project structure, there were several factors to consider when coding the interviews and life histories of the women participating in the project. For instance, would we use one common codebook, or would each City Research Team (CRT) create a codebook, each of which would be amalgamated into one comprehensive common codebook? The same question had to be considered in relation to a demographic dataset of the participants. As it concerned coding, we also needed to ask: Would each CRT code their data independently? Or would coding happen collaboratively from the outset? Coming to a decision on each of these questions was not straightforward, especially given that the CRTs arrived at the stage of processing their data at different times. Further, although many of the researchers had prior experience with coding qualitative data, none had experience of working in a comparative project, and not all had experience of working in NVivo.

Next, we outline the steps we followed to allow the inputting of data for coding into NVivo.

Step 1: Building the basis for the coding structure

Given the different times at which CRTs finished interviewing, we were not all ready to start coding at the same time. It was agreed that the common coding structure would be started by one CRT (Georgetown) with extensive experience of qualitative research. The principal investigator, Linda Peake, had worked for a number of decades on various research projects with the organisation Red Thread, from which members of the Georgetown CRT were drawn. The team was well qualified to start thinking through the data. The work started with drafting a table of thematic codes, which captured the narrative details and complexity of themes. We came to refer to this listing of thematic codes as the common codebook. Given that this coding process started in 2020, before our NVivo specialist was in place, the data were coded manually.

This process started with a three-week-long exercise in which six members of the Georgetown CRT (Susan Collymore, Karen de Souza, Joy Marcus, Linda Peake, Vanessa Ross, and Wintress White) sat and read through three of the in-depth interviews and life histories. We printed out each interview, with a wide margin along the right-hand side of each page. We talked through the issues and concerns that were emerging and selected three thematic codes that were issues in the lives of most of the women interviewed: housing, finances, and violence. To start the (parent) coding for housing, we each read through the same interview and the same life history, page by page, and marked in coloured pen every sentence in which housing was mentioned. Sometimes, a whole page was coloured in, sometimes, a paragraph, sometimes, half a sentence. At the end of this exercise, we looked at our results to see how similarly we had coded. Once we were convinced that we had captured each reference to housing, we read through the highlighted text to see the different ways in which the text referred to housing and began to develop child codes. In relation to housing, the data spoke to issues with tenure (ownership, renting, and squatting), the number of people living in the house, the condition of the house, the environment of the house, as well as the interviewee's housing history. We then discussed the categories within each child code. Again, we each read through each page of the same interview and the same life history and coded each section of coloured text in relation to the child codes for housing, writing in the margin of the text letters and numbers referring to the parent code (H for housing), child code (H1, H2, H3, and so on), as well as further numbers for each subcode category within the child code (H1i, H1ii, H1iii, and so on).

By the end of the first week, we had coded one interview and one life history for parent and child codes for housing, finances, and violence. At the end of this first week, we also reduced our team from six to four. Among the four of us remaining, we had coded parent and child codes in the same way over 95 percent of the time, whereas the other two women had only matched codes approximately 70 percent of the time. Our practice had been to read only one page, code the text, and then stop to compare codes, discussing why we had chosen certain text to code. This way, we were able to maintain quality control, keeping track of how close our codes were and whether text merited inclusion in the code.

By the end of the second week, as our familiarity with the process allowed us to work faster, we had worked through two more interviews and also identified 21 more parent codes (listed alphabetically): appearance, average day, beliefs, education, environment, family, the future, Georgetown, government, happiest and saddest times, health, hunger, identity, land, leisure, memories, mobility/transportation, religion, Sophia (the neighbourhood), technology, and work. During this week we also continued to add child codes and subcodes and to amend existing codes. By the end of the third week, we had finished the coding of the three interviews and three life histories. At this point, we had had four of us working on each transcript. After three intensive weeks, we now felt confident that we could reduce this number to two and still have the same high level of confidence in the consistency of the coding.

Step 2: Developing a common codebook

Over the course of the coding period, there was a consultation process starting with four virtual workshops, conducted over the course of two months (July–August 2021), followed by monthly meetings over a period of six months. Each of the four workshops was approximately two to three hours long and involved the participation of sixteen people across twelve time zones, from morning in North America to midnight in East Asia. The focus of the workshops was to discuss the thematic parent and child codes in the Georgetown common codebook. This was no small undertaking, as we had over twenty parent codes devised by the Georgetown CRT to work through, with each CRT adding new parent and child codes from their own coding processes, as well as amending the existing codes. It was a slow, painstaking process (the first workshop was spent discussing just one parent code), with numerous emails exchanged between each session.

Discussion revolved around how to transform the existing Georgetown-specific codebook into a comparative tool that did not merely generalise from the particular but that would also enable the specificities of each city to emerge and shape the analysis in an inclusive manner. The workshop facilitators (Biftu Yousuf and GenUrb postdoctoral fellow Elsa Koleth) tried to encourage the discussion to move beyond each specific research site to incorporate each city. This involved making decisions about (a) what it was relevant and necessary to draw from the data for the purpose of comparison, and (b) whether all the codes were legible across the city contexts. Members of each CRT were able to discuss the specificities of their data in a relational context, highlighting differences, convergences, and gaps. The outcome was an iteratively revised coding structure that continued to be revised after the workshops in response to ongoing communications from CRTs.

Creating a common codebook from the Georgetown codebook involved adding seven new parent codes (to give a total of 31) and adding to the child codes (to give a total of 118) while also rearranging and renaming codes so that the result would make sense in a comparative context. This process also necessitated the Georgetown CRT returning to and recoding their data once the common codebook was finalised.

At the conclusion of the four consultation workshops, the monthly meetings discussed and reviewed progress with coding through summary reports prepared by Biftu. Participants shared ideas on emergent coding patterns and resolved outstanding technical and non-technical issues.

In their coding of their data, each CRT adopted similar quality-assurance practices to the Georgetown CRT to ensure that when there was more than one person coding data, there was consistency across interviews and across codes in the coding process. This systematic approach to the coding process is why we have a high degree of confidence in a set of coded data created by five different research teams.

Step 3: Merging the data

Biftu oversaw the merging of data from the five CRTs for both the Windows and macOS systems. To adapt NVivo for a feminist project of this scope and size, Biftu raised 26 issues with QSR, which resulted in many email exchanges and troubleshooting sessions over the phone and via video platforms between May 2021 and April 2023. The data were merged in an iterative process, which began with uploading the Windows and macOS central databases (the transcripts from the interviews and life histories) into the NVivo Collaboration Cloud for team members to download and work on. Once team members completed their coding activity on any given day, they uploaded their copy back into the cloud for it to be merged back into the respective central databases. This was a recursive and laborious process repeated for over a year (from 2021 to 2022) until coding was finalised.

For this part of the coding process to happen seamlessly, it was crucial that the merging be conducted daily. This ensured that at the start of each subsequent day, the most up-to-date version of the central databases – which reflected all the coding activity to date – was available for team members to resume coding. The twelve time zones across which the CRTs were based meant that Biftu, for over a year, had to merge the individual coding activity into the central databases between 4:00 a.m. and 5:00 a.m. Toronto time daily, including weekends.

Biftu developed a comprehensive NVivo workflow instruction guide for both Windows and macOS users and routinely updated these documents to reflect new developments. Team members followed these guides carefully and thereby contributed to successful workflow. We also extensively documented everything we were doing, including keeping an active log of our meetings, which yielded over 70 pages of written text as a record of our deliberations.

By Linda Peake and Biftu Yousuf

The following section turns to a parallel task you need to consider as you are conducting your coding: that of writing memos and making annotations.

Section 4. Memos and annotations

Both memos and annotations offer users a place to record brief or comprehensive project-related comments, but they serve different purposes, as outlined in Table 18.1.

What are memos in NVivo? Memos are sites where you can store your insights into, interpretations of, and ideas and analytical thoughts about the data. They provide a space for you to keep your insights separate from but linked to the data, helping you to increase the transparency and reliability of your research findings. Memos can evolve into the story you tell about your research. There are different ways in which you can use them:

- to keep track of what you are doing, finding, learning, or (not) understanding or of what you are feeling, thinking, and reflecting on as you are reading and reviewing your documents;
- to write reflective notes to record your discoveries and ideas directly in NVivo, i.e. using memos to 'talk to yourself' as you seek to make sense of the data.

Table 18.1 Comparing memos and annotations in NVivo

Memos	*Annotations*
Standalone documents for text, tables, or images.	Textual notes embedded within data files that can be linked to select content.
Most useful for capturing reflective thoughts about a topic.	Most useful for making notes about a specific phrase or marking content for follow-up.
Can be coded.	Cannot be coded.
Search, find, and retrieve operations can be performed on memos.	Some annotations may be searched, depending on the text they contain.

Memos are useful since they can be exported as a Word document, and you can also create links between memos and various items in NVivo. There is more information on creating memos on the NVivo websites for Windows and Mac.

Annotations allow you to record comments, reminders, and observations about specific content in data files (Word documents, PDFs, pictures, audio and video files, and datasets). Unlike memos, annotations are embedded in the original data files. Annotations are visible in the repository of coded content, referred to as 'coding references' in NVivo. You can use annotations to interact with the data. There are illustrations of annotations on the NVivo websites for Windows and Mac.

In GenUrb example 18.3, I outline how memos and annotations were used alongside the coding process in the GenUrb project.

GenUrb example 18.3: Using memos and annotations

Within the NVivo program, we used both memos and annotations, although doing so required us to overcome NVivo's technical barriers in relation to sharing a project with both PC and Mac users.

Memos, which can be virtual or associated to specific codes in NVivo, functioned as a shared site for CRTs to make notes on emergent patterns and themes in their data, as well as to highlight any amendments and additions they wanted to make to the coding structure. The ability to link memos to codes was a key tool for the comparative analysis because (a) it gave teams a way to communicate their reflections on place-specific issues to each other and to other members of GenUrb; and (b) it ensured that the coding structure reflected what was emerging from the data in an inductive manner.

Annotations were used by the CRT members as they coded their data largely to explain and contextualise terms in their transcripts (such as terms specific to the language they were working in or specific cultural phenomena).

Using both functions for comparative purposes, allowing incoming insights to be merged into the central GenUrb database, required overcoming limitations in the NVivo software's functionalities related to the incommensurability of Windows and macOS. NVivo's functionalities at the time prevented the seamless merging of memos and annotations within and across platforms. When I came up against the limitations of the software, I had to invent new solutions using NVivo in tandem with other tools available to us.

For example, with the aim of keeping memos and annotations as collaborative tools, I had to find a way to overcome the merging limitation. After testing several possible solutions, we opted to move memos outside NVivo. To ensure that memos remained a joint resource that honoured the comparative intent of the project, memos were written and stored on OneDrive, with access granted to all team members.

It was not possible to implement the same solution with annotations, given that they are not standalone comments. A practical solution, which was less onerous for CRTs and allowed them to continue their work smoothly, was to collate the annotations outside NVivo for the Mac users because it is not possible to merge annotations created in projects on this platform. This was not an issue with NVivo Windows because annotations are automatically imported when these project copies are merged. Any time annotations were made as part of a team member's coding activity on the Mac platform, they would alert me by leaving a comment in the 'notes' section of the NVivo Collaboration Cloud. I would then manually copy and store the annotations and associated metadata, such as the name of the file, annotator, and version number, on the shared OneDrive documentation we developed for annotations.

Both memos and annotations were added back into the central NVivo GenUrb database once all the coding was complete. This was an arduous process for annotations because they are static in the sense that they are directly tied to specific locations within the data and had to be re-embedded accurately in the associated data files. Memos, however, already comprised individual documents that could be quickly linked to data files by way of association to parent codes.

By Biftu Yousuf

The following section addresses how to explore data in NVivo.

Section 5. Advancing to data exploration

In addition to thematic codes, GenUrb also classified data according to 'cases' (where each 'case' represented a woman interviewed in the research). This section outlines how to work with cases to classify descriptive data and use NVivo tools to explore coded data.

(i) Working with cases and attributes to classify and store descriptive data

What are cases? Cases are a way of recording descriptive information to support the comparisons you want to make. In NVivo, cases are codes that represent the 'units of analysis' in your project – for example, the people you interviewed or the communities you studied. Your project might have cases for people, organisations, places, or other entities that you want to analyse and compare. Just like thematic codes, you need to code data files into cases to be able to query the data. There are instructions on how to create and work with cases and the necessary steps to set the attribute values for cases on the NVivo Windows and Mac websites (see the book's website www.routledge.com/9781032668680).

Cases are like containers that hold all the information related to the units of analysis in your project. When you open a case, you can see all the information that has been coded there. For example, you could review everything that a particular participant said or everything about a particular community.

In NVivo cases are classified in order to:

- group multiple data files that relate to the same entity – for example, you can code an in-depth interview transcript and life history for the same person (case);
- and to re-organise a data file that houses data for multiple participants – for example, for a mother and daughter who both participated in a life history, you can code an in-depth interview transcript to each of the two speakers, either by coding an entire transcript to each of the two participants' cases or by coding their individual responses to their specific case.

Cases have attributes that refer to the characteristics of these entities. Case attributes can include, for example, demographic data such as age and gender or geographical location. You can use cases and their associated attributes to search for differences and similarities in the data based on a wide range of demographic characteristics and beyond. You can also analyse and compare cases based on these attributes to answer meaningful questions (such as land ownership or homeownership among cases in one geographical site or across geographical sites). Cases allow you, then, to:

- query demographic data – for example, you can use crosstabulations to compare participants' opinions on an issue according to their age or level of education;
- and visualise findings – for example, you can create a chart to represent the composition of the group of participants interviewed at a particular site in terms of their age or number of children.

In GenUrb example 18.4, we outline how we developed case attributes for the project.

GenUrb example 18.4: Developing case attributes for GenUrb data

To keep a record of all interviewees in the GenUrb project, I drafted a demographic table of participant characteristics, which would become a 'case classification sheet' in NVivo. In one of our consultation workshops, we discussed the common characteristics shared by participants that could be placed into meta groupings. This resulted in two groupings:

- personal characteristics, with variables for place, age, ethnicity, language and dialect, number of languages and dialects, religion, social class, and education;
- household and family characteristics, with variables for relationship status, household size, family type, household head, and number of children in the household.

This process involved abstracting from specific to general characteristics, but to abstract the particularities of participant data into broader generalisation was not a straightforward task and required much discussion. For example, understandings of what constituted a family differed across the CRTs.

The goal of this workshop was to create a collaborative space for teams to deliberate on the attributes and attribute values that were abstracted from the participant characteristics. The second goal was to ensure that all team members arrived at a shared understanding of the metagroupings and their definitions. Consulting the CRTs helped to ensure that the categories for the NVivo classification sheet were jointly developed to be sufficiently broad to allow comparison but also sufficiently specific to capture the nuances of each research site. Surprisingly, given the amount of discussion that coding sessions evoked, we were able to accomplish the task of finalising a demographic table for the purpose of transposing it into NVivo in just one meeting.

By Biftu Yousuf

(ii) Using NVivo tools to explore data

In addition to making memos and annotations, NVivo is equipped with various tools that can help you probe and explore data. These tools vary across NVivo versions and by operating systems. The NVivo websites for Windows and Mac provide detailed information on how to query and visualise your research.

Visualisations are ways to explore and present the data using charts, diagrams, maps, and other visualisation techniques. They can help you to:

- explore trends, find patterns, and make sense of what is happening in the data files;
- develop in-depth thinking through the process of building knowledge about your research topic;
- and illustrate your research by exporting image files to include in your report/presentation.

Queries help you to explore data using text searches, word frequencies, crosstabulations, and various forms of coding queries. They help you investigate hunches as you progress through your research. You can:

- find and analyse words or phrases in your files and codes;
- ask questions and find patterns based on your coding and check for coding consistency among team members;
- and create either quick and simple queries to get a sense of what is happening in the data or build detailed queries for a more focused analysis.

Notes (which include memos and annotations) can be used to manage ideas, record discoveries, and organise conceptual and theoretical knowledge. Notes are generated during the research process, and they can be linked to data in NVivo.

In GenUrb example 18.5, Araby Smyth, GenUrb postdoctoral fellow, outlines the approach she took to exploring data in NVivo.

GenUrb example 18.5: Exploring place ecologies of economic precarity using NVivo

One aspect of my work with the GenUrb project is exploring and analysing the interview data collected by the CRTs to develop a sense of the topographies and topologies of debt and finance in the lives of the women interviewed. Topographies are fixed locations where people encounter and incur debts, such as banks, credit unions, pawnshops, and retail stores. Chris Harker (2020, p. 4) defines debt topologies as 'a particular type of spatial relations, which tie indebted residents to people and institutions that have lent them money'. For example, some of the participants in the GenUrb project have debt with their employers, while others borrow from microcredit institutions. Each debt is unique in how it operates within and beyond neighbourhood, city, and national boundaries and how it impacts the lives of the people who manage it.

I began working on the GenUrb project in 2021 – about five years after it began. I was not, therefore, involved in creating the interview guide, conducting interviews, or, for the most part, data coding. The first thing I did to familiarise myself with the data was to read through the interview guide and codebook. At the same time, I assisted the Shanghai CRT with coding their data. Then, I read through the interview text coded 'Debt' and 'Finances'. From this I began to get a sense of just how diverse the women's economic lives were.

I then adopted a more systematic approach to my exploration of the data. For example, I used word frequency queries, looked at intersecting codes, and searched for text in the transcripts in relation to coding. Word frequency queries were helpful for learning about the diverse ways that participants experienced debt and finance. The frequency of words such as 'donate', 'gift', 'give', 'help', 'share', and 'support' suggested that women are doing much more than borrowing and lending money through formal financial institutions and that economic exchanges were weighted with social meaning. Looking specifically at the intersection of the codes for debt (or finance) and health was also helpful for getting to specific instances when women described going into debt to pay for healthcare and the mental and physical stresses they experienced as a result of being in debt. Searching for specific terms such as 'bank', 'credit', and 'pawn' was useful for quickly assessing the names and locations of financial institutions and services that the women utilised.

Through these queries, alongside meetings with all the CRTs, I made an Excel spreadsheet that itemised the debt topographies and topologies of each city. It includes the financial institutions women use (banks, microcredit, cooperatives, pawnshops, and money-wiring outlets); the community economy networks they engage in (including informal savings schemes, mutual aid, favours, in-kind exchange, and gifting); and their assets, regular expenses, and sources of income. There are details about why women need to borrow money, where and from whom they borrow, and where and how they manage debts. Following Harker's (2020) theorisation of debt topologies and topographies, we have a clearer picture of how the women's economic practices are both relational and fixed in urban space.

By Araby Smyth

It is important to note that in the GenUrb project engaging NVivo to explore the data was an iterative process that involved constant negotiation and consultation. NVivo was used as a tool to help manage, organise, and search through the large quantity of data we had and to put into operation the decisions taken by the GenUrb members about which aspects of the data we wanted to explore. This process was conducted alongside the constant acquisition of contextual information from CRT members in numerous analysis meetings and via email communication.

Having outlined the process of how NVivo can be used, the following section briefly outlines how to avoid some of the pitfalls of using NVivo.

Section 6. Complexities and practicalities of using NVivo

Finally, computer-assisted qualitative data analysis software is not without its limitations. Digital technology can be a powerful tool for collaborative research, given its ability to assist the organisation, exploration, and preliminary analysis of qualitative data. NVivo users are nonetheless cautioned to avoid some common pitfalls.

- *Not having a plan for NVivo*. You need to analyse the requirements of your research project based on its purpose, question(s), and method(s) as the starting point for analysing data.
- *Counting on NVivo to solve all research problems*. Think first. Software comes second!
- *Creating a coding framework without a clear strategy*. Identify your coding method and rationale before you start the coding process. Which approach will you take to coding and why? Will you implement broad-brush coding, line by line coding, or focused coding? Or will you combine coding approaches?
- *Over-coding*. Have your research question(s) nearby when coding as a reminder of what questions your study seeks to answer. Clearly define the coding levels required for your project: Jackson and Bazeley (2019) recommend no more than three (parent, child, and grandchild).
- *Data fragmentation resulting from decontextualisation*. Add codes to each segment of text with enough surrounding context so that you can derive meaning from individual passages when working with content in the coding references. For example, you can code the sentence before and after the specific passage selected for coding.
- *A lack of understanding of the role played by cases and classifications*. Familiarise yourself with cases and classifications, and explore the different ways you might effectively organise, code, and analyse data by leveraging these tools.
- *Not connecting the dots by bringing it all together in NVivo*. It helps to write memos to facilitate analysis and the development of theory. Use these memos to link together sections of text and codes when you are building themes across the data.
- *Reporting too many findings*. It is best to identify key themes to report and keep the storyline succinct and memorable.
- *Isolating modes of analysis*. It is possible to use digital and manual approaches in a complementary manner, selecting the aspects of each mode that best fit your purpose.
- *Overreliance on NVivo to store data securely*. Remember to save and back up your work regularly and methodically, as software is not infallible and can crash.

Despite these potential pitfalls, having a clear data management plan, a design framework for project setup in NVivo, and specified technical and exploratory strategies can help users recognise and circumvent potential errors.

The final section reflects on the GenUrb experience of employing NVivo in a feminist research project.

Section 7. Reflections on using NVivo for feminist research

Feminist analysis needs to account for how power circulates through all stages of research, including the contexts in which data are managed and processed. This standpoint prompted me to encourage reflection on whether using a codebook that originated from one CRT as the basis for comparative analysis would be compatible with a feminist commitment to minimise power disparities in team-based research. The consultation workshops served to ensure that other teams' voices were heard and that their work was incorporated into a common codebook for the project. As the NVivo specialist, I followed a feminist ethic of care in designing an NVivo database and workflow in a way that would facilitate the co-creation of knowledge at all stages. Using a single codebook that emerged out of one research site as the basis for building a comparative coding framework required and received further thought for these reasons.

The approaches taken to develop a workflow were driven by the feminist ethic to create spaces of inclusivity, which was made possible by centring 'team-based collaboration' as the guiding philosophical principle. One of the main ways we enacted this principle was through the collaborative work practices that we developed. These involved lengthy and detailed deliberation on issues and challenges, problem-solving, testing ideas, and constantly revising the database. It was through the process of documentation that critical reflexivity was exercised about how technology could be used in a way that served both feminist and comparative purposes.

Steps were taken that developed into collective strategies for creating a comparative coding framework, building the NVivo database, and implementing the coding process. The strategies were inherently experimental: there were no precedents to follow for how to build an NVivo database for a feminist project of this size and nature. We had to adapt the technology significantly – often pushing at its limits – to fit our purposes and ensure that the process was responsive to the specific and emergent needs of the project.

We engaged several methodological principles to redistribute power and to enable a more counter-hegemonic approach to feminist knowledge creation. One obvious practice was to build a comparative framework that was grounded and open to revision at all stages of data analysis. These principles enabled an evolving comparative coding framework that drew out contextual specificities by making visible the nuances of geographically distinct case studies.

Our commitment to an inclusive methodological design ensured a 'relational accountability' (Wilson, 2008, pp. 97–125) – a commitment to fulfil relational obligations in research that foregrounds communication and interaction among team members. The core principles we embodied through this process included trust, respect, cooperation, and humility. The integrity of our iterative and reflexive comparative framework for the co-creation of feminist knowledge also required us to uphold practices of transparency and verifiability. We were able to accomplish these through extensive documentation of in-depth consultations, deliberations, and decision-sharing, as well as writing NVivo summary reports and developing a robust digital archive to house these records for the project (see GenUrb example 18.2 The coding process). These practices enabled us to create a repository for 'situated knowledge' that will live beyond the life of the project.

All these activities involved an ongoing process of discussion, critical reflection, problem-solving, testing solutions, and re-purposing the technology and workflow in ways that aligned with the feminist and comparative goals of the project.

Section 8. Summary

This chapter has provided a practical guide to engaging with NVivo, reinforcing the view that software supports a dynamic and embodied process of qualitative data analysis. Researchers have the autonomy to build NVivo databases, beginning with the layout and setup of the project and continuing with the organisation, coding, exploration, and analysis of data. The linear framework provided is a template that users can draw on to help them through these procedures. The chapter also illustrated how a feminist ethics of care guided how GenUrb engaged NVivo. This commitment inspired an inclusive design for analysis that aligned with the feminist goals of the project. The number of in-depth deliberations, collaborative discussions, and decision-sharing practices involved in taking analysis forward show how our NVivo process was able to flatten power hierarchies and enable a counterhegemonic approach to feminist knowledge creation. Even though NVivo revealed many sources of incommensurability in the research, including political landscapes, competing attitudes toward technology, and software limitations, it played a crucial role in cultivating solidarity to advance the project's goals.

Part VI

Feminist approaches to knowledge mobilisation

19 Knowledge mobilisation in a feminist project

Araby Smyth, Linda Peake, and Jenna Blower

Learning objectives
After completing this chapter, readers should understand

- the professional purpose of knowledge mobilisation;
- the need for a knowledge-mobilisation plan;
- and the role of knowledge mobilisation in feminist research projects.

Further, readers should be able to

- identify different types of knowledge-mobilisation activities in a feminist research project;
- engage critically with institutional knowledge-mobilisation processes;
- and identify strategies to monitor and evaluate knowledge mobilisation in their own work.

This chapter explains the importance in contemporary feminist research of knowledge mobilisation, the co-creation of knowledge, and making an impact within and beyond academia. It expounds on the role of knowledge mobilisation in different stages of the research process, including prior to, during, and afterwards, drawing on insights developed in GenUrb, which has engaged in processes of mutual learning between feminist academics and grassroots women in diverse, global urban contexts and is committed to disseminating academic knowledge. Approaches to dissemination are engaged, including storytelling, which is accessible to advocacy and practitioner communities and to the general public and useful for urban policy formulation. The chapter is rooted in a feminist understanding that all knowledge emerges from uneven power dynamics and thus provides the reader with strategies across all phases of a research project for how to break down the barriers that exist between researchers and participants, communities, universities, and project partners.

Section 1. Knowledge mobilisation

The purpose of **knowledge mobilisation** is to connect the research process and results to other people and organisations within and beyond academia to 'generate useful knowledge' (Bowen and Graham, 2013, p. S5). As such, it goes beyond creating the traditional project website. Knowledge mobilisation is a process including 'a wide range of activities relating to the production and use of research results, including knowledge synthesis,

DOI: 10.4324/9781032668727-26

dissemination, transfer, exchange, and co-creation or co-production by researchers and knowledge users' that occurs at all stages of the research process (SSHRC, 2012a). Examples of knowledge mobilisation include:

- reaching an academic audience through academic publications and conferences;
- reaching the general public through **storytelling**, for example, producing blogs, podcasts, zines, or videos;
- reaching specific audiences, such as NGOs, civil society organisations, or policy-shapers, through
  - workshops or webinars;
  - writing for specific venues, for example, newspaper articles or policy briefs and memos;
  - press releases or social media;
  - and community events such as art exhibits, public forums, or performances.

Depending on where you are in the world, you may have heard knowledge mobilisation referred to as 'knowledge dissemination', 'knowledge translation', 'knowledge exchange', or 'research impact'. These variations are related to different national research agency policies and they also reflect the debates surrounding the meaning of 'useful knowledge' and informing ever-evolving standards of ethics and research practices. For instance, in the UK the Economic and Social Research Council (ESRC) 'has moved from using a linear conceptualisation of KT [knowledge transfer] to using "knowledge exchange (KE)", a term that emphasises more interaction and conversation' (Bannister and Hardill, 2015, pp. 4–5). In Canada, the term 'knowledge mobilisation' is preferred to that of 'knowledge dissemination'. The latter emphasises a one-way process by which research findings are provided to specific audiences, whereas the former implies a two-way dialogue between researchers and research users, which promotes a culture of partnership to promote co-production of knowledge to strengthen the development of practice, social innovation, and policy (SSHRC, 2012b). 'Research impact' describes the difference research makes and is the intended outcome of knowledge mobilisation. In GenUrb, our understanding of knowledge mobilisation has shifted from being primarily about researchers sharing findings with audiences to researchers entering into ongoing conversations with others.

The importance of knowledge mobilisation in the evaluation of research agendas has increased in recent years, emerging out of the recognition of a 'knowledge transfer gap', the failure to transfer information effectively to a specific audience. Sarah Bowen and Ian Graham (2013) argue that researchers should conceptualise knowledge transfer not merely as a technical issue but also as a political one, understanding that knowledge is power. Hence, knowledge mobilisation requires consideration not only of specific techniques and practices but also of the sociopolitical issues driving research questions and of how knowledge can be used (for example, to inform policy or community-based change). This understanding of knowledge mobilisation encourages more engaged scholarship between researchers, participants, and partners who each bring their own perspectives, experiences, and skills to enrich collaborative inquiries.

While knowledge-mobilisation practices vary by locale, it is now firmly entrenched in academic research cultures. In the social sciences, there is an increased expectation from public agencies that researchers will demonstrate relevance and knowledge mobilisation. This expectation is illustrated here for Canada, Singapore, South Africa, and the UK (see links to these national research agencies on the book's website www.routledge.com/9781032668680).

- The formalisation of knowledge mobilisation through the Social Sciences and Humanities Research Council of Canada (SSHRC) can be dated back to their *Strategic plan 2006–2011*, one component of which focused on effective knowledge sharing across diverse sectors of society, including community groups, policy makers, legislatures, business, and the media (Bennet and Bennet, 2013).
- The Social Science Research Council (SSRC) in Singapore has established funding criteria based on the potential impact of the research in Singapore (Social Science Research Council Singapore, 2020).
- In line with the goals of knowledge mobilisation, the National Research Foundation (NRF) in South Africa includes in its mission a statement to support knowledge creation that improves 'the quality of life of all South Africans' (National Research Foundation, 2020).
- In the United Kingdom, the Research Excellence Framework (REF), which has been adopted as a tool to evaluate academic performance, identifies the quality of 'impact' as a key component in this evaluation (Bannister and Hardill, 2015). According to the REF, impact can have 'an effect on, change or benefit to the economy, society, culture, public policy or services, health, the environment or quality of life, beyond academia' (UK Research and Innovation, n. d.).

Reflection exercise 19.1 asks you to investigate how your university and funding agencies related to your research field may define and evaluate knowledge mobilisation.

Reflection exercise 19.1: Crafting a knowledge-mobilisation strategy

Universities and research funding agencies frequently provide guidelines for crafting an effective knowledge-mobilisation strategy. For example, many universities have research support offices that help faculty and students prepare applications for funding. Many universities now have a web page on creating effective knowledge-mobilisation plans, and York University's Research Office has a knowledge-mobilisation specialist who offers feedback on and assistance with preparing grant applications (York University, n d.-a). Search (online or in person) for the research support office at your university (or one suggested in this exercise) and visit the websites of the funding agencies to which you might submit your research proposal (alternatively: review the websites of the SSHRC, REF, NRF, and SSRC).

Review how the university and specific funding agencies that you might apply to define and evaluate knowledge mobilisation. Take some notes and, based on these, ask yourself the following questions (modified from Carleton University [2014]).

- What knowledge will my research generate?
- Who is this knowledge generated with? Who is it generated for? In other words, who cares about this issue?
- Who will use the knowledge generated by my research, and how will they use it? In other words, who are the audiences for my research?
- How might I best connect the knowledge generated from my research with prospective audiences?

While the institutional incorporation of knowledge mobilisation into the social science research process is relatively recent, feminist researchers have long been concerned with the mobilisation of knowledge and social impact as a key research outcome. The next section brings together feminist knowledge production and the institutionalisation of knowledge-mobilisation requirements to articulate a feminist approach to contemporary knowledge-mobilisation practices.

Section 2. Feminist approaches to knowledge mobilisation

Within the social sciences, many epistemological traditions based on shared knowledge production preceded the formalisation of knowledge mobilisation – specifically, critical scholarship in feminist, anti-racist, anti-capitalist, and sustainability studies addressed similar concerns. As the feminist sociologist Ann Oakley (2000, p. 3) remarked, 'The goal of an emancipatory (social) science is . . . to develop the most reliable and democratic ways of knowing . . . in order to bridge the gap between ourselves and others'. Research approaches, such as participatory action research and feminist ethnographies, that emphasise collaborative relationships between the investigator and the community of study have also presupposed an emphasis on knowledge mobilisation as part of the research process (Stacey, 1988; Dyck, 1993; Buch and Staller, 2014; Katz and Warren, 2014; Warren, Katz and Heynen, 2019). Accordingly, feminist approaches to knowledge mobilisation may put less emphasis on 'deliverables' and more on building reciprocal long-term relationships with people and places and on building a civic role for the academy. Thus, incorporating a feminist perspective into knowledge-mobilisation strategies necessitates engaging in knowledge mobilisation as a process of reflexive and collaborative co-production in which there is active interrogation of '*who* currently gets to define *which forms of knowledge* will count as useful *for whom*' (Moss, 2013, p. 238).

Feminist knowledge mobilisation considers how knowledge that emerged from material and historical legacies of colonisation, extraction, and political domination continue to have an impact on how feminist knowledge circulates. It seeks not only to account for power structures that privilege certain accounts of knowledge but also to engage in decolonising the knowledge produced by including perspectives and voices that have been historically excluded from research (Marx Ferree and Verloo, 2016). Feminist and other critical scholars engaging with knowledge production in the global South have sought to distance themselves from the instrumental use of knowledge by some development policy organisations in which the academic is portrayed as the all-knowing expert, the purveyor of specialist knowledge, versus the research participant, who supposedly holds no significant insights into their own life (Williams, 2015). Such a hierarchical positioning, separating knowledge producers from users, runs counter to feminist notions of the co-production of knowledge.

This does not mean that feminist scholars have a romantic view of feminist knowledge mobilisation as a panacea for the building of non-hierarchical relations. Even in feminist projects, it is common for engagement beyond the academy to be tentative and not necessarily to lead to specific impacts. Thus, a feminist approach not only recognises the complex embodied politics of knowledge mobilisation but also the production of knowledge as often messy, not straightforward, and as proceeding in a non-linear manner (Hardill and Mills, 2015).

Most recently, during the COVID-19 pandemic, feminist scholars have had to contend with the move to online modes of knowledge mobilisation. This shift is expected to continue, with knowledge mobilisation adopting hybrid modes of face-to-face and online interventions (see Chapter 5, Section 3 Practising feminist urban research in a time of ecological crisis). Online gatherings of feminist networks have taken the form of virtual conferences using Zoom or other videoconferencing technology or more informal exchanges through Facebook, Telegram, WhatsApp groups, X (formerly Twitter), email, websites, or blogs (Crossley, 2015). Such online communities are said to expand 'feminist notions of knowledge production and community – where collaborative conversation, kinship, writing differently, and collective research is generated' (Black *et al.*, 2019, p. 123). The proliferation of virtual spaces to foster inclusion and effective knowledge mobilisation across networks of feminist communities has also led to addressing knowledge mobilisation in relation to gendered online issues such as sexual harassment and other care-related concerns (see Chapter 20, Section 4 Challenges faced using social media in feminist scholar-activism). Hybrid modes of knowledge mobilisation will likely become more popular because of the opportunities they offer researchers to engage communities and build networks across distance and time zones.

Reflection exercise 19.2 asks you to investigate the contributions to knowledge mobilisation from feminist scholar-activists in your field.

Reflection exercise 19.2: Feminist scholar-activism and knowledge mobilisation

Feminist scholars and activists, among others, were practising knowledge mobilisation long before the rise of the institutional focus on academic knowledge mobilisation.

- How have feminist scholar-activists in your field built meaningful relationships between themselves and the communities they work with?
- How has the feminist praxis of knowledge sharing prefigured knowledge-mobilisation practices?
- How might institutionalised practices of knowledge mobilisation, online or offline, create challenges for feminist researchers?

In the following section, we address the practical concerns of incorporating knowledge-mobilisation strategies into feminist research.

Section 3. Incorporating knowledge mobilisation into a feminist research project

Researchers should implement strategies for effective knowledge mobilisation from the inception of a research project. Aligning knowledge-mobilisation activities with specific knowledge users and producers, for example, is a necessary first step for effective knowledge-mobilisation strategies. Researchers should also be prepared to rethink knowledge-mobilisation activities throughout the research and beyond to ensure not only that project findings can be translated from one context to another and reach their intended audiences but also that they remain relevant to any changes that take place in the research project's objectives and in the context in which the research takes place. GenUrb example 19.1 considers how achievable GenUrb's knowledge-mobilisation plan was in the light of COVID-19.

GenUrb example 19.1: How knowledge mobilisation changed over the course of the research

In GenUrb, we expected that our schedule of knowledge-mobilisation activities would evolve as the partnership grew and engaged with the knowledge produced and as new modes of communication technology emerged. With hindsight, it is interesting that our expectations of change were all positive: in practice, they were largely negative, primarily because of COVID-19 and national politics. COVID-19 seriously undermined our ability to ensure that our knowledge mobilisation plan was achievable.

In our SSHRC application, we outlined how our knowledge-mobilisation plan provided a close fit with the project objectives of:

- engaging in innovative public-education initiatives with grassroots women in diverse urban contexts via processes of social and mutual learning;
- disseminating knowledge that would be accessible and useful to urban, national, and global policymakers, advocacy and practitioner communities, and the public;
- and developing a Gendered Urban Global South Knowledge Network that would build research capacity and train a new generation of scholars.

The first set of knowledge-mobilisation activities was cut short by COVID-19, which prevented any face-to-face contact (over differing periods of time in the different city locations). The second was not a particular success, either, because of the different national political contexts in which we worked, in which feminist research became increasingly difficult to undertake, and again partly because COVID-19 made contacting policymakers increasingly time-consuming and difficult and, in some cases, such as in Shanghai and Delhi, impossible. We consider the third set of knowledge-mobilisation activities to be a largely successful and ongoing process. Overall, and in the context of COVID-19, if we were to grade ourselves on our knowledge-mobilisation plan meeting project objectives, we would give ourselves a C+.

By Linda Peake

Figure 19.1 outlines how knowledge mobilisation can be addressed in different stages of the research process: pre-research, research, and post-research.

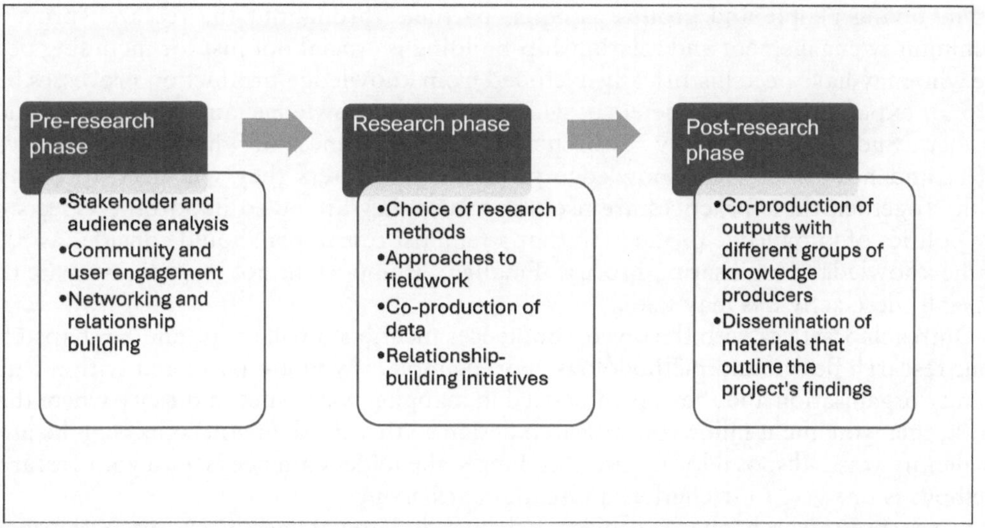

Figure 19.1 Knowledge-mobilisation strategies in different phases of the research project
Source: Araby Smyth

(i) Pre-research phase

In the initial stages of research, the goals of knowledge mobilisation should be developed to be inclusive of people who may have been historically excluded from knowledge-production processes to enhance the impact of the research and its relevance for diverse communities. Not only do research projects now include extensive knowledge mobilisation plans, but there has also been a fundamental shift in the way partnerships outside academia are developed, such that researchers and practitioners are commonly considered equal partners who collaborate at all stages of the research process (see Chapter 9, Section 1 Research partnerships). This may include collaboration in developing research questions, selecting a methodology, participating in data collection, and reporting and disseminating findings.

Universities are also increasingly employing knowledge-mobilisation specialists as brokers to facilitate relationships between academic communities and non-academic partners. If your university has such an employee, you should contact them for any guidance they can give. For example, at York University, the Knowledge Mobilisation Unit aims 'to broker relationships between York's researchers/students and non-academic research partners (primarily from the community sector and regional/municipal governments) so that York's research can have an impact on public policy and professional practice' (Phipps and Morton, 2013, p. 256).

Researchers should conduct analysis of who their stakeholders and audiences are, as well as engaging with partners, communities, and other users through brokering and networking (Briggs *et al.*, 2015). Stakeholder and audience analysis helps the researcher anticipate the needs and expectations of the people involved in and affected by the research project as well as those the research aims to reach. Community and user engagement incorporates input from people who the research outcomes will affect. Meanwhile, brokering and networking, or relationship-building, are essential for involving people and groups as equal partners throughout the research process. Community engagement and relationship-building is crucial not just for including people who may have been historically excluded from knowledge-production processes but also for expanding the parameters of what counts as 'knowledge' and who can produce 'theory'. Such practices allow researchers to gain awareness of who the research will affect and how, and what knowledge producers and users they will need to engage with. Together, these practices are useful because they attend to important aspects of the politics of knowledge production that a feminist researcher should consider as part of the knowledge mobilisation process. Finally, it is important not to underestimate the time all these activities may take.

Outreach practices with the diverse audiences mentioned will shape and be shaped by your research design and methodology. For example, if you are partnered with a community organisation and they are interested in mapping waste sites in the city where they work, then you might tailor your research design so that the data you collect can be used to identify waste-disposal locations. Matching stakeholder data needs with your research methods is one way to strengthen a research partnership.

In the pre-research phase, project websites and social media are often the most important sites for knowledge mobilisation where images of field sites, researchers, and participants can draw the viewer in. Ethical issues may arise here because permission is needed to use images of people. In the GenUrb case not all the researchers could put their photos on the project website given the hostility of some governments to feminist research.

GenUrb example 19.2 shows part of the knowledge-mobilisation plan submitted as part of the funding application. It identifies the different activities and outputs that were intended for different groups of knowledge producers and users. In practice, events occurring during the research period, such as national elections and the global pandemic, made it impossible to carry out all these activities and resulted in some of them being transferred online.

GenUrb example 19.2: Knowledge-mobilisation activities and outputs for different knowledge producers and users

Knowledge producers and users	*Knowledge mobilisation activities and outputs*	*Rationale*
Grassroots women research participants	Visual representations – infographics, photos, graphic pamphlets – social media, plays, short videos, podcasts, and exhibits	Visual media can reach communities where literacy rates are low and reading is not an everyday activity.
Policy-shapers engaging with SDGs 5 and 11	Policy briefs and memos, social media, workshops, and conferences	Without a targeted knowledge-mobilisation strategy of briefs and memos to reach policy-shapers, it is unlikely that they will access the outputs of the research.

Academic researchers	Website, books, edited books, refereed articles, and theme issues of (open access, international, peer-reviewed) journals, workshops, conferences, social media, and podcasts	It should be noted that while books, journal articles, and conferences are still important for knowledge mobilisation, academic outputs are diversifying to incorporate the use of a wide range of technologically enhanced outputs that can be published in real time.
General public	Website, open access publications, conferences, exhibits, blogs, and podcasts	There needs to be a wide range of knowledge-mobilisation strategies to access different constituencies.

By Linda Peake (from GenUrb knowledge-mobilisation plan, SSHRC application)

GenUrb example 19.3 illustrates the importance of knowledge-mobilisation activities even at very early stages of the research.

GenUrb example 19.3: The GenUrb opening ceremony in Shanghai

On July 31, 2018, the Shanghai City Research Team (CRT) organised an arts and cultural event at Yuwatai Gallery (a gallery near the research neighbourhood). The event was arranged as an 'opening ceremony' for the GenUrb project in Shanghai and included invited government officials, journalists, feminist scholars, a local organisation, and students, as well as the women participating in the research project.

The event was opened with a speech by the owner of the Yuwatai Gallery, who provided the venue space for us to show their public support for the research. This was followed by further opening remarks by Shanghai CRT team leads Tsung-Yi Michelle Huang and Penn Tsz Ting Ip. This event was based on a mental-mapping technique, structured as a game, with myself as the moderator. The game involved the women participants developing mental maps of their neighbourhood. In this activity the CRT collaborated with We and Equality, a Shanghainese non-profit organisation. We also recruited twelve student helpers through We and Equality to record the comments made by research participants during the mental-mapping activity. Adding a sense of ritual to the event and marking the official start of the research project with support from local patrons was calculated to increase the research participants' trust in the researchers and the research process.

Of the twenty women engaging in the research, only eight participated in this event. This low number was partially due to the low level of trust that some women participants had in the researchers, given that this event was held at the start of the research and they had not had time to get to know us well. It was also due to the extremely high number of women participants from the community we were working in who were physically ill from years of working in noxious factory environments: ten had cancer and five had heart disease. Despite this low turnout, this event was very influential in building trust in the research process among the research participants while also serving to make a range of stakeholders aware of the research project, increasing its legitimacy in the eyes of the participants.

By Penn Tsz Ting Ip

(ii) Research phase

Knowledge mobilisation continues through the choice of research methods, approaches to fieldwork, the co-production of data, and the building of relationships with and among related networks and communities. You may want to think about how your choice of research methods might help build relationships with stakeholders and people engaged in the research. How you approach fieldwork is also important for maintaining those relationships over time. For example, you may consider volunteering regularly for community organisations that focus on advocacy or building awareness around the themes you are researching.

The co-production of knowledge and the promotion of relationship-building initiatives can be achieved in various ways. In GenUrb knowledge mobilisation activities during the research phase included, for example, designing and holding workshops while in the field and employing these workshops to nurture research relationships and to further explore issues that arose during interviews. In GenUrb example 19.4 the Cochabamba CRT engaged participants with storytelling workshops to share life stories, express themselves, build connections, and network. GenUrb example 19.5 discusses how workshops were used by the Ramallah CRT to address issues that arose during the interviews. Knowledge-producing activities such as these that stem from and foster accountability and reciprocity are necessary for creating less exploitative and more respectful relations between researchers and participants.

GenUrb example 19.4: Storytelling workshops in Cochabamba

Near the end of the in-depth interview process, the Cochabamba CRT held a workshop to bring all the women who had participated in the GenUrb interviews together. The workshop provided the opportunity:

- to go over the goals of the project, receive feedback, and answer questions;
- for the women to share their stories with each other, because a key point that emerged from the interviews was a sense of isolation among the women;
- to ensure an iterative research process and introduce the idea of story sharing as part of the research before we undertook the life histories;
- and to foster feminist solidarity.

Evoking collective memory through storytelling is extremely common in Latin American cultures and has been invoked especially among working-class and Indigenous peoples of Bolivia involved in social protests (Farthing and Kohl, 2013). However, the women who participated in the GenUrb project were not accustomed to sharing their personal stories. What Karen de Souza (from Red Thread and the Georgetown CRT) had said about the COVID-19 diary-writing process in Georgetown rings true here: that the women did not have time to dwell on their emotions or experiences – they just had to get on with everyday life (see Razavi *et al.*, 2023). This workshop therefore provided a space to introduce the idea of life histories but also for the participants to build connections, network, nurture friendships, express themselves, and share their stories. Encouraging the women of Quechua background to share their stories helped ground the GenUrb research on urban place-making in personal experiences, and the stories women shared during the workshop provided deep insights into their lives and helped encourage solidarity.

The workshop only required a meeting space and a few materials. We decided to use a mandala as a visual guide for its symbolism of the circle of life. Using newspaper and magazine clippings, markers, paint, and paper, we created our own mandalas of life (*mandalas de la vida*) to illustrate our most important life events: the happy and sad moments, from childhood to present, concluding with our hopes for the future (Razavi, 2020a; see Document 19.1 *Mandala of Life Storytelling Workshop Guide* on the book's website www.routledge.com/9781032668680). The visual art component was useful to share emotions and memories that cannot easily be expressed verbally, partially due to the social stigmas around difficult issues, such as intimate partner violence.

The storytelling workshop also nurtured feminist solidarity by providing an opportunity for the CRT and participants to connect and build trust. Both participants and researchers took part in the exercise, allowing us to narrow somewhat the distance between us. We aimed to facilitate connection and exchange by providing space for mutual learning of multiple stories and ways of knowing, and, through the stories that were shared, to strengthen participants' voices. I found the mandala activity to be a useful practice to reflect on my own positionality as a researcher from the global North engaging with women of Quechua background in the global South. Being confronted with issues of positionality and representation leads us to focus on the 'ethical commitment to deconstruct the role of the researcher in perpetuating West/Other asymmetries and an imperative to develop ways to counter epistemic violence' (Weatherall, 2020, p. 474).

We did face some challenges in using this method. First, the only magazines and newspapers that we could find had images of white models that were not representative of the women in the workshop. Second, I grappled with the personal and potentially extractive nature of the workshop, as I was confronted with the tension of being a scholar from the global North. Third, we worried about the possible impact of this exercise and whether there were appropriate mental health resources available to the participants. To address these challenges, we clearly explained the GenUrb project; we obtained informed consent from all the participants; and we collectively established what would count as appropriate behaviour during the workshop. Both Sonia [Pardo Burgoa] and I, as facilitators, participated fully in the workshop, sharing our own life histories. We also provided information about health services via CETM. We tried our best to ensure that participants left the session feeling good rather than sad or dwelling on past traumas, but this kind of activity can generate strong emotional responses. However, as noted, we ensured the women were fully informed about the nature of the research and obtained their consent. They themselves made the decision to take part, and their time input was recognised with an honorarium.

By Nasya S. Razavi

GenUrb example 19.5: Using workshops to develop relationships in Ein Qiniya, Ramallah

In the in-depth interviews that we conducted with 25 women in Ein Qiniya, Ramallah-Jerusalem North, a widely expressed concern was that of the women's disconnection from health services and inability to respond swiftly to accidents. Interviewees often became emotional when they described how it takes at least twenty minutes to reach the closest health centre – when a vehicle is available, which is often not the case. We therefore organised a half-day workshop in March 2019 to engage in first-aid training with a professional healthcare worker from the Palestinian Red Crescent Society. The training was attended by 22 women, with several bringing their children along. The feedback from participants was that they found it really useful, especially in relation to learning about what they should do and not do in accidents and emergencies.

In June 2019, we addressed another one of the participants' strong concerns, that of social disconnection. By contrast with the highly structured nature of the workshop on first-aid training, we decided that this would be a looser and more intuitive workshop – a sweets-baking class with an experienced chef. This three-hour meeting was attended by eighteen women and over twenty children. The chef, Haifa Qawasmi, provided three recipes based on what the women were interested in learning: cinnamon rolls, chocolate rolls, and pancakes. She adjusted the ingredients so that the items were affordable and easy to make, gave out recipes, and conversed with participants about other dishes as well as their living conditions and routines. The feedback was again positive.

In August 2019, we moved to another issue that participants regularly brought up, creating an income-generating cooperative. The third workshop involved screening several short films and a discussion with Amal Juma', a veteran feminist trainer and advocate for women's rights with the Women's Affairs Technical Committee. It was attended by eleven women and 25 children (we pre-organised clowns to entertain the children in order to give the women space to engage). One film was about the excruciating working conditions of women agricultural labourers on colonial farms in the Jordan Valley, and three were about women-initiated cooperatives in agriculture, food processing, and lending in Battir, alShuyoukh, and Bethlehem.

After each film Amal asked the women about their impressions as an entry point to pass on information about available support organisations and pathways should they decide to create a cooperative for themselves in Ein Qiniya.

A fourth workshop, on reproductive health, was organised for March 2020 but was cancelled because of the COVID-19 pandemic. In April 2021, we followed up the question of creating women's cooperatives through a multi-session workshop on creating a collective garden. This was held at the Um Sleiman organic farm, Sakiya, a progressive academy located on ten hectares of land within Ein Qiniya that combines research, agriculture, and art. Due to a lack of funding and Mai's departure to Germany to complete her graduate studies, the work was suspended from our side, but Sakiya has kept this project on their agenda.

By Natasha Aruri and Mai Al-Battat

(iii) Post-research phase

Knowledge mobilisation in the dissemination phase includes the production of a range of outputs, adapted to reach multiple audiences, outlining the project's findings. To ensure project findings are wide reaching and accessible over time, activities at this stage may include open access publications (such as this book) and the creation of digital resources to be hosted on a variety of web platforms (Briggs *et al.*, 2015). Visual materials (e.g. documentaries, photographic exhibits) are particularly effective when making presentations about the research.

Ethical considerations also necessitate making research outputs intelligible and relevant to research participants. The co-production of such outputs is key so that project findings address the communities involved and are presented in useful and accessible ways. Feminist researchers engage alternative writing strategies, producing outputs that can be accessible to those who do not engage with the academy or for those who do not read or write daily, for whom a wide range of outputs may be designed, including, for example, pictorial pamphlets, radio programmes, and plays. Hence, it is important to maintain relationship-building initiatives among related networks and communities even after data collection and analysis is complete so that the research materials disseminated are in line with participants' wishes.

One important way in which outputs may be adapted to different audiences is in the way they tell 'their story'. Many researchers are trained in and accustomed to presenting research findings in descriptive and analytical styles of academic writing to be published in peer-review journals or university presses, in which many academics adopt a storyline that speaks in the voice of the 'impartial arbiter of social truths' (Bannister and O'Sullivan, 2015, p. 89). By contrast, storytelling requires narrative. As the sociologist Catherine Riessman (2006, p. 189) states 'The "truths" of narrative accounts are not in their faithful representations of a past world, but in the shifting connections they forge among past, present, and future'. Jon Bannister and Anthony O'Sullivan (2015, p. 88) suggest that part of the reluctance by scholars to engage in storytelling may be the absence of 'explicit criteria to assess the acceptability of the analytical content of a narrative'. This reluctance is, however, slowly being overcome as more feminist scholars explore existing traditions of storytelling outside the academy. The feminist geographers

Pavithra Vasudevan, Margaret Marietta Ramírez, Yolanda González Mendoza, and Michelle Daigle (2023, p. 1729) define storytelling as

> how people make sense of the world and our place in it, an iterative process of interpreting reality through observation and the exchange of ideas. Storytelling, as a means of theorizing the structural from embodied experience, is a fundamental methodological intervention of Indigenous, Black, and third world and women of color feminist traditions.

More specifically, the feminist planner Leonie Sandercock (2003, quoted in Bannister and O'Sullivan, 2015, p. 88) outlines the hallmarks of good narrative. She claims that in their analytical form stories have 'certain key properties: a temporal or sequential framework; an element of explanation or coherence; some potential for generalisability; a plot structure and protagonists, and "moral tension" '. Some of these qualities are captured in the GenUrb-based research narrated in 'Tiwa's morning', in which Grace Adeniyi-Ogunyankin and Linda Peake (2021b) use storytelling in the form of fictional ethnography to share 'a day in the life' narrative based on extensive interviews with women in Ibadan, Nigeria. Rather than share pieces of several women's lives, they amalgamate these to tell the story of how one woman struggles to make ends meet in the neoliberal city (see also Razavi, 2020b; Adeniyi-Ogunyankin, 2024). Grace continues this work in Adeniyi-Ogunyankin (2024) with the story of Mummy Ayesha.

GenUrb example 19.6 discusses how GenUrb approached knowledge mobilisation throughout the project.

GenUrb example 19.6: The success of initiatives taken to engage target audiences and participants in GenUrb's knowledge mobilisation plan

The importance the GenUrb project attached to knowledge mobilisation was highlighted by the creation of a knowledge-mobilisation team, a postdoctoral fellow position attached to this team and the allocation of 14 percent of the budget to knowledge-mobilisation activities. Furthermore, policymakers were included on the steering committee to ensure their participation in all GenUrb activities. CRT meetings were also to be held once every two months during years two to five to engage in storytelling, as well as facilitating critical listening and feedback sessions which ensured that co-production of knowledge with research participants was taking place and being fed back into steering committee meetings.

The knowledge-mobilisation team was to report to the steering committee meetings, and the agenda of those meetings and of the annual meetings included a dedicated slot for knowledge mobilisation. The knowledge mobilisation team was to:

- develop an engaged knowledge-mobilisation strategy addressing the objectives of the research and linked to the research activities within the CRTs, including the nurturing of emerging scholars and graduate students to provide support for a new generation of scholars with expertise in knowledge mobilisation;
- work closely with York University's Knowledge Mobilisation Unit at all stages of the research to ensure extensive outreach is being pursued;

- and reach out to all members of the project with appropriate (translated) and distinct outputs to foster continuous knowledge exchange and debate and to enable end users, other researchers, and the public to provide feedback, comments, and questions to influence the research.

While much of this was achieved (we did have a knowledge-mobilisation plan!) it was not always easy to keep knowledge-mobilisation activities on our agenda. The knowledge-mobilisation team was formed at the start of the project, but it was overshadowed by the research startup, an extremely time-consuming and arduous process. With packed steering committee agendas, knowledge mobilisation sometimes received scant attention, and it became increasingly difficult to secure the participation of policymakers at steering committee meetings. By the time of the COVID-19 pandemic, the policymakers had disappeared altogether, as had the two monthly CRT meetings. By the time knowledge mobilisation was seriously on the horizon, COVID-19 was with us and the knowledge-mobilisation team had disbanded, its role taken over by the principal investigator, the project manager, and postdoctoral fellows. Our connection to the York University Knowledge Mobilisation Unit remained unexplored. Knowledge-mobilisation activities – both academic and non-academic – have nonetheless taken place and continue to take place in terms of knowledge production and mobilisation, capacity-building and connectedness.

By Linda Peake

Use Reflection exercise 19.3 to review what you have learned about incorporating knowledge mobilisation into a feminist research project.

Reflection exercise 19.3: Incorporating knowledge mobilisation into a feminist research project

- Given that knowledge mobilisation requires dedicated time and resources, what cost-effective and time-efficient knowledge mobilisation activities would you include in your research?
- What information about possible research audiences will inform your knowledge mobilisation strategy?
- To what extent is it important for knowledge mobilisation to last beyond the life of your research project?

Section 4. Summary

This chapter has outlined how knowledge mobilisation is now a necessary part of contemporary research projects that seek to co-create knowledge with other knowledge producers and to make an overall impact within and beyond academia. Incorporated at all stages of the research process, as in the GenUrb project, knowledge-mobilisation activities aim to foster an engaged form of scholarship to ensure research findings benefit a wide audience. The chapter outlines a range of knowledge-mobilisation strategies and activities. Storytelling is one approach that is increasingly favoured by feminist scholars and that holds great promise for knowledge mobilisation.

20 Feminist engagement with social media

Mantha Katsikana

Learning objectives
After completing this chapter, readers should understand

- different uses of social media for scholarly networking and collaboration;
- the commonly used social media tools and their uses for feminist scholars;
- and challenges and ethical issues in the use of social media.

Further, readers should be able to

- integrate the use of social media in their feminist academic practice;
- assess which social media tools to engage with in their work;
- and navigate ethical challenges in the use of social media.

In this chapter, the uses of social media for feminist scholar-activism are explored, along with the challenges and possibilities their use entails. The chapter offers an overview of social media tools and platforms that have been used in feminist activism. It discusses social media as spaces for political engagement, solidarity, knowledge mobilisation, and resistance to systems of oppression that emerge both online and offline, within and beyond the academy. The chapter also explores challenges that feminist scholar-activists face in using social media, concerning: the politics of social media engagement for feminist activist research; online privacy, surveillance, and censorship; the relationship between online activism and on-the-ground organising; and the risk of exposure to online violence.

Section 1. Social media, feminist activism, and the web

During the last four decades the widespread use of the personal computer and the development of internet technologies have changed how people communicate; how they form and maintain relationships and communities; how they consume, produce, and share knowledge; and how they navigate the world. As a number of humanitarian crises – such as political uprisings and the recent COVID-19 pandemic – have shown, the widespread use of **social media** brings both possibilities and challenges. It provides valuable tools to navigate collectively across geographical contexts, but as internet technologies develop and the possibilities for connectivity and content creation increase, oppression in cyberspace is a growing problem.

Social media are internet-based data networks of communication, where users create and manage their own profiles, upload data in the form of text, image, sound, and

DOI: 10.4324/9781032668727-27

video, and link to content that is already online. Social media networks can be hosted on websites or mobile applications, and their functions of hosting data and connecting users evolve rapidly, with users 'moving' from one network to the next, depending on their communication needs, preferred formats, and available devices. Currently, the dominant and most popular social media are X (formerly Twitter), Facebook, Instagram, TikTok, and YouTube.

Feminists have long made innovative use of networked communication for activist engagement. Although their contributions may have received little mainstream attention, feminists have sought to counter harassment and create more community spaces online. Place-based information-sharing and social infrastructures, such as bulletin boards, printed material, newsletters, and physical archives (McKinney, 2015), have given rise to online equivalents such as virtual discussion groups using the ListServ software (Hyman, 2003), while the use of blogs and wikis as community resources has proliferated alongside the widespread use of social media (Siles, 2011; Gieseking, 2020). Online feminist activism has created a toolbox of critical practices against oppression, injustices, and corporate capitalism and its colonial biases, across and within vastly different contexts.

Feminist online activism is a set of heterogenous practices utilised by geographically dispersed communities which work together online towards liberation from gender-based and other forms of oppression. Feminist activist communities are reproduced and sustained online, using the internet as a tool for connecting collective action in grassroots organisations or social movements across local, national, regional, and transnational scales. Feminist and queer activists, scholars, educators, and community organisers have used social media in ways that reclaim online 'spaces' as platforms for consciousness-raising, disseminating information, sharing knowledge, creating solidarity networks, organising political resistance, and supporting individual and collective activism. They engage in making visible the experiences of those who have historically been excluded from knowledge-production processes. Social media offer possibilities for decolonising knowledge production in both academic and public realms, for example, by highlighting narratives and practices that are often perceived as universal (for example, around the experience of being a 'woman') and revealing the differences created by the different ways women and marginalised communities experience oppression globally. They can expose legacies of colonisation that are often obscured in mainstream historical and popular narratives. Social media can also aid in the decolonising of transnational activist organising by facilitating connection, forging of solidarities, and enabling information sharing by marginalised and Indigenous communities.

In the context of feminist scholar-activism, these possibilities have created a rich toolbox for the collaborative production and mobilisation of knowledge and the creation of solidarity networks that can respond to crises within and beyond the neoliberal university (see Chapter 4, Section 1 The academy and feminist scholar activism). Despite these developments, there are obstacles to accessing infrastructure, computers, and mobile applications that result from and perpetuate technological unevenness between places (for example, between rural and urban areas, between the global North and global South) (Graham *et al.*, 2014). Such inequalities are rooted in histories of colonisation, geopolitical agendas, and globalisation (Thorat, 2019). In addition, the incorporation of online technologies into everyday governance and surveillance and the commodification of personal data raise many questions about the ethics and challenges of navigating the internet. They also deepen divides across gender, race, class, disability, and physical location (see Amoore, 2020). Such difficulties make it more challenging to decolonise knowledge production through online tools. They shatter the illusion of the internet and social media as universal tools for democracy where anyone can express their opinion and reach out to others, forcing feminist scholar-activists to reflect on the geographies of privilege in which social media engagement unfolds.

The following section offers an overview of social media tools and associated platforms that have been used in feminist activism, along with their purposes and the opportunities they offer. It focuses in particular on podcasts.

Section 2. A guide to social media tools and their uses

As social media technology and trends evolve, the hierarchy of platforms changes, with location, socio-political, and cultural contexts affecting the choice of platforms preferred by users. Thus, while social media sites and applications (e.g. Facebook and Instagram), blogs and microblogs (e.g. Blogger and X), and content communities (e.g. YouTube and TikTok) appear to have global reach, the use of other platforms, particularly those focusing on instant messaging (e.g. WhatsApp, Telegram, Viber, Line, WeChat, and Kakao Talk) is restricted to particular countries or regions. An app may have been produced for a particular national context; may only support a specific language; or may only cater to the needs of specific communities. In some cases, access to the internet can be controlled and/or subject to surveillance by the state, as in China (see Su and Shyong, 2019; Wang, 2020; Miao, Jiang and Pang, 2021) or with RuNet in Russia (see Asmolov and Kolozaridi, 2017; Sivetc, 2021). However, users tend to develop different networked communication practices in countries where there is no affordable access to an internet connection or to hardware (e.g. personal computers, laptops, tablets, and phones) and no national internet regulation, local infrastructure, or governance.

Keeping in mind the continuous expansion and shifting usage patterns of social media, Table 20.1 focuses on a range of tools used within social media in their current form, either as integrated functions or as shared/embedded information hosted elsewhere on the web. It presents each social media tool, its function, and examples of sites.

Table 20.1 Tools provided by social media

Tool	*What does it do?*	*Sites*
Blog	Blogs consist of long-format data entries including text, images, audio, and video (embedded [i.e. linked from another platform] or hosted [i.e. directly uploaded on a website]), along with discoverable tags describing the entry contents.	Blogger, WordPress, Tumblr
Microblog	Microblogs consist of short text entries (for example the 280-character capacity per post on X [formerly Twitter]). Users have the ability to share and reshare text, images, video, links, hashtags, and mentions.	X, Mastodon, Hive
Podcast	Podcasts are digital audio files, available for download or streaming by users on their devices. Podcast content includes but is not limited to interviews, panel discussions, lectures, and audio documentaries. Podcasts can be produced as standalone content or as a series of episodes.	Soundcloud, Spotify, Apple Podcasts

(*Continued*)

Table 20.1 (Continued)

Tool	*What does it do?*	*Sites*
Image/video hosting	Image/video hosting allows the user to upload, store, and share images and videos from their devices. While most social media apps support image or video hosting, there are also websites that can handle larger file sizes or higher-resolution media that are used as archives and galleries.	Facebook, Instagram, X/Twitter, YouTube, Vimeo, Flickr, Internet Archive
File hosting/ cloud	File hosting enables users to host and share large files or many files on a site. Users can access, download, and edit files.	Dropbox, SharePoint, Google Drive
File sharing	File sharing allows for documents (such as articles, books, textbooks, etc.) to be uploaded and shared with other users in anonymous online libraries or through personal profiles on academic professional networking sites. For example, some non-profit organisations allow members to share archival material, as in a public library. Yet other sites operate on a shadow (piracy) basis and are therefore blocked in certain countries. Shadow libraries, for example, enable pirate file sharing of typically pay-walled content.	Commercial, for-profit sites, such as ResearchGate or Academia.edu, or non-profit sites like the Internet Archive; shadow libraries including aaaaarg and Library Genesis
Instant messaging	Instant messaging allows for real-time exchange of text, links, documents, images, and sound or video files between two users or a group of users in a dedicated private interface. Instant messengers can be standalone applications or integrated into social media platforms. Some offer voice and video calls.	WhatsApp, Telegram, Viber, Line, WeChat, and Kakao Talk
Livestreaming	Livestreaming allows for the real-time broadcasting of events, such as talks, lectures, or concerts, through a social media platform. It can be combined with synchronised live chats in which audience and broadcasters react to the broadcasted content.	Facebook, YouTube, Instagram, TikTok, Twitch, Clubhouse
Teleconferencing	Teleconferencing allows multiple users to share the same video call, accompanied by chat and the possibility of live captions, file sharing, recording, automatic production of transcripts, etc. Teleconferencing platforms usually cater to larger numbers of users than instant messenger video-calling.	Zoom, Skype, Google Meet, Microsoft Teams
Crowdsourced databases	Crowdsourced databases and geodatabases (for mapping) can be developed through users either submitting information to the creator and manager of a database (through submission forms and emails) or adding it directly to cloud-based databases (stored online and accessible to anyone with the link).	Google Docs, OpenStreetMap, and Queering the Map

One tool that is popular with feminist activist-scholars, particularly for knowledge mobilisation, is podcasting. This has flourished with the use of smartphones, access to simple audio-editing software, and the growth of streaming platforms. Reflection exercise 20.1 explores the process of conceptualising and designing a podcast for activist purposes. Since this exercise stresses the activist context, it may be useful first to complete the following exercises in Chapter 4: Reflection exercise 4.1 Situating the self in feminist activist praxis and Reflection exercise 4.2 Connecting with communities, creating solidarities.

Reflection exercise 20.1: Designing a podcast

This exercise helps you brainstorm ideas that could turn into a podcast for feminist scholar-activist purposes. Useful tips are provided for finding target audiences, obtaining financial support, engaging with communities, and giving space to under-represented and marginalised groups through ethical content creation and feminist collaboration.

Concept design

Podcast production starts with 'putting together' the podcast, a process that includes defining what will be its focus, purpose, and target audience. The structuring of a podcast will also depend on the theme, episode length, episode frequency, and number of people participating. Asking the following questions is therefore a vital part of designing a podcast.

- What is the theme/subject?
- What is the purpose?
- Who is the target audience?
- What is the length and frequency?
- Who is participating, and what are their main roles/tasks?
- Which language will be used?
- Who is responsible for research?
- Who is responsible for production?
- What is the desired mode of dissemination?

Answering these questions with reference to your project will help you initiate a brainstorming process on which the podcast can be based. Elements such as the theme, purpose, and target audience can also be used in the public description and presentation of the podcast for promotional purposes and in introducing the podcast to possible guests and collaborators.

Ethics

Building on feminist scholar-activism, it is important to reflect on whether the information in the podcast could harm participants or communities (see Chapter 7, Reflection exercise 7.1 Values in feminist research). Following the practice of reflexivity, participants and creators should think about how they are positioned within larger structures of power and oppression as well as how this positioning might affect the content of the podcast. For example, a podcast that is part of a broader knowledge-mobilisation strategy supported by an institution will probably need to have been covered already by the relevant institutional ethics procedures (see Chapter 8, Section 2 Ethics policies). Acquiring participants' consent can take the form of providing a script or questions in advance as a basis for review and approval by all participants. If possible, compensation for the time and energy of participants should be considered.

Funding

While making a podcast can be a very creative process it also requires a significant amount of labour, energy, and time. Compensating precariously employed and/or marginalised participants for their time is therefore very important, especially when university funds can be accessed for knowledge-mobilisation purposes. Including a podcast in the knowledge-mobilisation strategy of an academic project may secure funding for participants, equipment, and editing. In grassroots contexts, skills, equipment, and funds are often crowdsourced in the form of informal networks, fundraising, or donations.

These steps are crucial for designing a podcast project with clear goals and a transparent, collaborative, creative process. The results of brainstorming can be used to secure funding, to approach participants, and to promote the podcast.

In the following section, various uses of social media for feminist activism are explored, highlighting their empowering and subversive potential in different geographical, socio-political, and cultural contexts.

Section 3. Social media as a critical toolbox for feminist scholar-activism

Social media can provide a space of political engagement for feminist scholar-activists in local, national, and transnational contexts through practices such as organising, grassroots campaigning, responding to emergencies, fundraising, and aiding marginalised voices to find expression. Feminist scholar-activists use social media to disseminate knowledge and reach broad audiences, to mobilise around academic labour issues, to identify power imbalances and discrimination, and to form communities of solidarity and advocacy (Devereaux, 2014; Board, 2020). The possibilities of communication offered by social media can add another dimension to feminist scholar-activist decolonial praxis, creating tools for nurturing less exploitative and more respectful relations between researchers and the subjects of research, as well as between white and racialised researchers in hierarchical research teams (see Chapter 2, Section 1 Decolonising knowledge production in feminist urban research). Examples of ways in which feminist scholar-activists have used social media are provided here.

- They can use social media to co-ordinate direct action and responses to issues of concern, by campaigning, fundraising, crowdsourcing, and developing mutual aid (Taiwo, Baird and Sanders, 2022).
- They can archive material, including newsletters, pamphlets, flyers, zines, magazines, books, and other print ephemera, related to past and present struggles and to the histories of organisations and movements (Corbman, 2014).
- Feminist and queer activists have been producing material to disseminate ideas as well as critiques of broader issues in society and within social movements. Critiques take different forms. They can be translated into different languages; employ different tools (podcasts, blogs, personal accounts, feminist memes, and humorous tweets) and convey messages, promote mobilisation, and generate discussion (Hohenstein, 2016).
- Social media can be used to disseminate information, knowledge, and resources that are useful to specific communities or fields. For example, they can provide information on access to healthcare, including reproductive rights, sexual health, and mental health. They can be used to set up online support groups; to share knowledge on navigating healthcare systems (e.g. gaining access to HIV-related healthcare or to lesbian and trans healthcare); and to host directories on community resources (such as harm reduction and services for vulnerable communities) (Daniels, 2009; Younas, Naseem and Mustafa, 2020).
- Social media have been used as tools for organising against oppression and raising awareness across the world. For example, in Canada Indigenous women's activism has been at the forefront of Indigenous communities' engagement with social media, with campaigns such as the REDress Project and Walking with Our Sisters highlighting the violence against Indigenous girls and women that has been enabled and naturalised through colonialism (Saramo, 2016). In a different context, Indigenous women's ecofeminist activism against resource extraction is exposing oppression and forging alliances between different groups in Amazonia and internationally (Sempértegui, 2021). In organising against authoritarian regimes, feminist activists have promoted oppositional ideas and generated critique, building solidarity and raising awareness across national and diaspora communities (Kasana, 2014; Batmanghelichi and Mouri, 2017).
- Social media can be adopted by those trying to create safe spaces and document experiences of oppression both online and offline. As the backlash against feminism has intensified during the last decade, feminist activists have used social media to combat misogyny, homophobia, harassment, bullying, and the spread of racist, sexist, homophobic, and abusive rhetoric online (Shaw, 2014; Megarry, 2018; Vickery and Everbach, 2018; Loney-Howes, 2020). From dedicated blogs to posts on personal accounts and those of feminist groups, particular forms of violence against women, marginalised individuals, and communities are being addressed, changing mainstream patriarchal narratives, and offering community care and support to survivors while often holding their abusers publicly accountable (Loney-Howes, 2020). For example, the recent case at Harvard in which the renowned anthropologist, Professor John L. Comaroff, returned to teaching after sexual harassment accusations by his students led to feminist organising to create safe spaces on the campus and online (Hamid and Schisgall, 2023).

The practices described here are all key aspects of feminist local and transnational organising but also of individuals' engagement with feminist praxis online. Table 20.2 shares how feminist scholar-activists build on them to incorporate social media into their feminist praxis within the academy.

Feminist scholar-activists engage in the practices identified to different extents and in the context of their own lived realities, their relationship with academic and grassroots communities, and their political goals and priorities. GenUrb example 20.1 describes GenUrb's social media engagement in the context of transnational feminist scholar-activism.

Table 20.2 Social media practices for feminist scholar-activism in the academy

Documenting lived experiences
Social media are used as a platform for feminist scholar-activists (and individuals from marginalised communities) to describe their lived experiences, often revealing the power dynamics and biases of the academy. This practice involves issues that are often not talked about openly or for which there are no face-to-face venues. It has been used to report precarious academic labour conditions and sexist, homophobic, or racist behaviours in the academy and to promote calls for further action (Greenhalgh, 2019; Talbot and Pownall, 2022).
Pedagogy
Social media have been integrated into pedagogic practices – for example, to engage with ideas, examples, and the sharing of research, often in interactive ways (Guillard, 2016). It has also been used to establish informal off-campus co-learning networks between students, scholars, and communities (for example, in remote teaching during the COVID-19 pandemic) (see Greenhow, Li and Mai, 2019; Khan *et al.*, 2021).
Conducting remote and participatory research
Social media are used for participant outreach and for conducting preliminary and remote research; coordinating participation and providing feedback; connecting fellow researchers; and crowdsourcing data (Sheftel and Zembrzycki, 2017). Facebook and X are often used for participant outreach, and features such as video and voice calls are useful for conducting remote interviews and focus groups (see Chapter 14, Section 1 (ii) Modes of interview).
Knowledge-mobilisation activities
Social media are used to share resources and knowledge in various formats beyond academic institutions, such as pedagogical material or data from research projects. Through social media, links to syllabi, bibliographies, online free courses, livestreamed public lectures and discussions, resource lists, open access publications, archives, and podcasts can be shared (see Chapter 19, Section 1 Knowledge mobilisation).
Networking and professional development
ResearchGate, Academia.edu, X, and LinkedIn are used for the promotion of research through personal profiles, accessing professional development networks, finding collaborators, sharing job and funding opportunities, and for promoting and planning events such as conference, workshops, and calls for papers or other forms of participation. Additionally, virtual writing groups can be created and coordinated through social media (Greenhalgh, 2019; Talbot and Pownall, 2022).
Community and solidarity building
Social media have been used to connect feminist scholar-activists within institutions; across academic communities in different disciplines, locations, or institutions; and with marginalised communities and individuals in and outside the academy. Solidarity is often practised by publicising of community needs and demands on campuses; co-ordinating community responses; and calls to action through threads and Facebook groups, group chats on instant messaging apps, phone zaps, or online petitions (Linder *et al.*, 2016). Such practices contribute to feminist decolonial praxis within and beyond the academy, through redistribution of university resources towards marginalised and Indigenous communities.

GenUrb example 20.1: Social media engagement

Overall, GenUrb's engagement with social media has been muted, a consequence of having City Research Teams (CRTs) located in countries in which the risks attached to conducting feminist research cannot be overemphasised. This has resulted in some CRTs being unable to engage in social media and GenUrb having to impose an embargo on several of our activities, effectively preventing us from employing social media to advertise the events.

GenUrb has employed the following social media engagements and practices (see the book's website for the book's audio clips and podcasts www.routledge.com/9781032668680).

- From the outset, we have had a Twitter (now X) account, @GenUrbNetwork. Where possible, we use it to inform people about activities such as conferences, conference sessions, or workshops in which we are participating. We also use this account to advertise outputs from the GenUrb project, primarily publications.
- Since the first year of the project, we have had a GenUrb website (www.yorku.ca/genurb) that serves the same function as Twitter/X but also provides context about the organisational structure of GenUrb, of our committees, and of the CRTs.
- Individual CRTs have also employed social media.
  - The Shanghai CRT have used WeChat, Weibo, and Douban (a social networking platform with content on books, films, and music) to promote their 2021 book, *Ordinary Women: Qualitative Research on Workers' New Villages in Shanghai* (published in Chinese by Shanghai Literature and Art Publishing House).
  - The Cochabamba CRT, with the help of a local Quechua graphic illustrator, produced a series of short videos and graphic stills in Quechua and Spanish that detailed best practices in relation to COVID-19 (such as washing hands and wearing masks). The videos were uploaded to YouTube to facilitate sharing via WhatsApp, the messaging app of choice in Bolivia (see Chapter 15, GenUrb example 15.6 *Relatos en voz alta* – multimedia translation in Cochabamba)
  - Sound clips have also been recorded by members of the CRTs, describing their work and explaining who they are (see Chapter 0, Audio clip 2.1 The City Research Teams core members). They have also provided soundscapes of the environments in which they work (see Chapter 13, Audio clip 13.3 Navigating understandings of place).
  - Most recently, the GenUrb project has undertaken the production of podcasts produced by the author of this chapter, Mantha Katsikana, a PhD student in geography at York University, with an extensive background in feminist and queer activism and creative arts. The podcasts feature interviews on feminist scholar-activism and serve the important purpose of disrupting the dominant Anglo-American voice of academia (see Chapter 4, Audio Clip 4.1, Differing relationships to activism). In bringing to light the different voices of GenUrb, they open windows into the national and cultural contexts in which we operate. It is these very contexts that also explain why our social media presence is less than stellar.

By Linda Peake

As indicated by GenUrb's experience using social media and also by feminist activists in general, online presence is entangled with the challenges of everyday lived experiences that can differ for feminists because of their gender, race, class, institutional affiliation, geographic location, and socio-political context.

A further issue raised by online presence is concerned with ethical citation when sharing material or content online from different sources. The sudden extreme popularity of certain social media posts and content (popularly described as 'going viral') can often lead to a failure to credit the original source or creator, leading to misinterpretation and to the (mis)appropriation of authors' and content creators' work and intellectual property. The social inequalities and injustices that transfer onto the online realm, together with the structural biases of the internet and of social media, mean that social media citation politics disadvantage marginalised communities and individuals disproportionally. The material and intellectual labour of these communities and individuals is often miscredited or not credited at all, as noted by Black activists and scholar-activists (Nash, 2020; Thompson, 2023). Moreover, as corporations and individuals see online engagement and sharing of social-justice content as a way to build a more 'inclusive' and 'socially conscious' profile for themselves online, appropriation of activist material has been widespread. In other cases, a broader, uncritical engagement with social media leads users to neglect the task of providing the original source.

By engaging in ethical citation online users can contribute to how social change is made possible and transmitted within social media. They can also use their own platform and online networks to publicise the work and demands of groups and individuals that have been historically excluded from public dialogue and knowledge production. Reflection exercise 20.2 aims to help you engage in ethical citation online, building on feminist citation politics and decolonising knowledge production (see Chapter 2, Section 1 Decolonising knowledge production in feminist urban research).

Reflection exercise 20.2: Ethical citation practices for social media engagement

An ethical citation praxis in the context of social media engagement includes crediting the original author or creator; recognising where other users have miscredited or entirely neglected to credit a source or when they have (intentionally or otherwise) shared material out of context; and identifying cases of plagiarism. This exercise contains suggestions and strategies for developing ethical citation practices.

- As a reader, reflect on your practices of sharing content and information. How often do you share material found online? How often do you repost material from other accounts? Are you always familiar with the background of the creator or the spectrum of information the original source was sharing?
- Before sharing a post or information found online, think about the original source. Crediting the original author or creator when sharing material online is the key to ethical citation. This can be achieved through directly posting from the profile or site of the original creator and choosing to include a preview or screenshot of the original post.

- If you spot an uncredited post on your social media feed, can you trace this back to the original source? If you are able to trace the original source, think about the ways you could triangulate the results, confirming the accuracy in more than one way (for example through reverse-searching).
- If you spot a viral post that you suspect or already know to be miscredited or uncredited, ask yourself whether it is possible to trace this back to the original creator. If so, try to post a comment providing the original source, or repost providing the correct citation for others to share too. This practice often takes the form of publicly exposing and correcting the miscrediting, addressing the user/account/corporation responsible, and demanding accountability and transparency. In cases where profit is being made, users can demand the compensation of the original creators through comments and reposts or by direct messaging the offending poster.
- Reflect on your own privileges and platform as a user of social media. What is your following? Are you part of online communities or private groups/forums? Do you have people in your online networks who could influence a bigger audience? Such contacts might provide an opportunity to increase the exposure of material related to social justice or coming from marginalised creators and sources or to connect online communities with creators whose material they would benefit from.
- Reflect on your personal choices in resharing material. Are you sharing material on an issue from creators who have a lived experience of the issue or from someone with influence but no direct experience? It is important to raise the profile of marginalised authors and creators.
- Addressing questions such as these can help ensure you follow ethical citation practices online, reflecting on your own privileges, responsibility, and capacity as a user of social media interested in social justice and feminist praxis. An ethical citation praxis can contribute to collective goals of social justice and to a pedagogy of turning social media into safer places for marginalised communities.

In the following section, the ethical and political challenges facing feminist scholar-activists in their use of social media are illustrated in greater depth.

Section 4. Challenges faced using social media in feminist scholar-activism

Feminist scholar-activists face ethical and political challenges emerging from the use of social media that have been proven to be very hostile spaces for marginalised groups, especially along the lines of gender, race, class, sexuality, and disability. Feminist scholar-activists need to take on board these challenges in a way that is compatible with their feminist ethics and takes their personal and collective wellbeing into account. We outline some relevant issues in what follows.

(i) The politics of social media engagement

The structure of social media, based on profiles and networked communication, has been criticised as glorifying the individual and the promotion of the 'personal brand'. Users can gain popularity by increasing the number of followers they have, building an audience, and promoting their views, selling themselves in the neoliberal marketplace (Ringrose, 2018; Novoselova and Jenson, 2019). Feminist scholar-activists are not immune from such marketing practices, since having an online audience and measurable engagement is increasingly seen as necessary for the advancement of one's career and especially so for early-career scholars (Duffy and Pooley, 2017). They often find themselves caught between using social media for professional development, on the one hand, and knowledge mobilisation aimed at broader audiences, on the other (Baer, 2016). To complicate issues further, while funding agencies and universities are interested in analytics, the dissemination of a scholar's work online through blogs and other forms of public engagement is not valued as valid academic work, even though it requires time, effort, and emotional energy on top of already-heavy academic workloads.

Not all feminist scholars can easily develop an online profile. Those working outside the hegemony of English-language and global North–based institutions face challenges in maintaining online engagement. Such challenges often result in feminist scholars and scholar-activists from the global South not being able to engage with social media to the same extent or with the same reach, showing that the realities of online presence are directly linked to colonial regimes of knowledge. Facing challenges in maintaining online engagement might also be the result of not having access to an internet connection and technological devices, or of local regimes of surveillance and censorship (see Chapter 4, Section 4 (i) Feminist scholar-activist research and transnational power dynamics). Members of these networks may be dealing with broader political conditions such as a hostility to feminist activism and research. In GenUrb example 20.2 we discuss how GenUrb members have had to adjust their knowledge-mobilisation and public-engagement strategies accordingly.

GenUrb example 20.2: On the ethics and politics of social media engagement

On several occasions and at various stages of the research, we have been confronted with ethical and political considerations that have worked against our having an online presence. While ethical issues have manifested primarily in relation to the physical safety of GenUrb researchers and participants, political concerns have been linked to governments in India, Israel, and China, that are to varying degrees antagonistic to feminist research and researchers (see Chapter 0, GenUrb example 0.1 The evolution of the GenUrb partnership). These issues first became apparent when we wanted to populate the website with information about and photographs of our CRT members. In Anglo-American academic contexts these are standard procedures: academics in the Western academy may be photo-shy and even more reluctant about photos being updated, but such activities have become normalised. In the non-Western academy, however, such procedures can risk identification by regimes that are less than supportive of feminist scholarship and of Western 'interference'. It follows that our website is unevenly populated across the CRTs.

There have also been occasions when we have been unable to employ social media to advertise GenUrb events. Our three-day online workshop in August 2021, with Chinese feminist urban scholars from China, Europe, and North America, had to be held in secret with closed sessions. The climate for feminist (and queer) scholars in China has significantly deteriorated in the period that GenUrb has been operating, and the safety of the feminist scholars joining from China had to be protected. Similarly, a five-day GenUrb workshop held in Ramallah, Palestine, in September 2022 also had to be conducted covertly. Publicising the activity could have resulted in an entry ban on international participants, Israel having passed more regulations aimed at further isolating Palestinians. In addition, the GenUrb field site in Ein Qiniya, on the outskirts of Ramallah, has been the subject of intense land grabs by affluent elites close to the ruling Palestinian Authority.

Outputs from these events are forthcoming, and their temporal distance from the event in which they were initiated serves as a sufficient safety valve for participants. They include an edited book with Routledge, tentatively entitled *Feminist Explorations of Urban China*, and a book published in Arabic, *Takhayali Ein Quiniya* (Imagine Ein Quiniya), which explores urban futurities through female residents' visual representations of how they hope Ein Quiniya will develop in years to come.

In the online promotion (and writing) of their book *Ordinary Women*, the Shanghai CRT had to embark on a process of self-censorship and not mention feminism. This practice was necessary – given book censorship in the People's Republic of China and the climate of anti-feminism promoted by the state – to protect the research participants, researchers, and government officials who supported their research from state censorship and unpredictable repercussions.

In GenUrb, safety is our bottom line, and engagement in social media works against, and indeed sometimes exacerbates, these safety considerations, inhibiting our online presence.

By Linda Peake

(ii) Online privacy, surveillance, and censorship

Concerns over privacy and the use of personal data by tech companies, institutions, and the state have become pressing issues for online feminist scholar-activists. User data (including pictures and location logs) have been employed for identification purposes by the authorities policing political uprisings and protests to monitor and profile marginalised individuals and protesters (Linabary, Corple and Cooky, 2020). Social media advertising has also been used to influence political outcomes (Isaak and Hanna, 2018). In oppressive regimes, breaches of online privacy and anonymity can pose a safety issue for feminists, activists, scholars, and public intellectuals. Some universities and institutions also monitor and censor scholars' and faculty's online activity on social media, resulting in scholars having a negative relationship with their employers or even having their employment terminated (Salaita, 2015; Flaherty, 2022).

At the same time, feminist activists often face censorship and shadow banning (platforms removing content or reducing its visibility, hindering the creator's ability to interact with their audience). Rooted in 'the male bias in platforms and practices' (Megarry, 2018, p. 1074), the development of algorithms and automation can make it very difficult for feminist activist material to reach target audiences. For instance, social media posts addressing reproductive rights or displaying gendered bodies in non-sexualised ways are often targeted by hate speech or inappropriate content by algorithms without any opportunity for critical feedback from their creators. In some cases, feminist creators and scholars might be deprived of their entire account, thus losing an archive of public engagement and work but also their connection with their audience.

(iii) Connecting online activism with on-the-ground organising

Online activism is often criticised as inauthentic and performative, since it frequently does not require direct action (Loney-Howes, 2020). Rather, it focuses on discussions around activist methods, campaigning, organising, coordinating, and providing updates, all activities that are rarely acknowledged as activist work, especially when performed by women (Guillard, 2016). Feminist scholar-activists have highlighted the amount of effort and demands of such 'soft' activism alongside the fact that on-the-ground activism is not always possible for marginalised individuals who cannot participate due to limited access in terms of time, energy, mobility, and ability. Feminist academics have argued that engagement in online scholar-activism should be included in an understanding of activism as it engages in knowledge mobilisation and informs connections and communities across a range of geographical and political contexts (Jeppesen and Sartoretto, 2023). Despite complexities of access, privilege, and systemic inequalities, online transnational feminist networks have made issues visible, raised awareness, and pushed for change in coordination with on-the-ground activist work (Matos, 2017).

(iv) Exposure to online violence and its cost

While feminist scholar-activists use social media their online visibility exposes them to misogynists, white supremacists, and others spreading hate campaigns and to witnessing the online harassment of others (Nakamura, 2014; Shaw, 2014; Sultana, 2018b; Vickery and Everbach, 2018; Mott and Cockayne, 2021). Feminist material online is targeted by harassers with online sexism, homophobia, and racism often taking the form of offensive content, enabled by anonymity and often framed as 'humour' (Kavanagh and Brown, 2020, p. 1381). Online harassment disproportionally affects Black women and women of colour (including trans women) as sexism and homophobia intersect with racism and other forms of oppression online (Board, 2020). Feminist activist and academic events online have been targeted in cyberattacks (see GenUrb example 20.3). The exposure to harassment may traumatise communities, and individuals and it also adds to the amount of emotional labour, effort, and energy online activism requires. To complicate matters further, although having a presence and disseminating knowledge online are increasingly expected of faculty members at many universities, the amount of labour required to exist online, exposed to the intersections of gender, race, class, and disability, is hardly acknowledged (Kavanagh and Brown, 2020).

GenUrb example 20.3: 'Porn bombing' at a GenUrb conference session

At the July 2022 International Geographical Union (IGU) conference (held in Paris and online), GenUrb held a virtual session entitled 'Theorising Feminist Urban Comparative Research'. We were used to conference organising that appeared to place little value on feminist sessions, so we were not surprised that the IGU scheduled each Gender and Geography Commission session at the same time each day, making them compete. What we were not prepared for, however, was being 'Zoom bombed': having our session taken over by particularly virulent pornography. One minute we were looking at each other's faces on the Zoom call and the next we were watching intimate shots in close-up of Black male bodies engaged in violent sex and a voice-over discussing pornographic acts with children. It was difficult to make sense of what was happening. Not only the unexpectedness of being caught up in such an encounter but also the dissonance between what we were viewing on screen and what we were listening to made comprehension difficult. Perhaps it was only a few seconds before the realisation dawned, but in that pre-cognitive moment the affective void was filled with anger and disgust. When the mind, frantically racing to catch up, came into sync, a feeling formed of being dragged from an underwater abyss. Suddenly I could hear confused and upset voices shouting, 'turn it off!', 'what can we do?', 'who do we call?'. It did not take long for the more technically minded among us to find a way of removing the intruder who had taken over the screen.

In the calm that ensued after such a raucous intervention, we noticed that others were lurking in the chat section of the Zoom call, their inappropriate questions marking them out as less-than-welcome visitors. After expelling them, we decided to close down the chat and not take any questions from the audience. We agreed, though, to continue with the presentations. In hindsight, this was probably not the best decision to have made. The deeply intense feelings of being violated deserved more of our attention. Our anger and other emotions needed an outlet, but their aftermath was, at this stage, unknown to us.

That same day, we agreed to inform Zoom about what had happened. Following their complaints procedure, however, required us to trace the IP address of the Zoom bomber, and this was not possible. We had to pass the complaint on to York University's internet technology services, who were able to pass our complaint on to Zoom.

We also wrote to the organising body of the conference in Paris. It was clear that whatever security measures they had in place were inadequate. The online link for joining the conference's Zoom sessions could be accessed by anyone, and it was not curtained behind any paywall for the conference. The Paris committee claimed never to have heard of 'Zoom bombing'.

Perhaps it was too much to expect that by 2022 the abuse to which women and feminists are exposed online would be known to international conference organisers, or that someone on the organising committee would have addressed questions of online conference security as a matter of course. We asked the Paris organising committee for our money back. Sympathetic to our situation, they willingly complied. We asked for conference security to be an issue that would, in future, be automatically addressed and ensured not only by local conference organising committees but by the IGU executive. We have been assured it will be.

Neither of these forms of redress, however, could touch the culprits, the misogynistic vandals who invaded our space and attempted to wreck our session. Whether such redresses will serve in the future to puncture the carapace of women- and feminist-hating porn bombing that online environments spew out remains to be seen. One thing I do know, however: a year later my anger at this event has not gone away. That can only be a force for good.

By Linda Peake

Section 5. Summary

This chapter has explored social media as a critical toolbox for feminist online engagement. It has provided a guide to popular social media tools and their uses, illustrated with examples, including how to design a podcast for the purposes of knowledge mobilisation. The online practices feminists have been using to build community, disseminate knowledge, organise, voice marginalised experiences, and contribute to feminist decolonial praxis more broadly have been explored. Attention was also given to the ways feminist scholar-activists from different contexts use social media for raising awareness about and organising against systemic oppressions in academia; for holding institutions accountable; for forging alliances and carving out spaces for knowledge production; and for disseminating knowledge and resources to marginalised communities. Finally, the chapter has highlighted the ethical and political challenges of engaging in feminist scholar-activism online in different socio-political contexts, along with various responses and coping strategies feminist activists adopt to deal with these challenges.

Links to online materials

The book *Doing feminist urban research: insights from the GenUrb project* references resources that are helpful to feminist urban scholars at every stage of the research process. This companion website contains links to resources referenced in the book, including activist networks, international organisations, universities, research agencies, professional organisations, data analysis and management tools, and much more.

Chapter 0 Introducing GenUrb, by Linda Peake, Araby Smyth, and Nasya S. Razavi

GenUrb https://genurb.apps01.yorku.ca/

GenUrb profile on X (formerly Twitter) https://twitter.com/GenUrbNetwork

Part I. The building blocks of feminist decolonising urban research

Chapter 1 Decolonising feminist knowledge production, by Elsa Koleth

Red Thread https://redthreadguyana.org/

Chapter 2 Feminist comparative urban research, by Linda Peake, Mel Mikhail, and Elsa Koleth

Slum Dwellers International https://sdinet.org/

Chapter 3 Feminist engagements with translation, by Wiley Sharp

No weblinks

Chapter 4 Feminist scholar-activism, by Mantha Katsikana

No weblinks

Part II. The context of 21st-century urban feminist research and policy

Chapter 5 Feminist urban research in the time of COVID-19, by Mel Mikhail

World Health Organization Coronavirus dashboard https://covid19.who.int/

UN Women's Women Count data hub https://data.unwomen.org/resources/covid-19-and-gender-monitor

Global Health 50/50 data tracker https://globalhealth5050.org/the-sex-gender-and-covid-19-project/the-data-tracker/

Chapter 6 Feminist urban policy and the Sustainable Development Goals, by Nasya S. Razavi and Linda Peake

Charter for Women's Right to the City www.hlrn.org/img/documents/WFW-%20%20Charter%20for%20Womens%20Right%20to%20the%20City%20-%202004.pdf
European Charter for Women in the City www.hlrn.org/img/documents/1994%20-%20European%20Charter%20for%20Women%20in%20the%20City.pdf
Inclusion and Right to the City. Exercising Women's Citizen Rights: The Women's Agenda for Rosario, Argentina www.edcities.org/en/wp-content/uploads/sites/2/2015/03/ana-falu-EN.pdf
Toronto Women's City Alliance www.twca.ca
Women's Agenda for the City in Latin America
www.sitiosur.cl/agenda-de-mujeres-por-la-ciudad-en-america-latina/
Women's Right to City Manifesto www.right2city.org/news/womens-right-to-the-city-manifesto/
World Forum of Women. (2004) Charter for women's right to the city. www.hlrn.org/img/documents/WFW-%20%20Charter%20for%20Womens%20Right%20to%20the%20City%20-%202004.pdf
United Nations Sustainable Development Goals, Targets, and Indicators https://sdgs.un.org/goals

Part III. Feminist research standards

Chapter 7 Feminist research ethics, by Linda Peake

No weblinks

Chapter 8 Professional standards in feminist research, by Araby Smyth

Social Science Research Council Singapore www.ssrc.edu.sg/
National Science Foundation USA www.nsf.gov/
Economic and Social Research Council www.ukri.org/councils/esrc/
The City Institute at York University www.yorku.ca/cityinstitute/
Urban Studies Foundation www.urbanstudiesfoundation.org/
University of Washington, Office of Graduate Student Equity & Excellence https://grad.uw.edu/equity-justice/gsee-graduate-student-equity-excellence/
American Association of Geographers www.aag.org/
Black Geographers www.blackgeographers.com/
Black Geographies Specialty Group https://blackgeographies.org/
Space, Sexualities and Queer Research Group https://ssqrg.org/
Society of Woman Geographers www.iswg.org/
Feminist Geographies Specialty Group https://feministgeographies.org/
Feminist Intersectional Solidarity Group www.cag-acg.ca/figs
GeoBrujas, Comunidad de Geógrafas www.facebook.com/geobrujas/
Tri-Council Policy Statement: Ethical Conduct for Research Involving Humans – TCPS 2 (2018) https://ethics.gc.ca/eng/policy-politique_tcps2-eptc2_2018.html
Tri-Agency Framework: Responsible Conduct of Research (2021) https://rcr.ethics.gc.ca/eng/framework-cadre-2021.html

Chapter 9 Partnerships and teamwork in feminist collaborations, by Araby Smyth

Social Sciences and Humanities Research Council https://www.sshrc-crsh.gc.ca/home-accueil-eng.aspx

Chapter 10 Data management in feminist research projects, by Mel Mikhail

Whose Knowledge https://whoseknowledge.org/

Tri-Agency Research Data Management Policy by Canada's Tri-Agency funding programs (Social Sciences and Humanities Research Council of Canada, Canadian Institutes of Health Research, and National Sciences and Engineering Research Council) https://science.gc.ca/site/science/en/interagency-research-funding/policies-and-guidelines/research-data-management/tri-agency-research-data-management-policy

Research Data Management Capacity Building Initiative by Social Sciences and Humanities Research Council of Canada www.sshrc-crsh.gc.ca/funding-financement/programs-programmes/data_management-gestion_des_donnees-eng.aspx

Research Data Lifecycle by the UK Data Service www.youtube.com/watch?v=-wjFMMQD3UA

Responsible Data Governance for M&E in Africa. Part 1: Overview of Data Governance by Responsible Data in M&E (RDiME) Alliance www.im-portal.org/help-library/responsible-data-governance-for-me-in-africa-part-1-overview-of-data-governance

Responsible Data Governance for M&E in Africa. Part 2: Practical Guidance on Responsible Data Governance in M&E by Responsible Data in M&E (RDiME) Alliance www.im-portal.org/help-library/responsible-data-governance-for-me-in-africa-part-2-practical-guidance-on-responsible

OpenDOAR Global Directory of Open Access Repositories https://v2.sherpa.ac.uk/opendoar/

arXIV open-access repository of electronic preprints and postprints https://arxiv.org/

Zenodo open repository developed under the European OpenAIRE program and operated by CERN https://zenodo.org/

How to Open Up Data by Open Data Handbook https://opendatahandbook.org/guide/en/how-to-open-up-data/

Scenarios (international property rights scenarios for data sharing) by UK Data Service https://ukdataservice.ac.uk/learning-hub/research-data-management/rights-in-data/scenarios/

Creative Commons, international nonprofit organisation devoted to educational access that has released several copyright licenses, known as Creative Commons licenses, free of charge to the public https://creativecommons.org/

Licenses by Open Data Commons, Open Knowledge Foundation https://opendatacommons.org/licenses/

File Formats by UK Data Service https://ukdataservice.ac.uk/learning-hub/research-data-management/format-your-data/file-formats/

Research Data Management by UK Data Service https://ukdataservice.ac.uk/learning-hub/research-data-management/

DMPTool, service of the California Digital Library, University of California https://dmptool.org/

ReNamer, batch file renamer https://renamer.com/
Bulk Rename Utility, free file renaming utility for Windows www.bulkrenameutility.co.uk/
Git distributed version control system https://git-scm.com/about
The Anonymisation Decision-Making Framework: European Practitioners' Guide by Mark Elliot, Elaine Mackey, and Kieron O'Hara https://msrbcel.files.wordpress.com/2020/11/adf-2nd-edition-1.pdf
Mukurtu content management system and preservation protocol https://mukurtu.org/about/
South Asian American Digital Archive www.saada.org/browse
Chicacan por mi raza digital memory collective https://chicanapormiraza.org/
Feminist Principles of the Internet https://feministinternet.org/
Feminist Data Manifest-No www.manifestno.com/
Figshare online open access repository https://figshare.com/
Registry of Research Data Repositories www.re3data.org/
Making a research project understandable: Guide for data documentation by Siiri Fuchs & Mari Elisa Kuusniemi, Helsinki University Library, Data Support https://zenodo.org/record/1914401#.Xp8DVshKjIU
Data Documentation Initiative https://ddialliance.org/
A Toolkit for Centering Racial Equity Throughout Data Integration by Actionable Intelligence for Social Policy at University of Pennsylvania https://aisp.upenn.edu/wp-content/uploads/2022/07/AISP-Toolkit_5.27.20.pdf
Data Security Procedures for Researchers by Poverty Action Lab www.povertyactionlab.org/sites/default/files/Data_Security_Procedures_December.pdf
FAIR (Findable, Accessible, Interoperable, and Reusable) Principles by GO FAIR Initiative www.go-fair.org/fair-principles/

Part IV. Feminist methodologies and research methods

Chapter 11 Feminist methodologies and methods, by Linda Peake and Mel Mikhail

No weblinks

Chapter 12 Feminist approaches to fieldwork, by Araby Smyth, Elsa Koleth, and Linda Peake

Lens in Bloom: A Photography Journal by Community X Women https://genurb.apps01.yorku.ca/wp-content/uploads/2020/06/GenUrb%C3%97Vox-Book_Final.pdf
takhayali.net

Chapter 13 Feminist geo-ethnography, by Araby Smyth and Linda Peake

No weblinks

Chapter 14 Feminist interviews, by Araby Smyth, Linda Peake, and Elsa Koleth

Red Thread https://redthreadguyana.org/

Part V. Feminist data analysis

Chapter 15 Feminist practices of translation and interpreting, by Carmen Ponce

Cochabamba CRT website for information on COVID-19 www.relatosenvozalta.com

Quechua graphic illustrator, Phuyu (Facebook: @phuyudibujando; Instagram: @ phuyu.no)

Chapter 16 Feminist approaches to qualitative data analysis, by Linda Peake and Elsa Koleth

No weblinks

Chapter 17 Software-aided analysis for feminist research, by Biftu Yousuf

SPSS Statistical Package for the Social Sciences www.ibm.com/spss
QSR International on how to manage NVivo projects with Windows and MacOS https://help-nv.qsrinternational.com/20/win/Content/projects-teamwork/work-with-projects-windows-mac.htm?Highlight=mac
NVivo User Help – Windows https://help-nv.qsrinternational.com/20/win/Content/welcome.htm#
NVivo User Help – MacOS https://help-nv.qsrinternational.com/20/mac/Content/welcome.htm

Chapter 18 Using NVivo in feminist research, by Biftu Yousuf

At the time of writing QSR (NVivo's parent company) has combined with Lumivero and the QSR 'user help website' is no longer available. Lumivero is still creating and publishing 'online help pages and user guides' for each NVivo release. These are available for versions 10 and beyond, but only versions 12 and later are currently supported by Lumivero. See the weblinks below to access help.
NVivo User Help – Windows https://help-nv.qsrinternational.com/20/win/Content/welcome.htm#
NVivo User Help – MacOS https://help-nv.qsrinternational.com/20/mac/Content/welcome.htm

Part VI. Feminist approaches to knowledge mobilisation

Chapter 19 Knowledge mobilisation in a feminist project, by Araby Smyth, Linda Peake, and Jenna Blower

Social Science Research Council Singapore www.ssrc.edu.sg/
National Research Foundation South Africa www.nrf.ac.za/
Research Excellence Framework (REF) https://re.ukri.org/research/ref-impact/
Economic and Social Research Council UK https://esrc.ukri.org/
The Storytellers by Social Sciences and Humanities Research Council of Canada www.sshrc-crsh.gc.ca/society-societe/storytellers-jai_une_histoire_a_raconter/index-eng.aspx
Mandala de la vida: Mandala of Life Storytelling Workshop Guide by Nasya S. Razavi https://genurb.apps01.yorku.ca/about/city-research-teams/cochabamba/mandala-de-la-vida-mandala-of-life-storytelling-workshop-guide/

Chapter 20 Feminist engagement with social media, by Mantha Katsikana

GenUrb's Twitter (now X) account @GenUrbNetwork
GenUrb's website www.yorku.ca/genurb
Cochabamba CRT website for information on COVID-19 www.relatosenvozalta.com
Queering the Map www.queeringthemap.com

Glossary

Term	*Description*
Accountability	An important element of situated knowledge that refers to the willingness or obligation to take ownership of one's choices and actions.
Activism	The activity of collectively working to achieve political, economic, social, or environmental change towards a perceived greater good that depends on (usually long-term) commitment to a cause or an ideal.
Black feminism	A movement that centres primarily on the intersectional racism, sexism, and classism experienced by Black women.
CAQDAS	Computer-assisted qualitative data analysis software refers to computer programs that support the qualitative research process, including transcription, analysis, and dissemination.
Care	The everyday practices of attending to the emotional, material, social, and spiritual needs of other human and non-human beings.
Coding	The process of systematically arranging collected data into meaningful and unified categories for analysis.
Colonialism	The violent means employed by a political entity to dominate people and to dispossess them of their lands and resources for its own economic advantage. It is characterised by establishing a permanent population (settler colonialism), exploitation of resources, and the establishment of a colonial bureaucracy. Even after colonisation ends, the violent and traumatic legacy of colonialism survives and has detrimental impacts on cultures, economics, and politics.
Comparison	A longstanding approach in urban studies that is currently undergoing a revival in interest. It is used to produce knowledge about two or more urban places in relation to one another, with scholarship bringing relational and transnational epistemologies to bear on the urban.
Data	Observations, usually quantitative or qualitative, about the social world. Primary data are those collected for a specific purpose by a researcher, whereas secondary data are sources of data (e.g. census data) previously collected by other researchers.
Data justice	An approach that places social justice at the centre of its engagement with data-driven technologies and transformations. It seeks fairness for individuals and communities adversely affected by processes of data production and analysis.

Data management plan	Living documents that outline how data will be managed during and beyond the life of the research, taking into account the collection, analysis, securing, preservation, archiving, dissemination, and disposal of data.
Decolonisation	The process by which colonised countries gain independence from a colonising power. It is an ongoing struggle that people engage in to dismantle the internalised material and symbolic structures of colonialism.
Deductive analysis	An approach to research that begins by developing testable hypotheses that relate to your research questions and are derived from existing social theories about phenomena in the world. Once all data are collected the hypothesis is tested to see if it can be held true or, if the testing challenges the hypothesis and the underlying theory, the hypothesis can be realised as false.
Difference	A form of relational connectedness in which the other is understood hierarchically in terms of social differences such as gender, class, race, ethnicity, sexuality, ability, and religion.
Dispossession	The capitalist practice of expropriating wealth from public and private realms, depriving individuals or communities of land, property, or other possessions.
Equity, diversity, and inclusion (EDI/DEI)	Equity, diversity, and inclusion are a set of practices dedicated to identifying, understanding, and transforming structural inequalities in the places where we learn and work. EDI is commonly associated with justice (JEDI) and decolonising practices (DEDI).
Embodiment	A feminist concept that focuses on the scale of the body and the body's relationships with the world. Embodiment addresses the physicality and materiality of the body as well as the emotional lived bodily experiences that shape how identities are constructed and performed.
Epistemology	A theory of knowledge. It involves the study of how and what we know and is thus concerned with the methods used to validate knowledge as well as the nature, origin, and scope of knowledge.
Ethics	These are the moral principles that govern a person's behaviour, or the conducting of an activity, as well as a moral philosophy that defines what is morally right and wrong, just or unjust, acceptable and unacceptable in relation to societal norms.
Ethnography	A form of inquiry that typically involves a researcher from one community travelling to another to conduct an in-depth study of people, paying attention to lived experience, social relations, and cultural practices in that community.
Everyday	The spaces, practices, and relationships that shape daily life, including routine interactions with things, places, and environments, and how this life is understood and resisted. The study of everyday life pays close attention to the relations of power and inequality that shape these interactions.
Femicide	A symptom of global patriarchy, femicide refers to the widespread misogynistic hate crime of the targeted killings of women-identified individuals.
Feminism	A series of social and political movements, some local, some global, that seek to question the term 'women' while also challenging the subordinate position of women and other marginalised groups.

Feminist activism	The heterogeneous practices that work toward liberation from gender-based oppression, often taking the shape of collective action by women in grassroots social movements.
Feminist ethics of care	A feminist normative ethical theory that uses a relational and contextual approach toward morality and decision-making. It prioritises care as a virtue with relationships between specific individuals seen as the basis for ethical behaviour.
Feminist geo-ethnography	An approach to research that produces place-based, relational, and embodied research findings that is attuned to scale and is disruptive of oppressive power relations in the everyday. It differs from feminist ethnography primarily in the length of time and number of repeated visits to the 'field'.
Feminist scholar-activism	The work undertaken collectively by feminist scholars in and beyond academia through the adoption of an activist ethos in their research, pedagogy, and service.
Femocrat	A female bureaucrat working in a public body mandated to address policy issues affecting women such as violence, pay equity, and safety.
Field	The setting in which the sphere of inquiry for your research project is located. The field is often thought of as a place that the researcher goes to, but it is more than a bounded physical setting – it is dynamic and cannot be disentangled from the everyday lives of the researcher and participants.
Fieldwork	A process whereby the researcher collects data in the world, often though not always with research participants, using chosen research methods at predetermined field site(s).
Focus groups	A research method that requires assembling a group of people who engage in a moderated discussion around a topic or research question, with an emphasis on the data being a result of the communication between research participants or group interaction.
Gatekeeper	An individual, group, or organisation that, for better or worse, controls or can influence access to potential research participants.
Gender	The socially constructed differences between men and women, boys and girls that refer to gendered characteristics, norms, behaviours, and roles. Traditionally understood as a dualism of femininity and masculinity, it is now regarded as a continuum with a focus on performativity – that is, as something that people 'do'.
Generative comparative strategies	A strategy for comparison based upon distinctiveness that is directly influenced not only by the agency of the urban but also by the subjectivity of the researcher.
Genetic comparative strategies	A strategy for comparison that investigates urban spatialities through differentiation and diversity. While the study of differentiation has been employed primarily by those adopting a materialities approach, diversity has been taken up theoretically by political economy approaches.
Geo-ethnography	A research approach that brings spatiality to ethnographic methods. In addition to considering place-as-locale, it situates place in the broader spatialities of power relations that flow through it, connecting it to other places and scalar circulations of power.
Geographic imaginary	These are our taken-for-granted spatial orderings of the world, which are tacitly valorised and circulated not only through reason (through language and visualisations, such as the gaze) but also through structures of feeling and affect. As such they are more than representations: they are also simultaneously embodied, sensed, and discursively constructed.

Globalisation	The process of increasing economic integration between people, companies, and governments driven by free-market capitalism. Globalisation is characterised by unprecedented levels of economic liberalisation, direct foreign investment, labour migration, communications, and technology exchange.
Global North	As a geopolitical term 'the global North' refers to a common history of development based on colonisation of the global South that has led to high levels of economic and industrial development in the global North. As a geographical term it is commonly understood as a grouping of many Western countries in Australasia, North America, and Europe, as well as Japan and South Korea in East Asia, that contain about one quarter of the world's population – hence the alternative description of the global North as 'the minority world'.
Global South	As a geopolitical term, 'the global South' refers to a common history of imperialism, slavery and colonisation by states from the global North that has led to underdevelopment and political marginalisation. As a geographical term it is commonly understood as a grouping of many countries in Asia, Africa, the Caribbean, Latin America, and Oceania, that contain three-quarters of the world's population – hence the alternative description of the global South as 'the majority world'.
Governance	A set of practices or systems that an organisation, collective, or other such body uses to steward a project or process and specifically one that usually involves complex social dynamics and decision-making.
Grassroots	Ordinary people living in poor rural or urban communities, susceptible to risks and vulnerabilities because of their economic, social, and political marginalisation.
Hegemony	The exercise of power allowing the dominance of one group over another, through the legitimation of norms and the naturalisation of hierarchy, allowing a particular set of ideas to become commonsensical and intuitive, inhibiting the dissemination of alternative ideas.
Hermeneutics	The theory and methodology of the interpretation of written, verbal, and non-verbal communication concerned with analysing the ways in which communication is understood. Understanding is the interpretive act of integrating particular things, such as words, signs, and events, into a meaningful whole. Understanding an object, word, or fact requires us to make sense of it within our own life context.
Imperialism	The expansion of political and economic control of one state over another. Imperialism is often associated with colonialism, although it may not involve territorial acquisition, focusing on expanding political and economic influence often by force or coercion.
Inductive analysis	An approach to research that begins with data collection to explore relationships in the data without (explicit) reference to a theoretical framing. Data analysis and coding are continuous throughout the research process.
Interpreting	Interpreting involves conveying the meaning of spoken or signed words into another language.
Intersectionality	An analytic framework that both highlights the interdependent nature of multiple axes of structural difference, such as race, class, gender, age, ability and sexuality, that have the potential to oppress or empower, and provides a tool by which these power imbalances can be addressed. Its origins are in critical race studies, with the specific oppressions faced by Black women in the USA.

Interviews	Conversations between researchers and research participants allowing the former to gather primary information that serves as data. Research interviews are a power-laden practice wrapped up in research ethics challenges that require consent, preparation, and trust between researcher and participant.
Knowledge mobilisation	The reciprocal and complementary flow and uptake of research knowledge between researchers, knowledge brokers, and knowledge users – both within and beyond academia – to create new knowledge and catalyse change. It is a two-way process and views knowledge as a process, not as a product.
Knowledge production	Modes of conceptualising and conducting research, communicating research findings, and evaluating impact among key stakeholders in academic disciplines and beyond.
Labour	The physical and mental, paid and unpaid work done by people to reproduce human life and produce the goods or services in an economy.
Life histories	A qualitative method, based on an unstructured interview, that allows a person to talk about their life in their own words, with a core focus on the participant's experiences and subjectivities, allowing a researcher to understand the meanings people ascribe to past and present events and experiences and to collect data that provide depth, meaning, and context to their wider social lives.
Metadata	Metadata is defined as structured information that describes, explains, or locates your data. It can be collected for any type of data and it allows for the easy retrieval, management, and use of data. The most common type is descriptive, which can include information such as title, author, date created and file size as well as confidential information about the source of the data.
Methodology	A theoretically informed framework of principles and procedures for conducting research.
Methods	Techniques for the quantitative and qualitative collection of data, determined primarily by the methodology and epistemology employed in the research.
Monitoring and evaluation	Assessment of the progress, goals, and impact of a research project. Monitoring includes the continuous assessment of the research project, often conducted with the aim of fine-tuning research design. Evaluation involves the examination of the degree of success of the research project in terms of its relevance, effectiveness, efficiency, and the impact of activities undertaken with relation to short- and long-term objectives.
Neoliberalism	A hegemonic political project to advance free-market capitalism through privatisation, deregulation, free trade, fiscal austerity, and the reduction of state spending on social programmes.
NVivo	NVivo is a qualitative-analysis software. It works as a digital research tool that can be used with both textual and non-textual data.
Ontology	A branch of philosophy that asks what exists. It involves making claims about the nature of being as well as related concepts such as existence, becoming, and reality.
Oral histories	A research-interview method that utilises a loosely structured format and is based on listening to individuals talk about their lives, producing richly detailed and intimate primary data.

Participant observation — A qualitative research method in which the researcher participates in the day-to-day activities of research participants, asking questions and keeping records of observations about these activities. The questions are designed to uncover the meanings behind behaviours and the rules and norms that the participants may encounter.

Place — A complex synthesis of geographical location, human experience, and the spatialisation of power relations. Place plays an integral role in the world, making things possible and allowing them to emerge and develop in particular ways.

Place identity — A concept from environmental psychology which proposes that a person's knowledge and feelings, as well as their identity, develop through experience of their environment.

Positivism — A research philosophy that approaches the study of society by relying on scientific evidence, such as statistics and experimental results, to reveal the true nature of how society operates. It adheres to the view that only 'factual' knowledge gained is trustworthy and holds that society, like the physical world, operates according to general laws.

Post-colonialism — The experience and legacy of colonialism and imperialism marked by its past and present effects. Its study encompasses not only the historical period or state of affairs representing the aftermath of Western colonialism but also of the ongoing effects of colonialism on a former colony.

Positionality — A methodological endeavour to reflect on one's own situatedness and position within a given cultural context and to expose the interpersonal power dynamics that underpin research activities.

Professional standards — In the academy this refers to a framework of practices, ethics, and behaviours that govern work and relationships within the university.

Praxis — The process by which an idea is put into practice, shaping and changing the world, through a commitment to human well-being, a search for 'truth', and respect for others.

Production — The process of producing products (goods and services), which requires land, labour, and capital. These 'means of production' explain the relationship between products, their ownership, and the labour needed to produce them.

Qualitative analysis software — Computer-based programs, such as ATLAS.ti, Dedoose, MAXQDA, and NVivo, that support the qualitative research process. They offer facilities for transcription, analysis, and dissemination.

Qualitative data — Qualitative data is non-numeric information, such as interview and focus-group transcripts, personal diaries, field notes, maps, photographs, and other printed materials, audio-visual recordings, and images.

Questionnaires — A research instrument that comprises a set of questions or other prompts that aim to collect information from a respondent, typically employed in a survey. It is usually a mix of close-ended and open-ended questions and can be administered face to face, on the phone, or online.

Race — Race is a social construction that has emerged from assumptions of phenotypical and genotypical differences between groups of people. Human bodies become racialised not through biology but through discursive processes of racialisation and the lived conditions of racism that result: race may be a concept, but it is not an abstract concept.

Reflexivity	The examination of one's own feelings, beliefs, judgements, and practices during the research process, with the aim of identifying how these may influence the research.
Research ethics	The normative principles that govern how researchers should conduct their research as well as how they analyse complex problems and issues as they arise. Research ethics aim to minimise harm, mitigate risk, and protect human dignity.
Research ethics policies	Sets of rules and regulations that govern the conduct of research to mitigate possible harms to participants, researchers, and broader communities. They are commonly defined by national grant agencies and/or by ethics committees in universities.
Research partnerships	The bringing together of people and/or organisations from differing areas of expertise as well as with different financial and intellectual resources to conduct research that addresses challenges relevant to diverse stakeholders.
Settler colonialism	A violent ongoing colonial system of power characterised by the genocide and repression of Indigenous people and by continuous settler occupation and exploitation of lands and resources.
Secondary data	Data that already exist, collected by people or organisations other than the person now using them for a separate purpose.
Social media	The creation and exchange of ideas and information in many formats that takes place on online platforms.
Social reproduction	Labour, primarily that of women, that makes possible the continuation of human life on a daily and generational basis and that takes place inside and outside the home, particularly in institutions of education, health, and social care, and that can be global in its reach, as in transnational care chains.
Storytelling	A method of collecting data and mode of disseminating research results, storytelling is the interactive activity of sharing narratives that connects people.
Sustainable Development Goals (SDGs)	The SDGs are a call to action adopted by United Nations member states (2015–2030), comprising 17 interconnected goals to address global challenges related to poverty, environmental degradation, inequality, and justice.
Transcription	The practice of converting spoken or signed language into written text.
Translation	Written transfer of text from one language to another that communicates the meaning of that text.
Urban governance	The processes, policies, and decisions organised and enacted by government institutions, individuals, civil society, and other stakeholders that determine a city's or town's social and economic development and public life.
Urban policy	Urban policy refers to a broad category of (usually geographically oriented) policies aimed at influencing the social, economic, and cultural development of urban areas and urban lives, promoting long-term productive, inclusive, and resilient urban development. Framed through contested meanings and contested politics, urban policies can be fragmented and diverse in practice or co-ordinated under a national urban policy.

Urbanisation	The process of a country becoming more urbanised through the demographic process of natural increase in the urban population but also through migration from rural to urban areas and the reclassification of rural settlements as (part of) towns or cities.
Visual methods	Methods that involve the collection of visual elements such as maps, drawings, paintings, photos, videos, and three-dimensional objects.

References

Abu-Lughod, L. (1990) 'Can there be a feminist ethnography?', *Women and Performance: A Journal of Feminist Theory*, 5(1), pp. 7–27.

Ackerly, B. and True, J. (2008) 'Reflexivity in practice: power and ethics in feminist research on international relations', *International Studies Review*, 10(4), pp. 693–707.

Actionable Intelligence for Social Policy. (2022) 'A toolkit for centering racial equity throughout data integration', *University of Pennsylvania*. Available at: https://aisp.upenn.edu/wp-content/uploads/2022/07/AISP-Toolkit_5.27.20.pdf (Accessed: 28 August 2023).

Adams, P. (2017) 'Place', in D. Richardson, N. Castree, M. Goodchild, A. Kobayashi, W. Liu and R. Marston (eds.) *The American Association of Geographers international encyclopedia of geography: people, the earth, environment, and technology*. Oxford: John Wiley and Sons, Ltd., pp. 1–12.

Adeniyi-Ogunyankin, G. (2024) 'Moments', *GeoHumanities*, 0(0), pp. 1–5.

Adeniyi-Ogunyankin, G. and Peake, L. (2021a) 'The importance of a gendered analysis of COVID-19', in G.J. Andrews, V. Crooks, J.R. Pearce and J.P. Messina (eds.) *COVID-19 and similar futures: pandemic geographies*. New York: Springer Press, pp. 341–348.

Adeniyi-Ogunyankin, G. and Peake, L. (2021b) 'Tiwa's morning', in M. Lancione and C. McFarlane (eds.) *Global urbanism: knowledge, power and the city*. New York: Routledge, pp. 116–123.

Agencia Andaluza de Cooperación Internacional para el Desarrollo (AACID). (2021) *Estudio comparativo sobre sistemas de monitoreo local de los ODS a nivel local y regional*. Available at: www.juntadeandalucia.es/organismos/aacid/servicios/publicaciones/detalle/80433.html (Accessed: 19 November 2023).

Agnew, J. A. and Duncan, J. S. (eds.) (1989) *The power of place: bringing together geographical and sociological imaginations*. Oxford: Routledge.

Ahmed, S. (2012) *On being included: racism and diversity in institutional life*. Durham: Duke University Press.

Ahmed, S. (2013) *Feeling depleted?* Available at: http://feministkilljoys.com/2013/11/17/feeling-depleted/ (Accessed: 19 September 2023).

Ahmed, S. (2021) *Complaint!* Durham: Duke University Press.

Al-Ali, N. (2020) 'COVID-19 and feminism in the global South: challenges, initiatives and dilemmas', *The European Journal of Women's Studies*, 27(4), pp. 333–347.

Alarcón, W., Benfield, D.M., Fukushima, A.I. and Maese, M. (2020) 'Guest editors' introduction: "world"-making and "world"-travelling with decolonial feminisms and women of color', *Frontiers: A Journal of Women Studies*, 41(1), pp. x–xxi.

Ali, S.H. and Keil, R. (2006) 'Global cities and the spread of infectious disease: the case of severe acute respiratory syndrome (SARS) in Toronto, Canada', *Urban Studies*, 43(3), pp. 491–509.

Al-Saleh, D. and Noterman, E. (2021) 'Organizing for collective feminist killjoy geographies in a U.S. university', *Gender, Place and Culture*, 28(4), pp. 453–474.

Altschuler, S. and Wald, P. (2020) 'COVID-19 and the language of racism', *Signs*. Available at: http://signsjournal.org/covid/altschuler-and-wald/ (Accessed: 19 September 2023).

American Association of Geographers. (2022) *2022 Summer series for students and recent graduates*. Available at: www.aag.org/events/2022-summer-series-for-students-and-recent-graduates/ (Accessed: 13 September 2023).

Amoore, L. (2020) *Cloud ethics: algorithms and the attributes of ourselves and others*. Durham: Duke University Press.

Ampofo, A.A. (2010) 'One who has truth – she has strength: the feminist activist inside and outside the academy in Ghana', in A.A. Ampofo and S. Arnfred (eds.) *African feminist politics of knowledge: tensions, challenges, possibilities*. Uppsala: Nordiska Afrikainstitutet, pp. 28–51.

Ampofo, A.A. and Arnfred, S. (eds.) (2010) *African feminist politics of knowledge: tensions, challenges, possibilities*. Uppsala: Nordiska Afrikainstitutet.

Andaiye. (2000) 'The red thread story: resisting the narrow interests of a broader political struggle', in S.F. Brown (ed.) *Spitting in the wind: lessons in empowerment from the Caribbean*. Kingston: Ian Randle Publishers, pp. 51–98.

Anderson, C. (2012) 'Lost in space? Ethnography and the disparate geographies of social process', *The Professional Geographer*, 64(2), pp. 276–285.

Anderson, C.M. (2020) *Urbanism without guarantees: the everyday life of a gentrifying west side neighborhood*. Minneapolis: University of Minnesota Press.

Andrews, K. (2021) 'Racism is the public health crisis', *The Lancet*, 397(10282), pp. 1342–1343.

Angelo, H. and Wachsmuth, D. (2015) 'Urbanizing urban political ecology: a critique of methodological cityism', *International Journal of Urban and Regional Research*, 39(1), pp. 16–27.

Antonini, R., Cirillo, L., Rossato, L. and Torresi, I. (eds.) (2017) *Non-professional interpreting and translation: state of the art and future of an emerging field of research*. Amsterdam and Philadelphia: John Benjamins Publishing Company.

Antrobus, P. (2005) 'Critiquing the MDGs from a Caribbean perspective', *Gender and Development*, 13(1), pp. 94–104.

Anwar, N.H. and Viqar, S. (2017) 'Research assistants, reflexivity and the politics of fieldwork in urban Pakistan', *Area*, 49(1), pp. 114–121.

Arondekar, A. (2005) 'Without a trace: sexuality and the colonial archive', *Journal of the History of Sexuality*, 14(1/2), pp. 10–27.

Aruri, N., Al-Battat, M. and Qawasmi, S. in conversation with Adami, E. ((forthcoming) 'Takhayali Ein Qiniya: Twenty women reimagining their village' in A. Downey (ed) *Nida Sinnokrot: Palestine is not a garden*. London: Sternberg Press.

Asher, K. and Ramamurthy, P. (2020) 'Rethinking decolonial and postcolonial knowledges beyond regions to imagine transnational solidarity', *Hypatia*, 35(3), pp. 542–547.

Asmolov, G. and Kolozaridi, P. (2017) 'The imaginaries of RuNet: the change of the elites and the construction of online space', *Russian Politics*, 2(1), pp. 54–79.

Baer, H. (2016) 'Redoing feminism: digital activism, body politics, and neoliberalism', *Feminist Media Studies*, 16(1), pp. 17–34.

Bagheri, N. (2014) 'Mapping women in Tehran's public spaces: a geo-visualization perspective', *Gender, Place and Culture*, 21(10), pp. 1285–1301.

Bain, A. and Payne, W.J. (2016) 'Queer de-participation: reframing the co-production of scholarly knowledge', *Qualitative Research*, 16(3), pp. 330–340.

Bandia, P.F. (2008) *Translation as reparation: writing and translation in postcolonial Africa*. London: Routledge.

Banerjea, N. (2015) 'Critical urban collaborative ethnographies: articulating community with Sappho for equality in Kolkata, India', *Gender, Place and Culture*, 22(8), pp. 1058–1072.

Banerjea, N., Boyce, P. and Dasgupta, R.K. (2021) *COVID-19 Assemblages: queer and feminist ethnographies from South Asia*. Abingdon and New York: Routledge India.

Banks, S., Armstrong, A., Carter, K., Graham, H., Hayward, P., Henry, A., Holland, T., Holmes, C., Lee, A., McNulty, A., Moore, N., Nayling, N., Stokoe, A. and Strachan, A. (2015) 'Everyday ethics in community-based participatory research', in J. Bannister and I. Hardill (eds.) *Knowledge mobilisation and the social sciences: research impact and engagement*. London and New York: Routledge, pp. 97–111.

Bannister, J. and Hardill, I. (2015) 'Knowledge mobilisation and the social sciences: dancing with new partners in an age of austerity', in J. Bannister and I. Hardill, (eds.) *Knowledge mobilisation and the social sciences: research impact and engagement*. London and New York: Routledge, pp. 1–8.

Bannister, J. and O'Sullivan, A. (2015) 'Knowledge mobilisation and the civic academy: the nature of evidence, the roles of narrative and the potential of contribution analysis', in J. Bannister and I. Hardill (eds.) *Knowledge mobilisation and the social sciences*. London and New York: Routledge, pp. 83–96.

Barbour, K. (2018) 'Embodied ways of knowing: revisiting feminist epistemology', in L. Mansfield, J. Caudwell, B. Wheaton and B. Watson (eds.) *The Palgrave handbook of feminism and sport, leisure and physical education*. London: Palgrave Macmillan, pp. 209–226.

Bashevkin, S. (2006) *Tales of two cities: women and municipal restructuring in London and Toronto*. Vancouver: UBC Press.

Basu, P. and De Jong, F. (2016) 'Utopian archives, decolonial affordances: introduction to the special issue', *Social Anthropology*, 24(1), pp. 5–19.

Batmanghelichi, K.S. and Mouri, L. (2017) 'Cyberfeminism, Iranian style: online feminism in post-2009 Iran', *Feminist Media Histories*, 3(1), pp. 50–80.

Baxi, P. (2014) 'Sexual violence and its discontents', *Annual Review of Anthropology*, 43(1), pp. 139–154.

Bayley, J.E. and Phipps, D. (2017) 'Building the concept of research impact literacy', *Evidence and Policy: A Journal of Research, Debate and Practice*, 15(4), pp. 597–606.

Bell, L. (2014) 'Ethics and feminist research', in S.N. Hesse-Biber (ed.) *Feminist research practice: a primer*, 2nd ed. Los Angeles: Sage Publications, pp. 73–106.

Benería, L. and Sen, G. (1981) 'Accumulation, reproduction, and "Women's role in economic development": Boserup revisited', *Signs*, 7(2), pp. 279–298.

Benhabib, S. (1992) *Situating the self: gender, community, and postmodernism in contemporary ethics*. New York: Routledge.

Bennet, A. and Bennet, D. (2013) *Knowledge mobilization in the Social Sciences and Humanities: moving from research to action*, 2nd ed. Frost: MQI Press Knowledge Series.

Benson, K. and Nagar, N. (2006) 'Collaboration as resistance? Reconsidering the processes, products, and possibilities of feminist oral history and ethnography', *Gender, Place and Culture*, 13(5), pp. 581–592.

Berkhout, S.G. and Richardson, L. (2020) 'Identity, politics, and the pandemic: why is COVID-19 a disaster for feminism(s)?', *History and Philosophy of the Life Sciences*, 42(49). Available at: https://doi.org/10.1007/s40656-020-00346-7 (Accessed: 23 October 2023).

Bernal, V. and Grewal, I. (2014) 'The NGO form: feminist struggles, states, and neoliberalism', in V. Bernal and I. Grewal (eds.) *Theorizing NGOs: states, feminisms, and neoliberalism*. Durham: Duke University Press, pp. 1–18.

Bhabha, H.K. (2012) *The location of culture*, 2nd ed. Abingdon: Routledge.

Bhambra, G.K. (2014) 'Postcolonial and decolonial dialogues', *Postcolonial Studies*, 17(2), pp. 115–121.

Bhan, G. (2019) 'Notes on a Southern urban practice', *Environment and Urbanization*, 31(2), pp. 639–654.

Bharath, S., Musaifer, S. and Nagar, N. (2019) 'Closing notes: retelling dis/appearing tales', in R. Nagar (ed.) *Hungry translations: relearning the world through radical vulnerability*. Urbana: University of Illinois Press, pp. 241–258.

Bhaskar, R. (2008) *A realist theory of science*. London and New York: Routledge.

Billo, E. and Hiemstra, N. (2013) 'Mediating messiness: expanding ideas of flexibility, reflexivity, and embodiment in fieldwork', *Gender, Place and Culture*, 20(3), pp. 313–328.

Black, A.L., Crimmins, G., Dwyer, R. and Lister, V. (2019) 'Engendering belonging: thoughtful gatherings with/in online and virtual spaces', *Gender and Education*, 32(1), pp. 115–129.

Blaut, J.M. (1993) *The colonizer's model of the world: geographical diffusionism and Eurocentric history*. New York and London: The Gilford Press.

Blidon, M. and Zaragocin, S. (2019) 'Mapping gender and feminist geographies in the global context', *Gender, Place and Culture*, 26(7–9), pp. 915–925.

Board, W.E. (2020) '"Intersectionality went viral": toxic platforms, distinctive Black cyberfeminism and fighting misogynoir – an interview with Kishonna Gray', *Westminster Papers in Communication and Culture*, 15(1), pp. 68–73.

Bondi, L. and Rose, D. (2003) 'Constructing gender, constructing the urban: a review of Anglo-American feminist urban geography', *Gender, Place and Culture*, 10(3), pp. 229–245.

Borgerud, C. and Borglund, E. (2020) 'Open research data, an archival challenge?' *Archival Science*, 20, pp. 279–302.

Borgman, C.L. (2012) 'The conundrum of sharing research data', *Journal of the American Society for Information Science Technology*, 63(6), pp. 1059–1078.

Boria, M., Carreres, Á., Noriega-Sánchez, M. and Tomalin, M. (eds.) (2020) *Translation and multimodality: beyond words*. London and New York: Routledge.

Boris, E. (2022) 'Vulnerability and resilience in the Covid-19 crisis: race, gender, and belonging', in A. Triandafyllidou (ed.) *Migration and pandemics. IMSCOE Research Series*. New York: Springer, pp. 65–85.

Boserup, E. [1970] (2007) *Woman's role in economic development*. London: Earthscan.

Boudreau, J.A., with Boucher, N. and Liguori, M. (2009) 'Taking the bus daily and demonstrating on Sunday: reflections on the formation of political subjectivity in an urban world', *City*, 13(2–3), pp. 336–346.

Bowen, S.J. and Graham, I.D. (2013) 'From knowledge translation to engaged scholarship: promoting research relevance and utilization', *Archives of Physical Medicine and Rehabilitation*, 94(1), pp. S3–S8.

Brabeck, M. (2021) 'Open science and feminist ethics: promises and challenges of open access', *Psychology of Women Quarterly*, 45(4), pp. 457–474.

Brajkovic, L. (2018) 'Human capital investment or academic marginalism? Understanding the influence of political economy on higher education in post-socialist Europe', *Policy Reviews in Higher Education*, 2(2), pp. 151–175.

Brandt, W. (1980) *North-South: a programme for survival; report of the independent commission on international development issues*. Cambridge: MIT Press.

Brenner, N. and Schmid, C. (2015) 'Towards a new epistemology of the urban?' *City: Analysis of Urban Change, Theory, Action*, 19(2–3), pp. 151–182.

Brice, S. (2023) 'Making space for a radical trans imagination: towards a kinder, more vulnerable, geography', *Environment and Planning D: Society and Space*. Available at: https://doi.org/10.1177/02637758231187449 (Accessed: 19 September 2023).

Brigden, N. and Hallett, H. (2021) 'Fieldwork as social transformation: place, time, and power in a violent moment', *Geopolitics*, 26(1), pp. 1–17.

Briggs, G., Briggs, A., Whitmore, E., Maki, A., Ackerley, C., Maisonneuve, A. and Yordy, C. (2015) *Questing your way to a knowledge mobilization strategy: doing research that is useful and used*. Available at: https://carleton.ca/communityfirst/wp-content/uploads/KMB-Questing-Your-Way-to-a-KMb-Strategy-Jun-29-2015.pdf (Accessed: 24 October 2023).

Brown, W. (2015) *Undoing the demos: neoliberalism's stealth revolution*. New York: Zone Books.

Brown, E., Campbell, B., Cloke, J., Cook, J., Batchelor, S. and To, L.S. (2022) 'Low carbon energy for development network (LCEDN)', in L. Hantrais (ed.) *How to manage international multidisciplinary research projects*. Cheltenham: Edward Elgar, pp. 137–149.

Browne, K. (2005) 'Snowball sampling: using social networks to research non-heterosexual women', *International Journal of Social Research Methodology*, 8(1), pp. 47–60.

Browne, K., Banerjea, N., McGlynn, N., Sumita, B., Bakshi, L., Banerjee, R. and Biswas, R. (2017) 'Towards transnational feminist queer methodologies', *Gender, Place and Culture*, 24(10), pp. 1376–1397.

Buch, E.D. and Staller, K.M. (2007) 'The feminist practice of ethnography', in S.N. Hesse-Biber and P.L. Leavy (eds.) *Feminist research practice: a primer*. Thousand Oaks: Sage Publications, pp. 187–221.

Buch, E.D. and Staller, K.M. (2014) 'What is feminist ethnography?', in S.N. Hesse-Biber and P.L. Leavy (eds.) *Feminist research practice: a primer*, 2nd ed. Thousand Oaks: Sage Publications, pp. 107–144.

Buckley, M. and Strauss, K. (2016) 'With, against and beyond Lefebvre: planetary urbanization and epistemic plurality', *Environment and Planning D: Society and Space*, 34(4), pp. 617–636.

Budworth, P. and Hall, S.M. (forthcoming) 'The everyday: social lives and relationships', in L. Peake, A. Datta, and G. Adeniyi-Ogunyankin (eds.) *International handbook on gender and cities*. Cheltenham: Edward Elgar Publishing.

Bulk Rename Utility. (2023) *Bulk rename utility*. Available at: www.bulkrenameutility.co.uk/ (Accessed: 23 October 2023).

Bundhoo, D. and Lynch, K. (2021) 'Pacing emotional labour of qualitative research in an intractable conflict environment', *Area*, 53(1), pp. 47–55.

Butcher, M. and Maclean, K. (2018) 'Gendering the city: the lived experience of transforming cities, urban cultures and spaces of belonging', *Gender, Place, and Culture*, 25(5), pp. 686–694.

Butcher, S. (2021) 'Differentiated citizenship: the everyday politics of the urban poor in Kathmandu, Nepal', *International Journal of Urban and Regional Research*, 45(6), pp. 948–963.

Cáceres, D.M., Silvetti, F. and Díaz, S. (2016) 'The rocky path from policy-relevant science to policy implementation – a case study from the South American Chaco', *Current Opinion in Environmental Sustainability*, 19, pp. 57–66.

Çağlar, A. and Glick Schiller, N. (2018) *Migrants and city-making: dispossession, displacement, and urban regeneration*. Durham and London: Duke University Press.

Çağlar, A. and Glick Schiller, N. (2020) 'Relational multiscalar analysis: a comparative approach to migrants within city-making processes', *Geographical Review*, 111(2), pp. 206–232.

Cahuas, M.C. (2023) 'Building a holistic intersectional feminist praxis in geography: lessons from community', *The Professional Geographer*, 75(4), pp. 648–654.

Caretta, M.A. (2015) 'Situated knowledge in cross-cultural, cross-language research: a collaborative reflexive analysis of researcher, assistant and participant subjectivities', *Qualitative Research*, 15(4), pp. 489–505.

Caretta, M.A. and Jokinen, J.C. (2017) 'Conflating privilege and vulnerability: a reflexive analysis of emotions and positionality in postgraduate fieldwork', *The Professional Geographer*, 69(2), pp. 275–283.

Carleton University. (2014) *Tool: so what the heck Is knowledge mobilization and why should I care?* Available at: https://carleton.ca/communityfirst/2014/so-what-the-heck-is-knowledge-mobilization-and-why-should-i-care/ (Accessed: 23 October 2023).

Carpenter, D. (2017) 'The quest for generic ethics principles in social science research', in R. Iphofen (ed.) *Finding common ground: consensus in research ethics across the social sciences (Advances in research ethics and integrity, volume I)*. Bingley: Emerald Publishing, pp. 3–17.

Casimiro, I.M. and Andrade, X. (2010) 'Critical feminism in Mozambique', in A.A. Ampofo and S. Arnfred (eds.) *African feminist politics of knowledge: tensions, challenges, possibilities*. Uppsala: Nordiska Afrikainstitutet, pp. 137–156.

Castro, O. and Ergun, E. (2017) 'Introduction: re-envisioning feminist translation studies. Feminism in translation, translations in feminism', in O. Castro and E. Ergun (eds.) *Feminist translation studies: local and transnational perspectives*. New York: Routledge, pp. 1–11.

Centre for Feminist Foreign Policy. (2022) *Make foreign policy feminist: a feminist foreign policy manifesto for Germany*. Available at: https://centreforfeministforeignpolicy.org/2022/11/15/make-foreign-policy-feminist-a-feminist-foreign-policy-for-germany/ (Accessed: 24 October 2023).

Césaire, A. [1972] (2000) *Discourse on colonialism*. New York: Monthly Review Press.

Chakravarty, D. (2015) 'On being and providing "data": politics of transnational feminist collaboration and academic division of labor', *Frontiers: A Journal of Women's Studies*, 36(3), pp. 25–50.

Chant, S. and McIlwaine, C. (2016) *Cities, slums and gender in the global south: towards a feminized urban future*. London: Routledge.

Chapman, E. (2023) 'Revocalising human geography: decolonial language geographies beyond the nation-state', *Progress in Human Geography*, 47(1), pp. 24–42.

Chatterjea, A., Wilcox, H.N. and Williams, A.L. (eds.) (2022) *Dancing transnational feminisms: Ananya Dance Theatre and the art of social justice*. Seattle: University of Washington Press.

Chen, M.Y. (2020) 'Feminisms in the air', *Signs*. Available at: https://signsjournal.org/covid/chen/ (Accessed: 23 October 2023).

Cheyfitz, E. (1991) *The poetics of imperialism: translation and colonization from the tempest to Tarzan*. Oxford: Oxford University Press.

Chicana por mi raza. (2023) *Chicana por mi raza*. Available at: https://chicanapormiraza.org/ (Accessed: 23 October 2023).

Chilisa, B. (2019) *Indigenous research methodologies*, 2nd ed. Thousand Oaks: Sage Publications.

Choudry, A. (2020) 'Reflections on academia, activism, and the politics of knowledge and learning', *The International Journal of Human Rights*, 24(1), pp. 28–45.

Chowdhury, E.H. (2009) 'Locating global feminisms elsewhere: braiding US women of color and transnational feminisms', *Cultural Dynamics*, 21(1), pp. 51–78.

Christou, A. (2016) 'Feminisms, crises and affect: women in academia contemplating publics and performativities', *Interface: A Journal for and About Social Movements*, 8(1), pp. 34–44.

Chua, L. and Mathur, N. (eds.) (2018) *Who are 'we'? Reimagining alterity and affinity in anthropology*. New York: Berghahn.

Cifor, M., Garcia, P., Cowan, T.L., Rault, J., Sutherland, T., Chan, A., Rode, J., Hoffmann, A.L., Salehi, N. and Nakamura, L. (2019) *Feminist data manifest-no*. Available at: www.manifestno.com/ (Accessed: 23 October 2023).

Canadian Institutes of Health Research (CIHR), Natural Sciences and Engineering Research Council of Canada (NSERC) and Social Sciences and Humanities Research Council (SSHRC). (2021a) *Tri-agency framework: responsible conduct of research*. Available at: https://rcr.ethics.gc.ca/eng/framework-cadre-2021.html (Accessed: 23 October 2023).

CIHR, NSERC and SSHRC. (2021b) *Tri-agency statement of principles on digital data management*. Available at: https://science.gc.ca/site/science/en/interagency-research-funding/policies-and-guidelines/research-data-management/tri-agency-statement-principles-digital-data-management (Accessed: 25 October 2023).

CIHR, NSERC and SSHRC. (2022a) *Tri-agency guide for financial administration*. Available at: www.nserc-crsng.gc.ca/interagency-interorganismes/TAFA-AFTO/guide-guide_eng.asp (Accessed: 17 November 2023).

CIHR, NSERC and SSHRC. (2022b) *Tri-council policy statement: ethical conduct for research involving humans*. Available at: https://ethics.gc.ca/eng/policy-politique_tcps2-eptc2_2022.html (Accessed: 29 October 2023).

CIHR, NSERC and SSHRC. (2022c) *Tri-agency research data management policy*. Available at: https://science.gc.ca/site/science/en/interagency-research-funding/policies-and-guidelines/research-data-management/tri-agency-research-data-management-policy (Accessed: 17 November 2023).

Clark, L., Birkhead, A.S., Fernandez, C. and Egger, M.J. (2017) 'A transcription and translation protocol for sensitive cross-cultural team research', *Qualitative Health Research*, 27(12), pp. 1751–1764.

Clark, T., Foster, L., Sloan, L. and Bryman, A. (2021) *Bryman's social research methods*, 6th ed. Oxford: Oxford University Press.

Clarke, M. (2018) *Global South: what does it mean and why use the term?* Available at: https://onlineacademiccommunity.uvic.ca/globalsouthpolitics/2018/08/08/global-south-what-does-it-mean-and-why-use-the-term/ (Accessed: 23 October 2023).

Clarke, K.M. (2022) *The geopolitics of Black bones that matter*. Available at: www.americananthropologist.org/geopolitical-lives/clarke (Accessed: 19 September 2023).

Clifford, J. and Marcus, G. (1986) *Writing culture: the poetics and politics of ethnography*. Berkeley: University of California Press.

Clifford, N., Cope, M. and Gillespie, T. (eds.) (2023) *Key methods in geography*, 4th ed. London: Sage Publications.

Cliggett, L. (2005) *Grains from grass: aging, gender, and famine in rural Africa*. Ithaca: Cornell University Press.

Coddington, K. (2017) 'Voice under scrutiny: feminist methods, anticolonial responses, and new methodological tools', *The Professional Geographer*, 69(2), pp. 314–320.

Code, L. (1991) *What can she know? Feminist theory and the construction of knowledge*. Ithaca: Cornell University Press.

Cohen, D. and Crabtree, B. (2006) *Using qualitative methods in healthcare research: a comprehensive guide for designing, writing, reviewing and reporting qualitative research*. Available at: www.qualres.org/index.html (Accessed: 23 October 2023).

Cohen, M. (1986) 'Women and the urban street food trade: some implications for policy', in *Gender and planning working paper series, no. 55*. London: Bartlett School of Architecture and Planning, University College London. Available at: www.ucl.ac.uk/bartlett/development/sites/bartlett/files/migrated-files/WP55_0.pdf (Accessed: 23 October 2023).

Colombia Diversa. (2020) *Los órdenes del prejuicio: los crímenes cometidos sistemáticamente contra personas LGBT en el conflicto armado colombiano*. Available at: https://colombiadiversa.org/colombiadiversa2016/wp-content/uploads/2020/07/LIBRO-WEB-1.pdf (Accessed: 19 November 2023).

Combahee River Collective [1979] (2017) 'A Black feminist statement', in K.-Y. Taylor (ed.) *How we get free: Black feminism and the Combahee river collective*. Chicago: Haymarket Books.

Connell, R. (2007) *Southern theory: social science and the global dynamics of knowledge*, 1st ed. London and New York: Routledge.

Connell, R. (2014) 'Rethinking gender from the South', *Feminist Studies*, 40(3), pp. 518–539.

Connolly, J., Henriquez, F. and Hursthouse, A. (2022) 'Antimicrobial resistance in the real world (AMR)', in L. Hantrais (ed.) *How to manage international multidisciplinary research projects*. Cheltenham: Edward Elgar Publishing, pp. 63–74.

Cope, M. (2019) 'Qualitative methods and geohumanities', in J. Ash, R. Kitchin and A. Leszczynski (eds.) *Digital geographies*. London: Sage Publications, pp. 96–105.

Corbman, R.F. (2014) 'A genealogy of the lesbian herstory archives, 1974–2014', *Journal of Contemporary Archival Studies*, 1(1). Available at: https://elischolar.library.yale.edu/jcas/vol1/iss1/1 (Accessed: 25 October 2023).

Craig, C.J., Turcotte, J.F. and Coombe, R.J. (2011) 'What is feminist about open access?: a relational approach to copyright in the academy', *Feminists@law*, 1(1), pp. 1–35.

Craine, J. and Gardner, C. (2016) 'Visual methodology', in I. Hay (ed.) *Qualitative research methods in human geography*, 4th ed. Oxford: Oxford University Press, pp. 274–287.

Crenshaw, K. (1989) 'Demarginalizing the intersection of race and sex: a Black feminist critique of anti-discrimination doctrine: feminist theory and anti-racist politics', *The University of Chicago Legal Forum*, 1(8), pp. 139–168.

Crenshaw, K. (1991) 'Mapping the margins: intersectionality, identity politics, and violence against women of color', *Stanford Law Review*, 43(6), pp. 1241–1299.

Crimmins, T., Posthumus, E. and Prudic, K. (2021) *Citizen scientists are filling research gaps created by the pandemic*. Available at: www.theconversation.com/citizen-scientists-are-filling-research-gaps-created-by-the-pandemic-152521 (Accessed: 25 October 2023).

Croese, S. and Duminy, J. (2023) 'Co-producing urban expertise for SDG localization: the history and practices of urban knowledge production in South Africa', *Urban Geography*, 44(3), pp. 538–557.

Crossley, A.D. (2015) 'Facebook feminism: social media, blogs, and new technologies of contemporary U.S. feminism', *Mobilization: An International Quarterly*, 20(2), pp. 253–268.

Cuomo, D. and Massaro, V.A. (2016) 'Boundary-making in feminist research: new methodologies for "intimate insiders" ', *Gender, Place and Culture*, 23(1), pp. 94–106.
Cupples, J. (2002) 'The field as a landscape of desire: sex and sexuality in geographical fieldwork', *Area*, 34(4), pp. 382–390.
Curley, A. and Smith, S. (2020) 'Against colonial grounds: geography on Indigenous lands', *Dialogues in Human Geography*, 10(1), pp. 37–40.
Dados, N. and Connell, R. (2012) 'The global South', *Contexts*, 11(1), pp. 12–13.
Daigle, M. (2018) 'Embodying relations of accountability in settler colonial contexts,' *Political Geography*, 66, pp. 201–203.
Daigle, M. and Ramírez, M.M. (2019) 'Decolonial geographies', in Antipode Editorial Collective (eds.) *Keywords in radical geography: Antipode at 50*. Hoboken, NJ: Blackwell, pp. 78–84.
Dalton, C.M., Taylor, L. and Thatcher, J. (2016) 'Critical data studies: a dialog on data and space', *Big Data and Society*, 3(1), pp. 1–9.
Daniels, J. (2009) 'Rethinking cyberfeminism(s): race, gender, and embodiment', *Women's Studies Quarterly*, 37(1/2), pp. 101–124.
Datta, A. (2016) 'The genderscapes of hate: on violence against women in India', *Dialogues in Human Geography*, 6(2), pp. 178–181.
Davis, A. (1981) *Women, race, and class*. New York: Random House.
Davis III, C.H., Harris, J.C., Stokes, S. and Harper, S.R. (2019) 'But is it activist?: interpretive criteria for activist scholarship in higher education', *The Review of Higher Education*, 42(5), pp. 85–108.
Davis, D.-A. and Craven, C. (2022) *Feminist ethnography: thinking through methodologies, challenges, and possibilities*, 2nd ed. New York and London: Rowman and Littlefield.
Davis, H. and Todd, Z. (2017) 'On the importance of a date, or, decolonizing the Anthropocene', *ACME: An International Journal for Critical Geographies*, 16(4), pp. 761–780.
Decker, A.C. and Baderoon, G. (2018) 'African feminisms: cartographies for the twenty-first century', *Meridians*, 17(2), pp. 219–231.
Della Porta, D., Cini, L. and Guzmán-Concha, C. (2020) *Contesting higher education: student movements against neoliberal universities*. Bristol: Bristol University Press.
de Leeuw, S. and Hunt, S. (2018) 'Unsettling decolonizing geographies', *Geography Compass*, 12(7), pp. 1–14.
Dempsey, S.E., Parker, P.S. and Krone, K.J. (2011) 'Navigating socio-spatial difference, constructing counter-space: insights from transnational feminist praxis', *Journal of International and Intercultural Communication*, 4(3), pp. 201–220.
De Silva, M. (2018) 'Making the emotional connection: transnational eldercare circulation within Sri Lankan-Australian transnational families', *Gender, Place and Culture*, 25(1), pp. 88–103.
Devereaux, S. (2014) *Why these Tweets are called my back*. Available at: https://thenewinquiry.com/why-these-tweets-are-called-my-back/ (Accessed: 23 October 2023).
Deville, J., Guggenheim, M. and Hrdličková Z. (2016) 'Introduction: the practises and infrastructures of comparison', in J. Deville, M. Guggenheim and Z. Hrdličková (eds.) *Practising comparison: logics, relations, collaborations*. Manchester: Mattering Press, pp. 17–41.
Dhamoon, R.K. (2015) 'A feminist approach to decolonizing anti-racism: rethinking transnationalism, intersectionality, and settler colonialism', *Feral Feminisms*, 4, pp. 20–37.
Dhamoon, R.K. (2020) 'Racism as a workload and bargaining issue', *Socialist Studies/Études Socialistes*, 14(1), pp. 1–22.
Digital Curation Centre UK. (n.d.). *Practical guidance and examples to help you develop your data management plan*. Available at: www.dcc.ac.uk/dmps (Accessed: 25 October 2023).
Dill, B.T. and Zambrana, R.E. (2009) *Emerging intersections: race, class, and gender in theory, policy, and practice*. New Brunswick, NJ: Rutgers University Press.
D'Ignazio, C. and Klein, L.F. (2020) *Data feminism*. Cambridge: MIT Press.
Dingwall, R., Iphofen, R., Lewis, J., Oates, J. with Emmerich, N. (2017) 'Towards common principles for social science research ethics: a discussion document for the Academy of Social

Sciences', in R. Iphofen (ed.) *Finding common ground: consensus in research ethics across the social sciences (Advances in research ethics and integrity, volume I)*. Bingley: Emerald Publishing, pp. 111–123.

Dos Ventos Lopes Heimer, R. (2021) '*Travelling Cuerpo-Territorios*: a decolonial feminist geographical methodology to conduct research with migrant women', *Third World Thematics: A TWQ Journal*, 6(4–6), pp. 290–319.

Doshi, S. and Ranganathan, M. (2017) 'Contesting the unethical city: land dispossession and corruption narratives in urban India', *Annals of the American Association of Geographers*, 107(1), pp. 183–199.

Duffy, B.E. and Pooley, J.D. (2017) '"Facebook for academics": the convergence of self-branding and social media logic on Academia.edu', *Social Media + Society*, 3(1). Available at: https://doi.org/10.1177/2056305117696523 (Accessed: 25 October 2023).

Duncan, P. (2016) 'Hot commodities, cheap labour: women of colour in the academy', in J. Hobson (ed.) *Are all the women still white? Rethinking race, expanding feminisms*. Albany: State University of New York, pp. 177–204.

Dupont, V., Jordhus-Lier, D., Sutherland, C. and Braathen, E. (eds.) (2015) *The politics of slums in the global South: urban informality in Brazil, India, South Africa and Peru*. London: Routledge.

Dyck, I. (1993) 'Ethnography: a feminist method?', *The Canadian Geographer/Le Géographe Canadien*, 37(1), pp. 52–57.

Dyck, I. (2005) 'Feminist geography, the "everyday", and local-global relations: hidden spaces of place-making', *The Canadian Geographer/Le Géographe Canadien*, 49(3), pp. 233–243.

Edwards, R. and Holland, J. (2013) *What is qualitative interviewing?* London: Bloomsbury Publishing.

Edwards, R. and Mauthner, M. (2012) 'Ethics and feminist research: theory and practice', in T. Miller, M. Birch, M. Mauthner and J. Jessop (eds.) *Ethics in qualitative research*. London: Sage Publications, pp. 15–31.

Ehrenreich, B. and Hochschild, A.R. (2003) *Global woman: nannies, maids, and sex workers in the new economy*. New York: Metropolitan Books.

Elwood, S.A. and. Martin, D.G. (2000) '"Placing" interviews: location and scales of power in qualitative research', *The Professional Geographer*, 52(4), pp. 649–657.

Emejulu, A. and Littler, J. (2019) 'We do not have to be vicious, competitive, or managerial', *Soundings: A Journal of Politics and Culture*, 73, pp. 73–86.

England, K. (1994a) 'From "social justice and the city" to women-friendly cities? Feminist theory and politics', *Urban Geography*, 15(7), pp. 628–643.

England, K. (1994b) 'Getting personal: reflexivity, positionality, and feminist research', *The Professional Geographer*, 46(1), pp. 80–89.

England, K. (2002) 'Interviewing elites: cautionary tales about researching women managers in Canada's banking industry', in P. Moss (ed.) *Feminist geography in practice: research and methods*. Oxford: Blackwell Publishers Ltd., pp. 200–213.

Engster, D. and Hamington, M. (2015) *Care ethics and political theory*. Oxford: Oxford University Press.

Erkut, S. (2010) 'Developing multiple language versions of instruments for intercultural research', *Child Development Perspectives*, 4(1), pp. 19–24.

Eshelman, D.J. (2007) 'Feminist translation as interpretation', *Translation Review*, 74(1), pp. 16–27.

Esquivel, V. (2016) 'Power and the sustainable development goals: a feminist analysis', *Gender and Development*, 24(1), pp. 9–23.

Esquivel, V. and Sweetman, C. (2016) 'Gender and the sustainable development goals', *Gender and Development*, 24(1), pp. 1–8.

Eyles, J. (1989) 'The geography of everyday life', in D. Gregory and R. Walford (eds.) *Horizons in human geography*. London: Macmillan Education UK, pp. 102–117.

Eyles, J. and Smith, D.M. (eds.) (1988) *Qualitative methods in human geography*. Totowa: Barnes and Noble.

Fanon, F. (1986) *Black skin, white masks*. Translated by C.L. Markmann. London: Pluto Press.

Fanon, F. [1963] (2004) *The wretched of the earth*. Translated by R. Philcox. New York: Grove Press.

Faria, C., Klosterkamp, S., Torres, R.M. and Walenta, J. (2020) 'Embodied exhibits: toward a feminist geographic courtroom ethnography', *Annals of the American Association of Geographers*, 110(4), pp. 1095–1113.

Faria, C. and Mollett, S. (2016) 'Critical feminist reflexivity and the politics of whiteness in the "field"', *Gender, Place and Culture*, 23(1), pp. 79–93.

Faria, C. and Mollett, S. (2020) '"We didn't have time to sit still and be scared": a postcolonial feminist geographic reading of "an other geography"', *Dialogues in Human Geography*, 10(1), pp. 23–29.

Farthing, L. and Kohl, B. (2013) 'Mobilizing memory: Bolivia's enduring social movements', *Social Movement Studies*, 12(4), pp. 361–376.

Feminist Alliance for Rights. (2020) *Call for a feminist COVID-19 policy*. Available at: http://feministallianceforrights.org/blog/2020/03/20/action-call-for-a-feminist-COVID-19-policy/ (Accessed: 25 October 2023).

Feminist Principles of the Internet. (n.d.) *All principles*. Available at: https://feministinternet.org/en/principles (Accessed: 25 October 2023).

Fenster, T. (2005) 'The right to the gendered city: different formations of belonging in everyday life', *Journal of Gender Studies*, 14(3), pp. 217–231.

Fernández, F. and Evans, J. (eds.) (2018) *The Routledge handbook of translation and politics*. London: Routledge.

Fernández, J.S., Hisatake, K. and Nguyen, A. (2020) 'Decolonial feminism as reflexive praxis: Lugones's "world"-travelling as stories of friendship in academia', *Frontiers: A Journal of Women Studies*, 41(1), pp. 12–34.

Finlay, J.M. and Bowman, J.A. (2017) 'Geographies on the move: a practical and theoretical approach to the mobile interview', *The Professional Geographer*, 69(2), pp. 263–274.

Flaherty, C. (2022) *Free speech for me, and thee?* Available at: www.insidehighered.com/news/2022/04/12/princeton-says-it-wont-censor-webpage-criticizing-professor (Accessed: 25 October 2023).

Flinders, M. (2020) *Fit for the future: research leadership matters. A review of research leadership in the Social Sciences*. London: Economic and Social Research Council.

Flor, L.S., Friedman, J., Spencer, C.N., Cagney, J., Arrieta, A., Herbert, M.E., Stein, C., Mullany, E.C., Hon, J., Patwardhan, V., Barber, R.M., Collins J.K., Hay, S.I., Lim, S.S., Lozano, R., Mokdad, A.H., Murray, C.J.L., Reiner Jr., R.C., Sorensen, R.J.D., Haakenstad, A., Pigott, D.M. and Gakidou, E. (2022) 'Quantifying the effects of the COVID-19 pandemic on gender equality on health, social, and economic indicators: a comprehensive review of data from March 2020 to September 2021', *The Lancet*, 399(10344), pp. 2381–2397.

Flores, J. and Garcia, S. (2009) 'Latina *testimonios*: a reflexive, critical analysis of a "Latina space" at a predominantly White campus', *Race Ethnicity and Education*, 12(2), pp. 155–172.

Foucault, M. (1972) *The archaeology of knowledge*. London: Tavistock.

Fraser, N. (2009) 'Feminism, capitalism and the cunning of history', *New Left Review*, 56, pp. 97–117.

Fuchs, S. and Kuusniemi, M.E. (2018) 'Making a research project understandable – guide for data documentation (version 1.2)', *Zenodo*. Available at: https://doi.org/10.5281/zenodo.1914401 (Accessed: 25 October 2023).

Gabizon, S. (2016) 'Women's movements' engagement in the SDGs: lessons learned from the women's major group', *Gender and Development*, 24(1), pp. 99–110.

Gagliano, J.C. (2021) 'Solidarity for whom? Interrogating solidarity in geographic research methodologies', *Geography Compass*, 15(8). Available at: https://doi.org/10.1111/gec3.12587 (Accessed: 25 October 2023).

Gago, V. (2020) *Feminist international: how to change everything*. Translated by L. Mason-Deese. London: Verso.

Gago, V. and Mason-Deese, L. (2019) 'Rethinking situated knowledge from the perspective of Argentina's feminist strike', *Journal of Latin American Geography*, 18(3), pp. 202–209.

Gallagher, M., Kanngieser, A. and Prior, J. (2017) 'Listening geographies: landscape, affect and geotechnologies', *Progress in Human Geography*, 41(5), pp. 618–637.

Garcia-Rabines, D. and Bencich, B. (2021) 'Community-based resistance strategies among a group of trans women in Lima, Peru during the COVID-19 pandemic', *Journal of Homosexuality*, 68(4), pp. 663–672.

García Ramón, M.D., Simonsen, K. and Vaiou, D. (2006) 'Guest editorial: does anglophone hegemony permeate *Gender, Place and Culture*?' *Gender, Place and Culture*, 13(1), pp. 1–5.

Gardner, P. (2006) 'Oral history', in V. Jupp (ed.) *The Sage dictionary of social research methods*. London: Sage Publications, pp. 206–208.

Garwood, J. (2006) 'Data', in V. Jupp (ed.) *The Sage dictionary of social research methods*. London: Sage Publications, pp. 57–58.

Geertz, C. (1973) *The interpretation of cultures*. New York: Basic Books.

Gender and Development Network. (2019) *Feminist responses to COVID-19*. Available at: www.gadnetwork.org/gadn-resources/feminist-responses-to-covid-19 (Accessed: 25 October 2023).

Germes, M. and de Araújo, S.H. (2016) 'For a critical practice of translation in geography', *ACME: An International Journal for Critical Geographies*, 15(1), pp. 1–14.

Ghaddar, J. and Caswell, M. (2019) '"To go beyond": towards a decolonial archival praxis', *Archival Science*, 19(2), pp. 71–85.

Gieseking, J., Mangold, W., Katz, C., Low, S. and Saegert, S. (2014) 'Section 3 place and identity: editors' introduction and suggestions for further reading', in J. Gieseking, W. Mangold, C. Katz, S. Low and S. Saegert (eds.) *The people, place, and space reader,* 1st ed. London: Routledge, pp. 73–76.

Gieseking, J.J. (2015) 'Useful in/stability: the dialectical production of the social and spatial lesbian herstory archives', *Radical History Review*, (122), pp. 25–37.

Gieseking, J.J. (2020) *A queer New York: geographies of lesbians, dykes, and queers*. New York: NYU Press.

Gieseking, J.J. (2023) 'Reflections on a cis discipline and industry', Association of American Geographers Annual Meeting, Denver, 26 March.

Gilbert, L.S. (2002) 'Going the distance: "closeness" in qualitative data analysis software', *International Journal of Social Research Methodology*, 5(3), pp. 215–228.

Gilligan, C. (1982) *In a different voice: psychological theory and women's development*. Cambridge: Harvard University Press.

Gilmore, R.W. (2002) 'Fatal couplings of power and difference: notes on racism and geography', *The Professional Geographer*, 54(1), pp. 15–24.

Gilmore, R.W. (2007) *Golden gulag: prisons, surplus, crisis, and opposition in globalizing California*, 1st ed. Oakland: University of California Press.

Gilmore, R.W. (2022) *Abolition geography: essays towards liberation*. Edited by B. Bhandar and A. Toscano. New York: Verso.

Giroux, H.A. (2014) 'Neoliberalism's war against the radical imagination', *Tikkun*, 29(3), pp. 9–14.

Glaser, B. and Strauss, A. (1967) *The discovery of grounded theory: strategies for qualitative research*, 1st ed. New Brunswick and London: Transaction Publishers.

Glick Schiller, N. (2012) 'A comparative relative perspective on the relationships between migrants and cities', *Urban Geography*, 33(6), pp. 879–903.

Glick Schiller, N. and Çağlar, A. (2009) 'Towards a comparative theory of locality in migration studies: migrant incorporation and city scale', *Journal of Ethnic and Migration Studies*, 35(2), pp. 177–202.

Gobo, G. (2008) *Doing ethnography*. Thousand Oaks: Sage Publications.

Godard, B. (1989) 'Theorizing feminist discourse/translation', *Tessera*, 6, pp. 42–53.

Godlewska, A. and Smith, N. (eds.) (1994) *Geography and empire*. Oxford: Blackwell.

Gökarıksel, B. (2012) 'The intimate politics of secularism and the headscarf: the mall, the neighborhood, and the public square in Istanbul', *Gender, Place and Culture*, 19(1), pp. 1–20.

Golańska, D. (2022) 'Slow urbicide: accounting for the shifting temporalities of political violence in the West Bank', *Geoforum*, 132, pp. 125–134.

Gottlieb, H. (2017) 'Semiotics and translation', in K. Malmkjaer (ed.) *The Routledge handbook of translation studies and linguistics*. London: Routledge, pp. 45–63.

Gough, K.V. (2012) 'Reflections on conducting urban comparison', *Urban Geography*, 33(6), pp. 866–878.

Gouws, A. (2012) 'Reflections on being a feminist academic/academic feminism in South Africa', *Equality, Diversity and Inclusion: An International Journal*, 31(5/6), pp. 526–541.

Graham, M., Hogan, B., Straumann, R.K. and Medhat, A. (2014) 'Uneven geographies of user-generated information: patterns of increasing informational poverty', *Annals of the Association of American Geographers*, 104(4), pp. 746–764.

Graham, E. (1997) 'Philosophies underlying human geography research', in R. Flowerdew and D.M. Martin (eds.) *Methods in human geography: a guide for students doing a research project*. London: Routledge, pp. 8–33.

Grande, S. (2018) 'Refusing the university', in E. Tuck and W. Wang (eds.) *Toward what justice? Describing diverse dreams of justice in education*. New York: Routledge, pp. 47–64.

Greenhalgh, T. (2019) 'Twitter women's tips on academic writing: a female response to Gioia's rules of the game', *Journal of Management Inquiry*, 28(4), pp. 484–487.

Greenhow, C., Li, J. and Mai, M. (2019) 'From tweeting to meeting: expansive professional learning and the academic conference backchannel', *British Journal of Educational Technology*, 50(4), pp. 1656–1672.

Grewal, I. and Kaplan, C. (1994) *Scattered hegemonies: postmodernity and transnational feminist practices*. Minneapolis: University of Minnesota Press.

Grosz, E. (1994) *Volatile bodies: toward a corporeal feminism*. London: Allen and Unwin.

Guillard, J. (2016) 'Is feminism trending? Pedagogical approaches to countering (Sl)activism', *Gender and Education*, 28(5), pp. 609–626.

Gunaratnam, Y. (2003) 'Messy work: qualitative interviewing across difference', in *Researching 'race' and ethnicity: methods, knowledge and power*, 1st ed. London: Sage Publications, pp. 79–108.

Gustafson, D.L., Parsons, J.E. and Gillingham, B. (2019) 'Writing to transgress: knowledge production in feminist participatory action research', *Forum, Qualitative Social Research*, 20(2), pp. 1–25.

Haddour, A. (2019) 'Tradition, translation, and colonization', in A. Haddour (ed.) *Frantz Fanon, postcolonialism, and the ethics of difference*, 1st ed. Manchester: Manchester University Press, pp. 197–242.

Hale, C. (2008) *Engaging contradictions: theory, politics, and methods of activist scholarship*. Berkeley: University of California Press.

Hall, S.M. (2019) *Everyday life in austerity: family, friends and intimate relations*. London: Palgrave Macmillan.

Halley, J., Kotiswaran, P., Rebouché, R. and Shamir, H. (2018) *Governance feminism: an introduction*. Minneapolis, MN: University of Minnesota Press.

Hallowell, N., Lawton, J. and Gregory, S. (2005) *Reflections on research: the realities of doing research in the social sciences*. Milton Keynes: Open University Press.

Ham, S. (2021) 'Explaining gender gaps in the South Korean labor market during the COVID-19 pandemic', *Feminist Economics*, 27(1–2), pp. 133–151.

Hamadeh, N., Van Rompaey, C., Metreau, E. and Eapen, S.G. (2022) *New World Bank country classifications by income level: 2022–2023*. Available at: https://blogs.worldbank.org/

opendata/new-world-bank-country-classifications-income-level-2022–2023 (Accessed: 25 October 2023).

Hamid, R.D. and Schisgall, J.F. (2023) *Judge allows most counts in Comaroff Harassment Lawsuit against Harvard to proceed*. Available at: www.thecrimson.com/article/2023/4/4/comaroff-lawsuit-proceeds/ (Accessed: 24 October 2023).

Hamilton, A.R. (2020) *The killing fields*. Available at: https://blackgeographer.medium.com/the-killing-fields-bba8111ac137 (Accessed: 25 October 2023).

Hammersley, M. (2006) 'Induction', in V. Jupp (ed.) *The Sage dictionary of social research methods*. London: Sage Publications, pp. 146–147.

Hammond, M. and Wellington, J. (2013) *Research methods: the key concepts*. London and New York: Routledge Key Guides.

Hammonds, E. (2020) 'A moment or a movement? The pandemic, political upheaval, and racial reckoning', *Signs*. Available at: http://signsjournal.org/covid/hammonds/ (Accessed: 25 October 2023).

Hansen, K.T. and Dalsgaard, A.L. (2008) *Youth and the city in the global South*. Bloomington: Indiana University Press.

Haraway, D. (1988) 'Situated knowledges: the science question in feminism and the privilege of partial perspective', *Feminist Studies*, 14(3), pp. 575–599.

Haraway, D. (1991) *Simians, cyborgs, and women: the reinvention of nature*. New York: Routledge.

Harcourt, W., Van den Berg, K., Dupuis, C. and Gaybor, J. (eds.) (2022) *Feminist methodologies: experiments, collaborations and reflections*. Cham: Springer International Publishing.

Hardill, I. and Mills, S. (2015) 'Enlivening evidence-based policy through embodiment and emotions', in J. Bannister and I. Hardill (eds.) *Knowledge mobilisation and the social sciences*. London and New York: Routledge, pp. 155–166.

Harding, S. (1986) 'From the woman question in science to the science question in feminism', in *The science question in feminism*. Ithaca: Cornell University Press, pp. 15–29.

Harding, S. (1991) *Whose science? Whose knowledge?* Milton Keynes: Open University Press.

Harding, S. (1993) 'Rethinking standpoint epistemology: what is "strong objectivity?"', in L. Alcoff and E. Porter (eds.) *Feminist epistemologies*. New York: Routledge, pp. 49–82.

Harker, C. (2020) *Spacing debt: obligations, violence, and endurance in Ramallah, Palestine*. Durham: Duke University Press.

Harney, S. and Moten, F. (2013) *The undercommons: fugitive planning and black study*. Wivenhoe: Minor Compositions.

Harrison, F. (2007) 'Feminist methodology as a tool for ethnographic inquiry on globalization', in N. Gunewardena and A. Kingsolver (eds.) *The gender of globalization: women navigating cultural and economic marginalities*. Santa Fe: School for Advanced Research Press, pp. 23–31.

Hart, G. (2018) 'Relational comparison revisited: Marxist postcolonial geographies in practice', *Progress in Human Geography*, 42(3), pp. 371–394.

Hartsock, N.C.M. (1983) 'The feminist standpoint: developing the grounds for a specifically feminist historical materialism', in S. Harding and M.B. Hintikka (eds.) *Discovering reality: feminist perspectives on epistemology, metaphysics, methodology and philosophy of science*. Dordrecht, Holland, Boston and London: D. Reidel Publishing, pp. 283–310.

Harvey, D. (2005) *A brief history of neoliberalism*. Oxford: Oxford University Press.

Hashimoto, Y. (2021) 'Racing the creative class: diversity, racialized discourses of work, and color-blind redevelopment', *Urban Geography*, 42(4), pp. 528–550.

Hatim, B. and Munday, J. (2004) *Translation: an advanced resource book*. New York: Routledge.

Haug, S. (2021) *What or where is the 'global South'? A social science perspective*. Available at: https://blogs.lse.ac.uk/impactofsocialsciences/2021/09/28/what-or-where-is-the-global-south-a-social-science-perspective/ (Accessed: 25 October 2023).

Haworth, B.T., Cassal, L.C.B. and de Paula Muniz, T. (2023) '"No-one knows how to care for LGBT community like LGBT do": LGBTQIA+ experiences of COVID-19 in the United Kingdom and Brazil', *Disasters*, 47(3), pp. 584–607.

Hawthorne, C. (2019) 'Black matters are spatial matters: Black geographies for the twenty-first century', *Geography Compass*, 13(11), pp. 1–13.

Hayden, D. (1981) *The grand domestic revolution: a history of feminist designs for American homes, neighbourhoods, and cities*. Boston: MIT Press.

Heeks, R. and Renken, J. (2018) 'Data justice for development: what would it mean?', *Information Development*, 34(1), pp. 90–102.

Heine, J. (2023) *The global South is on the rise – but what exactly is the global South?* Available at: https://theconversation.com/the-global-south-is-on-the-rise-but-what-exactly-is-the-global-south-207959 (Accessed: 25 October 2023).

Heintz, J., Staab, S. and Turquet, L. (2021) 'Don't let another crisis go to waste: the COVID-19 pandemic and the imperative for a paradigm shift', *Feminist Economics*, 27(1–2), pp. 470–485.

Held, V. (2005) *The ethics of care: personal, political, and global*. Oxford: Oxford University Press.

Henry, F., Dua, E., James, C.E., Kobayashi, A., Li, P., Ramos, H. and Smith, M. (2017) *The equity myth: racialization and Indigeneity at Canadian universities*. Vancouver and Toronto: UBC Press.

Herbert, S. (2000) 'For ethnography', *Progress in Human Geography*, 24(4), pp. 550–568.

Hesse-Biber, S.N. (2014) *Feminist research practice*, 2nd ed. Thousand Oaks: Sage Publications.

Hill Collins, P. (1991) *Black feminist thought: knowledge, consciousness, and the politics of empowerment*. London: Routledge.

Hill Collins, P. (2000) *Black feminist thought: knowledge, consciousness, and the politics of empowerment*. London: Routledge.

Hill Collins, P. (2019) *Intersectionality as critical social theory*. Durham and London: Duke University Press.

Hoang, L. (2019) 'Twin privatization in Vietnam higher education: the emergence of private higher education and partial privatization of public universities', *Higher Education Policy*, 32(3), pp. 359–380.

Hohenstein, S. (2016) 'The heroine and the meme: participating in feminist discourses online', *Current Objectives of Postgraduate American Studies*, 17(1). Available at: https://doi.org/10.5283/copas.250 (Accessed: 25 October 2023).

Hong, G.K. (2015) *Death beyond disavowal: the impossible politics of difference*. Minneapolis and London: University of Minnesota Press.

hooks, b. (1981) *Ain't I a woman: Black women and feminism*. Boston: South End Press.

hooks, b. (2004) *The will to change: men, masculinity, and love*. New York: Atria Books.

Horn, P. and Grugel, J. (2018) 'The SDGs in middle-income countries: setting or serving domestic development agendas? Evidence from Ecuador', *World Development*, 109, pp. 73–84.

Hosein, G.J. (2011) 'Caribbean feminism, activist pedagogies and transnational dialogues', *Feminist Review*, 98(1), pp. 116–129.

Houston, S.D., McLean, D.J., Hyndman, J. and Jamal, A. (2010) 'Still methodologically becoming: collaboration, feminist politics and "Team Ismaili"', *Gender, Place and Culture*, 17(1), pp. 61–79.

Howitt, R. and Stevens, S. (2016) 'Cross-cultural research: ethics, methods, and relationships', in I. Hay (ed.) *Qualitative research methods in human geography*, 4th ed. Oxford: Oxford University Press, pp. 45–75.

Hudson, B. (1977) 'The new geography and the new imperialism: 1870–1918', *Antipode*, 9(2), pp. 35–41.

Hyman, A. (2003) 'Twenty years of Listserv as an academic tool', *The Internet and Higher Education*, 6, pp. 17–24.

Inter-University Consortium for Political and Social Research. (2023) *Sample data management plan for depositing data with ICPSR*. Available at: www.icpsr.umich.edu/web/pages/datamanagement/dmp/plan.html (Accessed: 25 October 2023).

Ip, P.T.T., Huang, T.M. and Wang, J. (2023) 'Gendering grassrootscapes: the sociospatial relations of lower working-class women dwelling in the socialist Workers' New Villages in post-reform Shanghai', *Journal of Urban Affairs*, pp. 1–21. https://doi.org/10.1080/07352166.2023.2247502

Irshad, I. and Yasmin, M. (2022) 'Feminism and literary translation: a systematic review', *Heliyon*, 8(3). Available at: https://doi.org/10.1016/j.heliyon.2022.e09082 (Accessed: 25 October 2023).

Isaak, J. and Hanna, M.J. (2018) 'User data privacy: Facebook, Cambridge Analytica, and privacy protection', *Computer*, 51(8), pp. 56–59.

Jackson, K. (2017) 'Where qualitative researchers and technologies meet: lessons from interactive digital art', *Qualitative Inquiry*, 23(10), pp. 818–826.

Jackson, K. (2021) ' "What should I do now?": navigating relational ethics as an early career researcher as illustrated by a qualitative interview study about women's drinking practices', *International Journal of Qualitative Methods*, 20. Available at: https://doi.org/10.1177/1609406920986044 (Accessed: 25 October 2023).

Jackson, K. and Bazeley, P. (2019) *Qualitative data analysis with NVivo*, 2nd ed. Los Angeles: Sage Publications.

Jacobs, J.M. (2012) 'Commentary: comparing comparative urbanisms', *Urban Geography*, 33(6), pp. 904–914.

James, A.L. and Tynes, B.A. (n.d.) 'Entitled IX', *Zora's Daughters*, S2, E10 [Podcast]. Available at: https://zorasdaughters.com/entitled-ix/ (Accessed: 25 October 2023).

James, C. and Chapman-Nyaho, S. (2017) 'We would never be hired these days: the precarious work situation of racialized and Indigenous faculty members', in F. Henry, E. Dua, C.E. James, A. Kobayashi, P. Li, H. Ramos and M. Smith (eds.) *The equity myth: racialization and indigeneity at Canadian universities*. Vancouver and Toronto: UBC Press, pp. 84–114.

Jamil, G. (2017) 'Real-life methods: feminist explorations of segregation in Delhi', in K. Kannabiran and P. Swaminathan (eds.) *Re-presenting feminist methodologies: interdisciplinary explorations*. London and New York: Routledge, pp. 91–112.

Jauk, D. (2017) 'Feminist scholar-activism goes global: experiences of "Sociologists for Women in Society" at the UN', *Brock Education: A Journal of Educational Research and Practice*, 27(1), pp. 79–100.

Jazeel, T. (2019) *Postcolonialism*. London: Routledge.

Jelin, E. (1977) 'Migration and labor force participation of Latin American women: the domestic servants in the cities', *Signs*, 3(1), pp. 129–141.

Jeppesen, S. and Sartoretto, P. (2023) 'Cartographies of resistance: counter-data mapping as the new frontier of digital media activism', *Media and Communication (Lisboa)*, 11(1), pp. 150–162.

Jiménez, A.C. (2017) 'Auto-construction redux: the city as method', *Cultural Anthropology*, 32(3), pp. 450–478.

Johnson, J.A. (2016) 'The question of information justice', *Communications of the Association for Computing Machinery*, 59(3), pp. 27–29.

Johnston, R. and Sidaway, J. (2016) *Geography and geographers: Anglo-American human geography since 1945*, 7th ed. London and New York: Routledge.

Johnston-Anumonwo, I. (2019) 'Mentoring across difference: success and struggle in an academic geography career', *Gender, Place and Culture*, 26(12), pp. 1683–1700.

Jokinen, J.C. and Caretta, M.A. (2016) 'When bodies do not fit: an analysis of postgraduate fieldwork', *Gender, Place and Culture*, 23(12), pp. 1665–1676.

Jones, A. and Jenkins, K. (2008) 'Rethinking collaboration: working the Indigene-colonizer hyphen', in N.K. Denzin, Y.S. Lincoln and L. Tuhiwai-Smith (eds.) *Handbook of critical and Indigenous methodologies*. Los Angeles: Sage Publications, pp. 471–486.

Joshi, S., McCutcheon, P. and Sweet, E.L. (2015) 'Visceral geographies of whiteness and invisible microaggressions', *ACME: An International Journal for Critical Geographies*, 14(1), pp. 298–323.

Kabeer, N. (2005) 'Gender equality and women's empowerment: a critical analysis of the third millennium development goal', *Gender and Development*, 13(1), pp. 13–24.

Kabeer, N., Razavi, S. and van der Meulen Rodgers, Y. (2021) 'Feminist economic perspectives on the COVID-19 pandemic', *Feminist Economics*, 27(1–2), pp. 1–29.

Kagal, N. and Latchford, L. (2020) 'Towards an intersectional praxis in international development: what can the sector learn from Black feminists located in the global North?', *Gender and Development*, 28(1), pp. 11–30.

Kaika, M. (2017) '"*Don't call me resilient again!*": the New Urban Agenda as immunology . . . or . . . what happens when communities refuse to be vaccinated with "smart cities" and indicators', *Environment and Urbanization*, 29(1), pp. 89–102.

Kapoor, P. (2017) 'Between fact and fiction: can there be a postcolonial feminist ethnography?', *Women's Studies International Forum*, 60, pp. 58–68.

Kasana, M. (2014) 'Feminisms and the social media sphere', *WSQ: Women's Studies Quarterly*, 42(3–4), pp. 236–249.

Kašić, B. (2016) '"Unsettling" women's studies, settling neoliberal threats in the academia: a feminist gaze from Croatia', *Women's Studies International Forum*, 54, pp. 129–137.

Katz, C. (1994) 'Playing the field: questions of fieldwork in geography', *The Professional Geographer*, 46(1), pp. 67–72.

Katz, C. (2001) 'On the grounds of globalization: a topography for feminist political engagement', *Signs,* 26(4), pp. 1213–1234.

Katz, C. (2004) *Growing up global: economic restructuring and children's everyday lives*. Minneapolis: University of Minnesota Press.

Katz, C. and Warren, G.C. (2014) *Gwendolyn Warren and Cindi Katz in conversation*. Available at: https://gcees.commons.gc.cuny.edu/gwendolyn-warren-cindi-katz-conversation/ (Accessed: 25 October 2023).

Kaufman, E.C. (2021) 'Staying with the trouble of collegiality, professionalism and care: fertilities in academia', *Environment and Planning C: Politics and Space*, 39(8), pp. 1737–1754.

Kavanagh, E. and Brown, L. (2020) 'Towards a research agenda for examining online gender-based violence against women academics', *Journal of Further and Higher Education*, 44, pp. 1379–1387.

Kelley, R.D. (2016) *Black study, Black struggle*. Available at: http://bostonreview.net/forum/robin-d-g-kelley-black-study-black-struggle (Accessed: 26 October 2023).

Kelly, K. (2020) 'COVID-19 requires an intersectional feminist policy response', in G.W. Muschert, K.M. Budd, M. Christian and R. Perrucci (eds.) *Social problems in the age of COVID-19: volume 2 – global perspectives*. Bristol: Policy Press, pp. 111–122.

Kelly, L. (2005) 'Inside outsiders', *International Feminist Journal of Politics*, 7(4), pp. 471–495.

Khan, M.N., Ashraf, M.A., Seinen, D., Khan, K.U. and Laar, R.A. (2021) 'Social media for knowledge acquisition and dissemination: the impact of the COVID-19 pandemic on collaborative learning driven social media adoption', *Frontiers in Psychology*, 12, pp. 1–13.

Kidman, J. (2020) 'Whither decolonisation?: Indigenous scholars and the problem of inclusion in the neoliberal university', *Journal of Sociology*, 56(2), pp. 247–262.

Kingsolver, A.E., Wagner, G., Barrera, R., Brown, B., Civitillo, J., Clark, D., Gerald, V., Goodrich, D., Hewlett, M., Hughes, M., Leader, J., Liger, L., Little Water, T., Loftlin, C., Parra-Medina, D., Rymer, D., Scott, C., Schumpert, C., Thompson, N. and Wilson, T. (2003) 'Teaching anthropological ethics at the University of South Carolina: an example of critical ethical dialogues across communities', in Fluehr-Lobban, C. (ed.) *Ethics and the profession of Anthropology: dialogue for ethically conscious practice*. Walnut Creek: Rowman Altamira, pp. 197–224.

Kinkaid, E. (2022) 'Trans in geography: outlines of a socio-political location', *Queer Geographies Post-Graduate Reading Group*. Available at: https://sites.google.com/view/edenkinkaid/recorded-presentations/trans-in-geography. (Accessed: 26 October 2023).

Kinkaid, E. (2023) 'Whose geography, whose future? Queering geography's disciplinary reproduction', *Dialogues in Human Geography*. Available at: https://doi.org/10.1177/20438206221144839 (Accessed: 25 October 2023).

Kinkaid, E., Parikh, A. and Ranjbar, A.M. (2022) 'Coming of age in a straight white man's geography: reflections on positionality and relationality as feminist anti-oppressive praxis', *Gender, Place and Culture*, 29(11), pp. 1556–1571.

Klein, P., Fatima, M., McEwen, L., Moser, S.C., Schmidt, D. and Zupan, S. (2011) 'Dismantling the ivory tower: engaging geographers in university – community partnerships', *Journal of Geography in Higher Education*, 35(3), pp. 425–445.

Kobayashi, A. (1994) 'Coloring the field: gender, 'race,' and the politics of fieldwork', *The Professional Geographer*, 46(1), pp. 73–80.

Kobayashi, A. and Peake, L. (1994) 'Unnatural discourse: "race" and gender in geography', *Gender, Place, and Culture*, 1(2), pp. 225–243.

Koch, N. (2020) 'Deep listening: practicing intellectual humility in geographic fieldwork', *The Geographical Review*, 110(1–2), pp. 52–64.

Kocsis, J. (2024) 'Research, rigour, and rape: facing the reality of gender-based violence in academic fieldwork', *Gender, Place & Culture*, 0(0), pp. 1–9.

Koleth, E., Peake, L., Razavi, N. and Adeniyi-Ogunyankin, G. (2023) 'Decolonising feminist explorations of urban futures', *Urban Geography*, 44(9), pp. 1843–1852.

Kong, L. and Qian, J. (2019) 'Knowledge circulation in urban geography/urban studies, 1990–2010: testing the discourse of Anglo-American hegemony through publication and citation patterns', *Urban Studies*, 56(1), pp. 44–80.

Koopman, S. (2008) 'Cutting through topologies: crossing lines at the school of the Americas', *Antipode*, 40(5), pp. 825–847.

Koopman, S. (2011) 'Alter-geopolitics: other securities are happening', *Geoforum*, 42(3), pp. 274–284.

Koopman, S. (2020) 'Building an inclusive peace is an uneven socio-spatial process: Colombia's differential approach', *Political Geography*, 83, 102252.

Koopman, S. (forthcoming) 'Putting more geo in Geography: tactics for better bridging language gaps in fieldwork, conferencing, and publishing'.

Krause, P., Szekely, O., Bloom, M., Christia, F., Daly, S., Lawson, C., Marks, Z., Milliff, A., Miura, K., Nielsen, R., Reno, W., Souleimanov E.A. and Zakayo, A. (2021) 'COVID-19 and fieldwork: challenges and solutions', *PS: Political Science and Politics*, 54(2), pp. 264–269.

Kushnir, I. (2020) 'The voice of inclusion in the midst of neoliberalist noise in the Bologna process', *European Educational Research Journal*, 19(6), pp. 485–505.

Kvale, S. (2013) *Doing interviews*. Los Angeles: Sage Publications.

Kwan, M.P. (2002a) 'Introduction: feminist geography and GIS', *Gender, Place and Culture*, 9(3), pp. 261–262.

Kwan, M.P. (2002b) 'Is GIS for women? Reflections on the critical discourse in the 1990s', *Gender, Place and Culture*, 9(3), pp. 271–279.

Kwan, M.P. and Ding, G. (2008) 'Geo-narrative: extending geographic information systems for narrative analysis in qualitative and mixed-method research', *The Professional Geographer*, 60(4), pp. 443–465.

Lapadat, J.C. (2000) 'Problematizing transcription: purpose, paradigm and quality', *International Journal of Social Research Methodology*, 3(3), pp. 203–219.

Lather, P. (2001) 'Postbook: working the ruins of feminist ethnography', *Signs*, 27(1), pp. 199–227.

Lau, T.C.W. (2019) 'Slowness, disability, and academic productivity: the need to rethink academic culture', in C. McMaster and B. Whitburn (eds.) *Disability at the university: a disabled students' manifesto*. New York: Peter Lang, pp. 11–19.

Lawhon, M. (2021) *Making urban theory: learning and unlearning through Southern cities*. London: Routledge.

Lawhon, M. and Truelove, Y. (2020) 'Disambiguating the southern urban critique: propositions, pathways and possibilities for a more global urban studies', *Urban Studies*, 57(1), pp. 3–20.

Le Blanc, A.M. (2017) 'Disruptive meaning-making: qualitative data analysis software and postmodern pastiche', *Qualitative Inquiry*, 23(10), pp. 789–798.

Lee, R. and Ahtone, T. (2020) *Land-grab universities*. Available at: www.hcn.org/issues/52.4/indigenous-affairs-education-land-grab-universities (Accessed: 26 October 2023).

Lefebvre, H. [1967] (1996) 'The right to the city', in E. Kofman and E. Lebas (eds.) *Writing on cities*. Oxford: Blackwell, pp. 63–184.

Le Galès, P. and Robinson, J. (eds.) (2023) *The Routledge handbook of Comparative Global Urban Studies*. London: Routledge.

Leurs, K. (2017) 'Feminist data studies: using digital methods for ethical, reflexive and situated socio-cultural research', *Feminist Review*, 115, pp. 130–154.
Lewis, D. (1973) 'Anthropology and colonialism', *Current Anthropology*, 14(5), pp. 581–602.
Li, D. and Li, Y. (2022) 'Preparing for RAE 2020 in Hong Kong: academics research, writing and publishing trajectories in a neoliberal governance landscape', *Studies in Higher Education*, 47(7), pp. 1410–1422.
Liebman, A., Rhiney, K. and Wallace, R. (2020) 'To die a thousand deaths: COVID-19, racial capitalism, and anti-Black violence', *Human Geography*, 13(3), pp. 331–335.
Linabary, J.R., Corple, D.J. and Cooky, C. (2020) 'Feminist activism in digital space: postfeminist contradictions in #WhyIStayed', *New Media and Society*, 22(10), pp. 1827–1848.
Linder, C., Myers, J.S., Riggle, C. and Lacy, M. (2016) 'From margins to mainstream: social media as a tool for campus sexual violence activism', *Journal of Diversity in Higher Education*, 9(3), pp. 231–244.
Linz, J. (2021) 'Where crises converge: the affective register of displacement in Mexico City's post-earthquake gentrification', *Cultural Geographies*, 28(2), pp. 285–300.
Linz, J., Smyth, A., Billo, E., Curran, W., Hawkins, R., Mullings, B., Mountz, A., Parizeau, K., Walton-Roberts, M., Whitson, R., Elledge, A., Faria, C., Whitesell, D., Al-Saleh, D., Noterman, E. and FLOCK Geography Collective. (2021) 'Calling all collectives: interviews with feminist geography collectives', in B. Gökariksel, M. Hawkins, C. Neubert and S. Smith (eds.) *Feminist geography unbound: discomfort, bodies, and prefigured futures*. Morgantown: West Virginia University Press, pp. 261–290.
Liu, H. (2020) *Redeeming leadership. An anti-racist feminist intervention*. Bristol: Bristol University Press.
Lofland, L. (1995) 'Analytical ethnography: features, failings, futures', *Journal of Contemporary Ethnography*, 24(1), pp. 30–67.
Loney-Howes, R. (2020) *Online anti-rape activism: exploring the politics of the personal in the age of digital media*. Bingley: Emerald Publishing.
Longhurst, R. (2005) 'Situating bodies', in L. Nelson and J. Seager (eds.) *A companion to feminist geography*. Malden: Blackwell Publishing, pp. 337–349.
Lorde, A. (1984) *Sister outsider: essays and speeches*. Berkeley: Crossing Press.
Lourenco, S.F. and Tasimi, A. (2020) 'No participant left behind: conducting science during COVID-19', *Trends in Cognitive Sciences*, 24(8), pp. 583–584.
Loyd, J., Veledenitsky, S., Diaz, I., Ibrahim, S., Giddings, C., Caldwell, K., Hawkins, R. and Bonds, A. (2022) 'Dear feminist collective: how does one take up slow scholarship (in the midst of crises)?', in S. Lovell, S. Coen and W. Rosenberg (eds.) *The Routledge handbook of methodologies in human geography*. London: Routledge, pp. 395–406.
Low, S.M. and Merry, S.E. (2010) 'Engaged anthropology: diversity and dilemmas', *Current Anthropology*, 51(2), pp. 203–226.
Lugones, M. (2007) 'Heterosexualism and the colonial/modern gender system', *Hypatia*, 22(1), pp. 186–219.
Lugones, M. (2010) 'Toward a decolonial feminism', *Hypatia*, 25(4), pp. 742–759.
Lunny, D.R. (2019) 'English hegemony, Anglo privilege and the promise of "Allo" lingual citational praxis in transnational feminisms research', *Feminist Review*, 121(1), pp. 66–80.
Mackenzie, S. (1989) *Visible histories: women and environments in a post-war British city*. Montreal: McGill-Queen's Press.
Mackey, E., Elliot, M. and O'Hara, K. (2016) *The anonymisation decision-making framework (ADF)*. Available at: https://ukanon.net/framework/ (Accessed: 26 October 2023).
Madge, C., Noxolo, P. and Raghuram, P. (2004) 'Bodily contours: geography, metaphor and pregnancy', *WGSG, Geography and Gender Reconsidered*, pp. 68–83.
Mahler, A.G. (2017) 'Global South', in E. O'Brien (ed.) *Oxford bibliographies in literary and critical theory*. New York: Oxford University Press.

Mahler, A.G. (2018) *From the tricontinental to the global South: race, radicalism, and transnational solidarity*. Durham: Duke University Press.

Mahtani, M. (2014) 'Toxic geographies: absences in critical race thought and practice in social and cultural geography', *Social and Cultural Geography*, 15(4), pp. 359–367.

Maira, S. and Sze, J. (2012) 'Dispatches from pepper spray university: privatization, repression, and revolts', *American Quarterly*, 64(2), pp. 315–330.

Mani, L. (1990) 'Multiple mediations: feminist scholarship in the age of multinational reception', *Feminist Review*, 35(1), pp. 24–41.

Manuel, G. and Posluns, M. (1974) *The Fourth World: an Indian reality*. Toronto: Collier-Macmillan Canada.

Markham, A. (2017) *Reflexivity: some techniques for interpretive researchers*. Available at: https://annettemarkham.com/2017/02/reflexivity-for-interpretive-researchers/ (Accessed: 26 October 2023).

Martínez, S. and Agüero, J. (2021) 'Cartography of Southern feminisms: contributions of decolonial feminisms and community feminisms', *International Social Work*, 66(3). Available at: https://doi.org/10.1177/00208728211001417 (Accessed: 2 October 2023).

Martínez-Gómez, A. (2017) 'Non-professional interpreting', in R. Antonini, L. Cirillo, L. Rossato and I. Torresi (eds.) *Benjamins translation library*. Amsterdam: John Benjamins Publishing Company, pp. 417–431.

Marx Ferree, M. and Verloo, M. (2016) 'Foreword gender inequality and institutional knowledge: the politics of knowledge transfer', in M. Bustelo, L. Ferguson and M. Forest (eds.) *The politics of feminist knowledge transfer: gender training and gender expertise*. Basingstoke: Palgrave Macmillan, pp. x–xvii.

Mason-Deese, L. (2020) *Translator's foreword to Feminist international: how to change everything, by V. Gago*. London: Verso, pp. vii–x.

Massey, D. (1993) 'Power-geometry and a progressive sense of place', in J. Bird, B. Curtis, T. Putnam, G. Robertson and L. Tuckner (eds.) *Mapping the futures: local cultures, global change*. Abingdon: Routledge, pp. 60–70.

Massey, D. (1994) *Space, place, and gender*. Minneapolis: University of Minnesota Press.

Massey, D. (2005) *For space*. London: Sage Publications.

Matos, C. (2017) 'New Brazilian feminisms and online networks: cyberfeminism, protest and the female "Arab Spring" ', *International Sociology*, 32(3), pp. 417–434.

Matrix. (1984) *Making space: women and the man made environment*. London: Pluto Press.

Matthews, S.A., Detwiler, J.E. and Burton, L.M. (2005) 'Geo-ethnography: coupling geographic information analysis techniques with ethnographic methods in urban research', *Cartographica: The International Journal for Geographic Information and Geovisualization*, 40(4), pp. 75–90.

Mattingly, D.J. and Falconer-Al-Hindi, K. (1995) 'Should women count? A context for the debate', *The Professional Geographer*, 47(4), pp. 427–435.

Mayers, L. (2021) 'Gendered "risk" and racialized inheritance: toward a feminist analysis of debt in US higher education', *Feminist Studies*, 47(3), pp. 729–752.

Mbete, S. and Goodman, A. (2021) ' "Perfect storm" of pandemic, poverty and jailing ex-president unleashes mass protest in South Africa', *Democracy Now!* Available at: www.democracynow.org/2021/7/15/south_africa_protests_jacob_zuma (Accessed: 26 October 2023).

McCall, L. (2005) 'The complexity of intersectionality', *Signs*, 30(3), pp. 1771–1800.

McClaurin, I. (2001) *Black feminist anthropology: theory, politics, praxis and poetics*. New Brunswick: Rutgers University Press.

McCray, J.A. (2023) *Reflections on leaving*. Available at: https://medium.com/@jordan.a.mccray/reflections-on-leaving-93fc34b07608 (Accessed: 26 October 2023).

McDowell, L. (1999) *Gender, identity and place*. Cambridge: Polity.

McFarlane, C. (2010) 'The comparative city: knowledge, learning, urbanism', *International Journal of Urban and Regional Research*, 34(4), pp. 725–742.

McFarlane, C. and Robinson, J. (2012) 'Introduction: experiments in comparative urbanism', *Urban Geography*, 33(6), pp. 765–773.
McGuirk, P. (2015) 'Geographical theorizing for a world of cities', *Progress in Human Geography*. Available at: https://doi.org/10.1177/0309132515603838 (Accessed: 26 October 2023).
McKinney, C. (2015) 'Newsletter networks in the feminist history and archives movement', *Feminist Theory*, 16(3), pp. 309–328.
McKittrick, K. (2013) 'Plantation futures', *Small Axe: A Caribbean Journal of Criticism*, 17(3–42), pp. 1–15.
McKittrick, K. (2021) *Dear science and other stories*. Durham: Duke University Press.
Megarry, J. (2018) 'Under the watchful eyes of men: theorising the implications of male surveillance practices for feminist activism on social media', *Feminist Media Studies*, 18(6), pp. 1070–1085.
Mei-Singh, L. (2021) 'Accompaniment through carceral geographies: abolitionist research partnerships with Indigenous communities', *Antipode*, 53(1), pp. 74–94.
Merleau-Ponty, M. (1962) *The phenomenology of perception*. Translated by C. Smith. London: Routledge and Kegan Paul.
Meyerhoff, E. and Noterman, E. (2019) 'Revolutionary scholarship by any speed necessary: slow or fast but for the end of this world', *ACME: An International Journal for Critical Geographies*, 18(1), pp. 217–245.
Miao, W., Jiang, M. and Pang, Y. (2021) 'Historicizing Internet regulation in China: a meta-analysis of Chinese Internet policies (1994–2017)', *International Journal of Communication*, 15, 2003–2026.
Mignolo, M. and Walsh, C. (2018) *On decoloniality: concepts, analytics, praxis*. Durham and London: Duke University Press.
Militz, E. (2023) 'Intimate geographies of virginal blood', *Cultural Geographies*, 30(1), pp. 123–139.
Miraftab, F. (2004) 'Can you belly dance? Methodological questions in the era of transnational feminist research', *Gender, Place and Culture*, 11(4), pp. 595–604.
Miraftab, F. (2016) *Global heartland: displaced labor, transnational lives and local placemaking*. Bloomington: Indiana University Press.
Moewaka Barnes, H. and McCreanor, T. (2022) 'Decolonising qualitative research design', in O. Flick (ed.) *Sage handbook of qualitative research design*. London: Sage Publications, pp. 210–233.
Mohanty, C. (1988) 'Under Western eyes: feminist scholarship and colonial discourses', *Feminist Review*, 30(1), pp. 61–88.
Mohanty, C.T. (2003a) '"Under Western eyes" revisited: feminist solidarity through anticapitalist struggles', *Signs*, 28(2), pp. 499–535.
Mohanty, C.T. (2003b) *Feminism without borders: decolonizing theory, practicing solidarity*. Durham: Duke University Press.
Mölders, T. (2019) 'Rethinking gender: feminist perspectives on sustainable development goals in the light of (re)productivity', *GAIA: Ecological Perspectives for Science and Society*, 28(2), pp. 95–99.
Mollett, S. and Faria, C. (2013) 'Messing with gender in feminist political ecology', *Geoforum*, 45, pp. 116–125.
Mollett, S. and Faria, C. (2018) 'The spatialities of intersectional thinking: fashioning feminist geographic futures', *Gender, Place and Culture*, 25(4), pp. 565–577.
Moraga, C. and Anzaldúa, G. (eds.) (1981) *This bridge called my back: writings by radical women of color*. New York: Kitchen Table: Women of Color Press.
Morse, J. (2015) 'Critical analysis of strategies for determining rigor in qualitative inquiry', *Qualitative Health Research*, 25(9), pp. 1212–1222.
Moser, C.O.N. (1996) *Confronting crisis: a summary of household responses to poverty and vulnerability in four poor urban communities*. Washington: World Bank.
Moser, C.O.N. and Peake, L. (1987) *Women, human settlements, and housing*. London: Tavistock Publications.

Möser, C. (2017) 'Gender travelling across France, Germany and the US: the feminist gender debates as cultural translations', in O. Castro and E. Ergun (eds.) *Feminist translation studies: local and transnational perspectives*. London and New York: Routledge, pp. 80–92.

Moss, G. (2013) 'Research, policy and knowledge flows in education: what counts in knowledge mobilisation?', *Contemporary Social Science*, 8(3), pp. 237–248.

Moss, P. (2002) 'Taking on, thinking about, and doing feminist research in geography', in P. Moss, K. Falconer-Al Hindi and H. Kawabata (eds.) *Feminist geography in practice: research and methods*. Malden, MA: Wiley-Blackwell, pp. 1–17.

Mott, C. and Cockayne, D. (2017) 'Citation matters: mobilizing the politics of citation toward a practice of "conscientious engagement" ', *Gender Place and Culture*, 24(7), pp. 954–973.

Mott, C. and Cockayne, D. (2021) 'Understanding how hatred persists: situating digital harassment in the long history of white supremacy', *Gender, Place and Culture*, 28(11), pp. 1521–1540.

Mountz, A., Bonds, A., Mansfield, B., Loyd, J., Hyndman, J., Walton-Roberts, M., Basu, R., Whitson, R., Hawkins, R., Hamilton, T. and Curran, W. (2015) 'For slow scholarship: a feminist politics of resistance through collective action in the neoliberal university', *ACME: An International E-Journal for Critical Geographies*, 14(4), pp. 1235–1259.

Mountz, A., Miyares, I., Wright, R. and Bailey, A.J. (2003) 'Methodologically becoming: power, knowledge and team research', *Gender, Place and Culture*, 10(1), pp. 29–46.

Mukurtu. (2023) *Mukurtu*. Available at: https://mukurtu.org/ (Accessed: 11 September 2023).

Müller, M. (2021) 'Worlding geography: from linguistic privilege to decolonial anywhere', *Progress in Human Geography*, 45(6), pp. 1440–1466.

Mullings, B. (1999) 'Insider or outsider, both or neither: some dilemmas of interviewing in across-cultural setting', *Geoforum*, 30(4), pp. 337–350.

Mullings, B. and Mukherjee, S. (2018) 'Reflections on mentoring as decolonial, transnational, feminist praxis', *Gender, Place and Culture*, 25(10), pp. 1405–1422.

Mullings, B., Peake, L. and Parizeau, K. (2016) 'Cultivating an ethic of wellness in geography', *The Canadian Geographer*, 60(2), pp. 161–167.

Mullings, L. (2000) 'African American women making themselves: Notes on the role of black feminist research', *Souls: A Critical Journal of Black Politics, Culture and Society*, 2(4), pp. 18–29.

Mulvey, L. (1975) 'Visual pleasure and narrative cinema', *Screen*, 16(3), pp. 6–18.

Münch, R. (2014) *Academic capitalism: universities in the global struggle for excellence*. New York: Routledge.

Muncie, J. (2006) 'Discourse analysis', in V. Jupp (ed.) *The Sage dictionary of social research methods*. London: Sage Publications, pp. 74–76.

Munday, J. (2016) *Introducing translation studies: theories and applications*, 4th ed. London: Routledge.

Murphy, E.R. (2015) 'An interview with Professor Paul Bandia', *Perspectives: Studies in Translatology*, 23(1), pp. 143–154.

Murray, A. and Spronk, S. (2019) 'Blended financing, Canadian foreign aid policy, and alternatives', *Studies in Political Economy: A Socialist Review*, 100(3), pp. 270–286.

Mustafa, D., Matson, P., Roberts, E. and Sharpe, J. (2024) 'Ecologies of sustainable development goals: a mid-term perspective', *International Development Planning Review*, 46(2), pp. 107–121.

Mwangi, N. (2019) ' "Good that you are one of us": positionality and reciprocity in conducting fieldwork in Kenya's flower industry', in L. Johnstone (ed.) *The politics of conducting research in Africa: ethical and emotional challenges in the field*. London: Palgrave Macmillan, pp. 13–33.

Nagar, R. (2013) 'Storytelling and co-authorship in feminist alliance work: reflections from a journey', *Gender, Place and Culture*, 20(1), pp. 1–18.

Nagar, R. (2014) *Muddying the waters: coauthoring feminisms across scholarship and activism*. Urbana, Chicago, and Springfield: University of Illinois Press.

Nagar, R. (2019) *Hungry translations: relearning the world through radical vulnerability*. Urbana: University of Illinois Press.

Nagar, R. (in consultation with Ali, F. and Sangtin Women's Collective). (2003) 'Collaboration across borders: moving beyond positionality', *Singapore Journal of Tropical Geography*, 24(3), pp. 356–372.

Nagar, R., Aslan, Ö., Hasan, N.Z., Rahemtullah, O.S., Upadhyay, N. and Uzun, B. (2016) 'Feminisms, collaborations, friendships: a conversation', *Feminist Studies*, 42(2), pp. 502–519.

Nagar, R., Davis, K., Butler, J., Keating, A.L., de Lima Costa, C., Alvarez, S.E., Altinay, A.G., Ergun, E. and Castro, O. (2017) 'A cross-disciplinary roundtable on the feminist politics of translation', in O. Castro and E. Ergun (eds.) *Feminist translation studies: local and transnational perspectives*. Routledge, pp. 111–136.

Nagar, R. and Geiger, S. (2007) 'Reflexivity and positionality in feminist fieldwork revisited', in A. Tickell, E. Sheppard, J. Peck and T.J. Barnes (eds.) *Politics and practice in economic geography*. London: Sage Publications.

Nagar, R., Lawson, V., McDowell, L. and Hanson, S. (2002) 'Locating globalization: feminist (re) readings of the subjects and spaces of globalization', *Economic Geography*, 78(3), pp. 257–284.

Nagar, R., Meier I. and Spathopoulou, I. (2023) 'Refusals, radical vulnerability, and hungry translations – a conversation with Richa Nagar', *Fennia: International Journal of Geography*, pp. 1–7. Available at: https://doi.org/10.11143/fennia.121797 (Accessed: 26 October 2023).

Nagar, R. and Raju, S. (2003) 'Women, NGOs and the contradictions of empowerment and disempowerment: a conversation', *Antipode*, 35(1), pp. 1–13.

Nagar, R. and Selmeczi, A. (2018) 'The labor of political theatre as embodied politics: a conversation', in S. Choi, A. Selmeczi and E. Strauss (eds.) *Critical methods for the study of world politics: creativity and transformation*. London: Routledge, pp. 84–100.

Nagar, R. and Shirazi, R. (2019) 'Radical vulnerability: hungering collectively for justice', in Antipode Editorial Collective (eds.) *Keywords in radical geography: Antipode at 50*. London: Wiley-Blackwell, pp. 236–242.

Nakamura, L. (2014) '"I WILL DO EVERYthing that am asked": scambaiting, digital show-space, and the racial violence of social media', *Journal of Visual Culture*, 13(3), pp. 257–274.

Naples, N. (2003) *Feminism and method: ethnography, discourse analysis and activist research*. New York: Routledge.

Nash, J.C. (2019) *Black feminism reimagined: after intersectionality*. Durham: Duke University Press.

Nash, J.C. (2020) 'Citational desires: on Black Feminism's institutional longings', *Diacritics*, 48(3), pp. 76–91.

Nash, M. (2023) 'Breaking the silence around blood: managing menstruation during remote Antarctic fieldwork', *Gender, Place and Culture*, 30(8), pp. 1083–1103.

National Research Foundation. (2020) *National research foundation*. Available at: www.nrf.ac.za/ (Accessed: 26 October 2023).

Naylor, L., Daigle, M., Zaragocin, S., Ramírez, M.M. and Gilmartin, M. (2018) 'Interventions: bringing the decolonial to political geography', *Political Geography*, 66, pp. 199–209.

Neely, A.H. and Lopez, P.J. (2022a) 'The differential impact of the COVID-19 pandemic in geography in the United States', *The Professional Geographer*, 74(3), pp. 430–439.

Neely, A.H. and Lopez, P.J. (2022b) 'Toward healthier futures in post-pandemic times: political ecology, racial capitalism, and black feminist approaches to care', *Geography Compass*, 16(2). Available at: https://doi.org/10.1111/gec3.12609 (Accessed: 26 October 2023).

Nijman, J. (2007) 'Introduction: comparative urbanism', *Urban Geography*, 28(1), pp. 1–6.

Niranjana, T. (1992) *History, post-structuralism, and the colonial context*. Berkeley: University of California Press.

Nixon, J. (ed.) (2017) *Higher education in austerity Europe*. London: Bloomsbury Academic, an Imprint of Bloomsbury Publishing Plc.

Nnaemeka, O. (2004) 'Nego-feminism: theorizing, practicing, and pruning Africa's way', *Signs*, 29(2), pp. 357–385.

Novoselova, V. and Jenson, J. (2019) 'Authorship and professional digital presence in feminist blogs', *Feminist Media Studies*, 19(2), pp. 257–272.

Novovic, G. (2021) 'Are SDGs counting what counts: feminist analysis of Agenda 2030's policy shrinking', *Global Governance*, 27(3), pp. 366–387.

Novovic, G. (2022) 'Can agenda 2030 deliver on "localization"? Policy limitations of agenda 2030 in the broader global governance system', *Development Policy Review*, 40(4). Available at: https://doi.org/10.1111/dpr.12587 (Accessed: 19 November 2023).

Noxolo, P. (2017) 'Decolonial theory in a time of the re-colonisation of UK research', *Transactions of the Institute of British Geographers*, 42(3), pp. 342–344.

Nuttall, S. (2021) 'The university now: what it will have been for what it is becoming', in J. Barnard-Naude (ed.) *Decolonising the neoliberal university: law, psychoanalysis and the politics of student protest*. New York: Birkbeck Law Press, pp. 49–59.

Nyambura, R., Valiana, S., Ajl, M. and Castel-Branco, R. (2021) 'African feminist and anti-capitalist responses to COVID-19: labor, health and ecological questions', *African EcoFeminisms*. Available at: https://ces.uc.pt/en/agenda-noticias/conversas-com-o-mundo/6-abr-2020/african-feminist-and-anti-capitalist-responses (Accessed: 27 September 2022).

O'Connell, R. and Brannen, J. (2022) 'Families and food in hard times (FFP)', in L. Hantrais (ed.) *How to manage international multidisciplinary research projects*. Cheltenham: Edward Elgar, pp. 126–136.

O'Connor, H. and Madge, C. (2017) 'Online interviewing', in N. Fielding, R. Lee and G. Blank (eds.) *The Sage handbook of online research methods*. Los Angeles: Sage Publications.

O'Toole, E., Feeney, L., Heard, K. and Naimpally, R. (2018) *Data security procedures for researchers*. Available at: www.povertyactionlab.org/resource/data-security-procedures-researchers (Accessed: 26 October 2023).

Oakley, A. (1981) 'Interviewing women: a contradiction in terms?', in H. Roberts (ed.) *Doing feminist research*. London: Routledge and Kegan Paul, pp. 30–61.

Oakley, A. (2000) *Experiments in knowing: gender and method in the social sciences*. New York: New Press.

Oakley, A. (2016) 'Interviewing women again: power, time and the gift', *Sociology*, 50(1), pp. 195–213.

Oeur, F.B. (2021) 'Fever dreams: W. E. B. Du Bois and the racial trauma of COVID-19 and lynching', *Ethnic and Racial Studies*, 44(5), pp. 735–745.

Ogundele, T. and Walton-Roberts, M. (2021) *Covid-19 and health worker infections: the need for disaggregated intersectional data*. Available at: www.genderandcovid-19.org/editorial/covid-19-and-health-worker-infections-the-need-for-disaggregated-intersectional-data/ (Accessed: 26 October 2023).

Okech, A. and Musindarwezo, D. (2019) 'Transnational feminism and the post-2015 development agenda: what can we learn from the experiences of FEMNET?', *Soundings*, 71, pp. 75–90.

Oldfield, S. (2023) *High stakes, high hopes: urban theorizing in partnership*. Athens: University of Georgia Press.

Oldfield, S. and Parnell, S. (2014) 'From the south', in S. Parnell and S. Oldfield (eds.) *The Routledge handbook on cities of the global South*. London: Routledge, pp. 1–4.

Olutayo, M.S. and Yalley, A.A. (2019) 'Feminists across borders: transnational feminism, knowledge production and university education in Nigeria', *International Journal of Humanities and Social Science*, 9(8), pp. 35–41.

Open Knowledge. (2023) *The open data handbook*. Available at: https://opendatahandbook.org/guide/en/ (Accessed: 26 October 2023).

Ordorika, I. (2022) 'Student movements and politics in Latin America: a historical reconceptualization', *Higher Education*, 83(2), pp. 297–315.

Oswin, N. (2020) 'An other geography', *Dialogues in Human Geography*, 10(1), pp. 9–18.

Pan American Health Organization and World Health Organization. (2020) 'Why data disaggregation is key during a pandemic', *PAHO: Pan American Health Organization*. Available at: iris.

paho.org/bitstream/handle/10665.2/52002/Data-Disaggregation-Factsheet-eng.pdf?sequence=19 (Accessed: 26 October 2023).

Pandey, A. (2009) 'Unwelcomed and unwelcoming encounters', in P. Ghassem-Fachandi (ed.) *Violence: ethnographic encounters*. New York: Berg Publishers, pp. 135–144.

Parker, B. (2011) 'Material matters: gender and the city', *Geography Compass*, 5(6), pp. 433–447.

Parnell, S. (2016) 'Defining a global urban development agenda', *World Development*, 78, pp. 529–540.

Parnell, S. and Robinson, J. (2012) '(Re)theorizing cities from the global South: looking beyond neoliberalism', *Urban Geography*, 33(4), pp. 593–617.

Parry, B. (2004) 'The institutionalization of postcolonial studies', in N. Lazarus (ed.) *The Cambridge companion to postcolonial literary studies*. Cambridge: Cambridge University Press, pp. 66–80.

Pascoe, S., Sanders, A., Rawluk, A., Satizábal, P. and Toumbourou, T. (2020) *Holding space for alternative futures in academia and beyond*. Available at: https://antipodeonline.org/2020/04/22/holding-space-for-alternative-futures-in-academia-and-beyond/ (Accessed: 26 October 2023).

Patrick, S. and Huggins, A. (2023) *The term 'global South' is surging. It should be retired*. Available at: https://carnegieendowment.org/2023/08/15/term-global-south-is-surging.-it-should-be-retired-pub-90376. (Accessed: 26 October 2023).

Peake, L. (1993) 'Race and sexuality: challenging the patriarchal structuring of urban social space', *Environment and Planning D: Society and Space*, 11(4), pp. 415–432.

Peake, L. (on behalf of Red Thread Women's Development Program). (2000) *Women researching women: Red Thread report on surveys of domestic violence and women's reproductive health in Guyana. Project TC-97-07-40-9-GY*. Georgetown, Guyana: Inter-American Development Bank.

Peake, L. (2009) 'Feminist and quantitative? Measuring the extent of domestic violence in Georgetown, Guyana', *Treballs de La Societat Catalana de Geografia*, 66, pp. 133–148.

Peake, L. (2015) 'The Suzanne Mackenzie Memorial Lecture: rethinking the politics of feminist knowledge production in geography', *The Canadian Geographer*, 59(3), 257–266.

Peake, L. (2017a) 'Feminism and the urban', in J. Short (ed.) *A research agenda for cities*. Cheltenham: Edward Elgar, pp. 82–97.

Peake, L. (2017b) 'Feminist methodologies', in D. Richardson, N. Castree, M. Goodchild, A. Kobayashi, W. Liu and R. Marston (eds.) *The AAG international encyclopedia of geography: people, the earth, environment and technology*, vol. 5. Oxford: John Wiley and Sons Ltd, pp. 2331–2340.

Peake, L. (2024) 'Urban feminist theory', in L. Peake, N.S. Razavi and A. Smyth (eds.) *Doing feminist urban research: insights from the GenUrb project*. London: Routledge.

Peake, L. and de Souza, K. (2010) 'Feminist academic and activist praxis in service of the transnational', in R. Nagar and A.L. Swarr (eds.) *Critical transnational feminist praxis*. Minneapolis: University of Minnesota Press, pp. 206–218.

Peake, L. and Mullings, B. (2016) 'Critical reflections on mental and emotional distress in the academy', *ACME: An International Journal for Critical Geographies*, 15(2), pp. 253–284.

Peake, L., Patrick, D., Reddy, R.N., Tanyildiz, G.S., Ruddick, S. and Tchoukaleyska, R. (2019) 'Placing planetary urbanization in other fields of vision', *Environment and Planning D: Society and Space*, 36(3), pp. 374–386.

Peake, L. and Rieker, M. (2013) 'Rethinking feminist interventions into the urban', in L. Peake and M. Rieker (eds.) *Rethinking feminist interventions into the urban*. London and New York: Routledge.

Peake, L. and Sheppard, E. (2014) 'The emergence of radical/critical geography within North America', *ACME: An International Journal for Critical Geographies*, 13(2), pp. 305–327.

Peake, L. and Trotz, A. (1999) *Gender, ethnicity and place: women and identities in Guyana*. London: Routledge.

Pearce, M. (2021) 'What shall we map next? Expressing Indigenous geographies with cartographic language', in U. Dieckmann (ed.) *Mapping the unmappable? Cartographic explorations with Indigenous peoples in Africa*. Bielefeld: Transcript Verlag, pp. 317–340.
Peck, J. (2015) 'Cities beyond compare?' *Regional Studies*, 49(1), pp. 160–182.
Pels, P. (2008) 'What has anthropology learned from the anthropology of colonialism?', *Social Anthropology/Anthropologie Sociale*, 16(3), pp. 280–299.
Pels, P. and Salemink, O. (1994) 'Introduction: five theses on ethnography as colonial practice', *History and Anthropology*, 8(1–4), pp. 1–34.
Pereira, C. (2010) 'Reflections of a feminist scholar-activist in Nigeria', in S. Arnfred and A. Adomako Ampofo (eds.) *African feminist politics of knowledge: tensions, challenges, possibilities*. Uppsala: Nordiska Afrikainstitutet, pp. 83–110.
Perez-Brumer, A. and Silva-Santisteban, A. (2020) 'COVID-19 policies can perpetuate violence against transgender communities: insights from Peru', *AIDS and Behavior*, 24(9), pp. 2477–2479.
Perez-Gonzalez, L. (2014) *Audiovisual translation: theories, methods and issues*. London: Routledge.
Phipps, D. and Morton, S. (2013) 'Qualities of knowledge brokers: reflections from practice', *Evidence and Policy: A Journal of Research, Debate and Practice*, 9(2), pp. 255–265.
Pink, S. (2006) 'Visual methods', in V. Jupp (ed.) *The Sage dictionary of social research methods*. London: Sage Publications, pp. 320–321.
Polster, C. and Newson, J. (2015) *A penny for your thoughts: how corporatization devalues teaching, research, and public service in Canada's universities*. Ottawa: Canadian Centre for Policy Alternatives. Available at: https://policyalternatives.ca/sites/default/files/uploads/publications/National%20 Office/2015/05/Penny_Thoughts_TOC_preface_intro.pdf (Accessed: 26 October 2023).
Ponder, C.S. (2021) 'Spatializing the municipal bond market: urban resilience under racial capitalism', *Annals of the American Association of Geographers*, 111(7), pp. 2112–2129.
Porter, C., Thandi Sulé, V. and Croom, N. (eds.) (2023) *Black feminist epistemology, research, and praxis: narratives in and through the academy*. London: Routledge.
Pratt, G. and Johnston, C. (2017) 'Crossing oceans: testimonial theatre, Filipina migrant labour, empathy and engagement', *GeoHumanities*, 3(2), pp. 279–291.
Pratt, G. and Rosner, V. (2006) 'The global and the intimate', *Women Studies Quarterly*, 34(1/2), pp. 13–24.
Pratt, G. and Yeoh, B. (2003) 'Transnational (counter) topographies', *Gender, Place and Culture*, 10(2), pp. 159–166.
Puar, J.K. (2012) ' "I would rather be a cyborg than a goddess": becoming-intersectional in assemblage theory', *Philosophia*, 2(1), pp. 49–66.
Pulido, L. (2008) 'FAQs: frequently (un)asked questions about being a scholar activist', in C.R. Hale (ed.) *Engaging contradictions: theory, politics, and methods of activist scholarship*. Berkeley: University of California Press, pp. 341–365.
Quataert, J.H. (2014) 'A knowledge revolution: transnational feminist contributions to international development agendas and policies, 1965–1995', *Global Social Policy*, 14(2), pp. 209–227.
Quaye, S.J., Shaw, M.D. and Hill, D.C. (2017) 'Blending scholar and activist identities: establishing the need for scholar activism', *Journal of Diversity in Higher Education*, 10(4), pp. 381–399.
Quijano, A. (2000) 'Coloniality of power, Eurocentrism, and Latin America', *International Sociology*, 15(2), pp. 215–232.
Rabaka, R. (2015) *The Negritude movement: W.E.B. Du Bois, Leon Damas, Aimé Césaire, Léopold Senghor, Frantz Fanon, and the evolution of an insurgent idea*. Lanham, MA: Lexington Books.
Radecki, J. and Schonfeld, R. (2020) ' "The impacts of COVID-19 on the research enterprise": a landscape review', *Ithaka S+R*. Available at: https://doi.org/10.18665/sr.314247 (Accessed: 26 October 2023).
Rafael, V. (1993) *Contracting colonialism: translation and Christian conversion in Tagalog society under early Spanish rule*. Durham: Duke University Press.

Rafael, V. (2015) 'Betraying empire: translation and the ideology of conquest', *Translation Studies*, 8(1), pp. 82–93.

Raghuram, P., Noxolo, P. and Madge, C. (2014) 'Rising Asia and postcolonial geography', *Singapore Journal of Tropical Geography*, 35(1), pp. 119–135.

Ramohai, J. (2019) 'A black woman's perspective on understanding transformation and diversity in South African higher education', *Transformation in Higher Education*, 4(a58), pp. 1–10.

Ramos, S. (2021) *COVID-19's impact felt by researchers: scientists, graduate students talk about conducting research during a pandemic.* Available at: www.apa.org/science/leadership/students/COVID-19-impact-researchers (Accessed: 26 October 2023).

Rapp, R. (1975) *Toward an anthropology of women.* New York and London: Monthly Review Press.

Rao, S. (2021) 'Beyond the Coronavirus: understanding crises of social reproduction', *Global Labour Journal*, 12(1), pp. 39–53.

Ravitch, S.M. and Carl, N.M. (2021) *Qualitative research: bridging the conceptual, theoretical and methodological*, 2nd ed. Los Angeles: Sage Publications.

Razavi, S. (2016) 'The 2030 Agenda: challenges of implementation to attain gender equality and women's rights', *Gender and Development*, 24(1), 25–41.

Razavi, N.S. (2020a) 'Mandala de la vida, mandala of life, storytelling workshop guide', *GenUrb, Centro de estudios y trabajo de la mujer.* Available at: https://genurb.apps01.yorku.ca/about/city-research-teams/cochabamba/mandala-de-la-vida-mandala-of-life-storytelling-workshop-guide/ (Accessed: 26 October 2023).

Razavi, N.S. (2020b) 'Women in cities: the nexus between SDG 5 and SDG 11', in W. Leal Filho, A. Azul, L. Brandli, A. Lange Salvia and T. Wall (eds.) *Gender equality, encyclopedia of the UN Sustainable Development Goals.* Available at: https://doi.org/10.1007/978-3-319-70060-1_153-1 (Accessed: 19 November 2023).

Razavi, N.S., Adeniyi-Ogunyankin, G., Balogun, D., Basu, S., Datta, A., de Souza, K., Koleth, E., Ip, P., Marcus, J., Miraftab, F., Mullings, B., Nmormah, S., Pardo Burgoa, S. and Peake, L. (2023) 'Everyday urbanisms in the pandemic city: a feminist comparative study of the gendered experiences of Covid-19 in southern cities', *Social and Cultural Geography*, 24(3–4), pp. 582–599.

Ren, J. and Luger, J. (2015) 'Comparative urbanism and the "Asian City": implications for research and theory', *International Journal of Urban and Regional Research*, 39(1), pp. 145–156.

Reynolds, K., Block, D. and Bradley, K. (2018) 'Food justice scholar-activism and activist scholarship', *ACME: An International Journal for Critical Geographies*, 17(4), pp. 988–998.

Richardson, D., Castree, N., Goodchild, M., Kobayashi, A., Liu, W. and Marston, R. (eds.) (2017) *The AAG international encyclopedia of geography: people, the earth, environment, and technology.* Malden and Oxford: John Wiley and Sons, Ltd.

Riessman, C.K. (2006) 'Narrative analysis', in V. Jupp (ed.) *The Sage dictionary of social research methods.* London: Sage Publications, pp. 186–189.

Rigg, J. (2007) *An everyday geography of the global South.* London: Routledge.

Ringrose, J. (2018) 'Digital feminist pedagogy and post-truth misogyny', *Teaching in Higher Education*, 23, pp. 647–656.

Roberts, S.M. (2004) 'Gendered globalization', in E. Kofman, L. Peake and L.A. Staeheli (eds.) *Mapping women, making politics: feminist perspectives on political geography.* New York: Routledge, pp. 127–140.

Robinson, D. (1997) *Translation and empire: postcolonial theories explained.* Manchester: St. Jerome Publishing.

Robinson, F. (2011) *The ethics of care: a feminist approach to human security.* Philadelphia, PA: Temple University Press.

Robinson, J. (2005) 'Urban geography: world cities, or a world of cities', *Progress in Human Geography*, 29(6), pp. 757–765.

Robinson, J. (2006) *Ordinary cities: between modernity and development.* London: Routledge.

Robinson, J. (2011) 'Cities in a world of cities: the comparative gesture', *International Journal of Urban and Regional Research*, 35(1), pp. 1–23.

Robinson, J. (2016a) 'Thinking cities through elsewhere: comparative tactics for a more global urban studies', *Progress in Human Geography*, 40(1), pp. 3–29.

Robinson, J. (2016b) 'Comparative urbanism: new geographies and cultures of theorizing the urban', *International Journal of Urban and Regional Research*, 40(1), pp. 187–199.

Robinson, J. (2022) *Comparative urbanism: tactics for global urban studies*. Chichester: Wiley.

Rodrigues, C.H. (2018) 'Translation and signed language: highlighting the visual-gestural modality', *Cadernos de Tradução*, 38(2), pp. 294–319.

Rodríguez, C. (2018) *Decolonizing academia: poverty, oppression and pain*. Halifax and Winnipeg: Fernwood Publishing.

Rosaldo, M. and Lamphere, L. (eds.) (1974) *Woman, culture, and society*. Stanford, CA: Stanford University Press.

Rose, G. (1997) 'Situating knowledges: positionality, reflexivities and other tactics', *Progress in Human Geography*, 21(3), pp. 305–320.

Rose, G. (2022) *Visual methodologies: an introduction to researching with visual materials*, 5th ed. London: Sage Publications.

Rosenberg, R. (2023) 'On surviving a cis discipline', *Environment and Planning D: Society and Space*, 41(4), pp. 600–605.

Roy, A. (2009) 'The 21st-century metropolis: new geographies of theory', *Regional Studies*, 43(6), pp. 819–830.

Roy, A. (2011) 'Slumdog cities: rethinking subaltern urbanism', *International Journal of Urban and Regional Research*, 35(2), pp. 223–238.

Roy, A. (2016) 'Who's afraid of postcolonial theory?', *International Journal of Urban and Regional Research*, 40(1), pp. 200–209.

Roy, A. (2020) '"The shadow of her wings": respectability politics and the self-narration of geography', *Dialogues in Human Geography*, 10(1), pp. 19–22.

Roy, A., Wright, W.J., Al-Bulushi, Y. and Bledsoe, A. (2020) '"A world of many Souths": (anti) Blackness and historical difference in conversation with Ananya Roy', *Urban Geography*, 41(6), pp. 920–935.

Ruddick, S., Peake, L., Tanyildiz, G.S. and Patrick, D. (2019) 'Planetary urbanization: an urban theory for our time?' *Environment and Planning D: Society and Space*, 36(3), pp. 387–404.

Ryle, G. (1968) *The thinking of thoughts: what is 'Le Penseur' doing?* Regina: University of Saskatchewan Press.

SAADA. (2023) *SAADA: South Asian American digital archive*. Available at: www.saada.org/ (Accessed: 27 October 2023).

Said, E. (1978) *Orientalism*. New York: Pantheon Books.

Salaita, S. (2015) *Uncivil rites: Palestine and the limits of academic freedom*. Chicago: Haymarket Books.

Saldaña, J. (2015) *The coding manual for qualitative researchers*. London: Sage Publications.

Sandercock, L. (2003) 'Out of the closet: the importance of stories and storytelling in planning practice', *Planning Theory and Practice*, 4(1), pp. 11–28.

Sandoval, C. (2000) *Methodology of the oppressed*. Minneapolis: University of Minnesota Press.

Sangtin Writers Collective and Nagar, R. (2006) *Playing with fire: feminist thought and activism through seven lives in India*. Minneapolis: University of Minnesota Press.

Santaemilia, J. (2017) 'A corpus-based analysis of terminology in gender and translation research: the case of feminist translation', in O. Castro and E. Ergun (eds.) *Feminist translation studies: local and transnational perspectives*. London: Routledge, pp. 15–28.

Santiago-Alarcon, D. and MacGregor-Fors, I. (2020) 'Cities and pandemics: urban areas are ground zero for the transmission of emerging human infectious diseases', *Journal of Urban Ecology*, 6(1). Available at: https://doi.org/10.1093/jue/juaa012 (Accessed: 26 October 2023).

Saramo, S. (2016) 'Unsettling spaces: grassroots responses to Canada's missing and murdered Indigenous women during the Harper government years', *Comparative American Studies: An International Journal*, 14(3–4), pp. 204–220.

Sayer, A. (2000) *Realism and social science*. London: Sage Publications.

Schensul, J.J. and LeCompte, M.D. (2012) *Essential ethnographic methods: a mixed methods approach*, 2nd ed. Lanham: AltaMira Press.

Schmid, C., Karaman, O., Hanakata, N.C., Kallenberger, P., Kockelkorn, A., Sawyer, L., Streule, M. and Wong, K.P. (2018) 'Towards a new vocabulary of urbanisation processes: a comparative approach', *Urban Studies*, 55(1), pp. 19–52.

Schneider, M., Lord, E. and Wilczak, J. (2020) 'We, too: contending with the sexual politics of fieldwork in China', *Gender, Place and Culture*, 28(4), pp. 519–540.

Scott, J. (2006) 'Content analysis', in V. Jupp (ed.) *The Sage dictionary of social research methods*. London: Sage Publications, pp. 40–41.

Sempértegui, A. (2021) 'Indigenous women's activism, ecofeminism, and extractivism: partial connections in the Ecuadorian Amazon', *Politics and Gender*, 17(1), pp. 197–224.

Sen, G. (2018) 'The SDGs and feminist movement building', *UN Women Discussion Paper No. 27*. Available at: www.unwomen.org/en/digital-library/publications/2018/12/discussion-paper-the-sdgs-and-feminist-movement-building (Accessed: 26 October 2023).

Sevenhuijsen, S. (1998) *Citizenship and the ethics of care: feminist considerations on justice, morality, and politics*. New York: Routledge.

Shamma, T. (2018) 'Translation and colonialism', in S.A. Harding and O.C. Cortés (eds.) *The Routledge handbook of translation and culture*. London and New York: Routledge.

Sharp, G. and Kremer, E. (2006) 'The safety dance: confronting harassment, intimidation, and violence in the field', *Sociological Methodology*, 36(1), pp. 317–327.

Sharp, J. (2005) 'Geography and gender: feminist methodologies in collaboration and in the field', *Progress in Human Geography*, 29(3), pp. 304–309.

Shaw, I.G.R., Dixon, D.P. and Jones III, J.P. (2010) 'Theorizing our world', in B. Gomez and J.P. Jones III (eds.) *Research methods in geography: a critical introduction*. Malden: Blackwell Publishing Ltd, pp. 9–25.

Shaw, A. (2014) 'The Internet is full of jerks, because the world is full of jerks: what feminist theory teaches us about the Internet', *Communication and Critical/Cultural Studies*, 11(3), pp. 273–277.

Sheftel, A. and Zembrzycki, S. (2017) 'Slowing down to listen in the Digital Age: how new technology is changing oral history practice', *The Oral History Review*, 44(1), pp. 94–112.

Sheppard, E., Leitner, H. and Maringanti, A. (2013) 'Provincializing global urbanism: a manifesto', *Urban Geography*, 34(7), pp. 893–900.

Sheppard, E. and Nagar, R. (2004) 'From east – west to north – south', *Antipode*, 36(4), pp. 557–563.

Shotwell, A. (2020) *The virus is a relation*. Available at: uppingtheanti.org/blog/entry/the-virus-is-a-relation (Accessed: 26 October 2023).

Siles, I. (2011) 'From online filter to web format: articulating materiality and meaning in the early history of blogs', *Social Studies of Science*, 41(5), pp. 737–758.

Simon, D., Palmer, H. and Riise, J. (eds.) (2020) *Comparative urban research from theory to practice: co-production for sustainability*. Bristol: Bristol University Press, Policy Press.

Simon Fraser University. (2022) *Responsible conduct of research (R60.01)*. Available at: www.sfu.ca/policies/gazette/research/r60-01.html (Accessed: 27 October 2023).

Simone, A. (2004) 'People as infrastructure: intersecting fragments in Johannesburg', *Public Culture*, 16(3), pp. 407–429.

Simone, A. (2017) 'The majority world and the politics of everyday living in Southeast Asia', in R. Burdett and S. Hall (eds.) *The Sage handbook of the 21st century city*. London: Sage Publication, pp. 399–415.

Simone, A. (2018) *Improvised lives: rhythms of endurance in an urban south*. Cambridge and New York: Polity.
Simone, A. (2020) Cities of the global South. *Annual Review of Sociology*, 46(1), pp. 603–622.
Simone, A. (2022) *The surrounds: urban life within and beyond capture*. Durham: Duke University Press.
Simone, A. and Pieterse, E. (2017) *New urban worlds: inhabiting dissonant times*. Cambridge and Malden: Polity.
Simpson, A. (2007) 'On ethnographic refusal: Indigeneity, "voice" and colonial citizenship', *Junctures: The Journal for Thematic Dialogue*, (9), pp. 67–80.
Simpson, L.B. (2017) *As we have always done: Indigenous freedom through radical resistance*. Minneapolis: University of Minnesota Press.
Sin, C.H. (2003) 'Interviewing in "place": the socio-spatial construction of interview data', *Area*, 35(3), pp. 305–312.
Singh, T. and Mathews, T.J. (2019) 'Facilitating queer of color feminist co-mentorship: reflections on an online archive of scholar-activism', *Gender, Place and Culture*, 26(12), pp. 1701–1720.
Sivetc, L. (2021) 'Controlling free expression 'by infrastructure' in the Russian Internet: the consequences of RuNet sovereignization', *First Monday*, 26(5). Available at: *https://doi.org/10.5210/fm.v26i5.11698* (Accessed: 27 October 2023).
Sletto, B., Wagner, A., Bryan, J. and Hale, C. (eds.) (2020) *Radical cartographies: participatory mapmaking from Latin America*. Austin: University of Texas Press.
Smiles, D. (2020) *Respectful engagement during fieldwork*. Available at: https://u.osu.edu/geographyblog/2020/05/13/respectful-engagement-during-fieldwork/ (Accessed: 27 October 2023).
Smith, D. (2005) *Institutional ethnography: a sociology for people*. Walnut Creek: AltaMira Press.
Smith, L.T. [1999] (2012) *Decolonizing methodologies: research and indigenous peoples*, 2nd ed. London: Zed Books Ltd.
Smith, M. (2006) 'Realism', in V. Jupp (ed.) *The Sage dictionary of social research methods*. London: Sage Publications, pp. 255–257.
Smith, S. (2020) *Intimate geopolitics: love, territory, and the future on India's northern threshold*. New Brunswick: Rutgers University Press.
Smyth, A. (2023) 'Proceeding through colonial past-presents in fieldwork: methodological lessons on accountability, refusal, and autonomy', *Antipode*, 55(1), pp. 268–285.
Social Science Research Council Singapore. (2020) *About Us*. Available at: www.ssrc.edu.sg/who-we-are/about-us (Accessed: 27 October 2023).
Sonn, C., Stevens, G. and Duncan, N. (2013) 'Decolonisation, critical methodologies and why stories matter', in G. Stevens, N. Duncan and D. Hook (eds.) *Race, memory and the Apartheid archive*. New York: Palgrave Macmillan.
Spivak, G.C. (1985) 'The Rani of Sirmur: an essay in reading the archives', *History and Theory*, 24(3), pp. 247–272.
Spivak, G.C. (1988) 'Can the subaltern speak?' in C. Nelson and L. Grossberg (eds.) *Marxism and the interpretation of culture*. Basingstoke: Macmillan, pp. 271–313.
Spivak, G.C. (1999) *A critique of postcolonial reason: toward a history of the vanishing present*. Cambridge: Harvard University Press.
Social Sciences and Humanities Research Council (SSHRC) (2012a) *Guidelines for effective knowledge mobilization*. Available at: www.sshrc-crsh.gc.ca/funding-financement/policies-politiques/knowledge_mobilisation-mobilisation_des_connaissances-eng.aspx (Accessed: 27 October 2023).
Social Sciences and Humanities Research Council (SSHRC). (2012b) *Partnerships*. Available at: www.sshrc-crsh.gc.ca/about-au_sujet/partnerships-partenariats/index-eng.aspx (Accessed: 27 October 2023).
Social Sciences and Humanities Research Council (SSHRC). (2020) *Tips for managing a partnership*. Available at: www.sshrc-crsh.gc.ca/about-au_sujet/partnerships-partenariats/tips_for_managing-conseils_pour_gerer-eng.aspx (Accessed: 27 October 2023).

Social Sciences and Humanities Research Council (SSHRC). (2021a) *Best practices in equity, diversity and inclusion in research*. Available at: www.sshrc-crsh.gc.ca/funding-financement/nfrf-fnfr/edi-eng.aspx#2 (Accessed: 27 October 2023).

Social Sciences and Humanities Research Council (SSHRC). (2021b) *SSHRC doctoral fellowships*. Available at: www.sshrc-crsh.gc.ca/funding-financement/programs-programmes/fellowships/doctoral-doctorat-eng.aspx#7 (Accessed: 27 October 2023).

Stacey, J. (1988) 'Can there be a feminist ethnography?', *Women's Studies International Forum*, 11(1), pp. 21–27.

Staeheli, L.A. and Lawson, V.A. (1994) 'A discussion of women in the field: the politics of feminist fieldwork', *The Professional Geographer*, 46(1), pp. 96–102.

Staeheli, L. and Mitchell, D. (2005) 'The complex politics of relevance in geography', *Annals of the Association of American Geographers*, 95(2), pp. 357–372.

Stahlberg, F. (2020) 'Neural machine translation: a review', *The Journal of Artificial Intelligence Research*, 69, pp. 343–418.

Stewart, A., Grabe, S. and Zheng, W. (2023) 'Women's movement activism in authoritarian states: lessons from the Global Feminisms Project', *Signs*, 49(2).

Stoler, A.L. (2002) 'Colonial archives and the arts of governance', *Archival Science*, 2(1–2), pp. 87–109.

Storper, M. and Scott, A.J. (2016) 'Current debates in urban theory: a critical assessment', *Urban Studies*, 53(6), pp. 1114–1136.

Strega, S. and Brown, L.A. (eds.) (2015) *Research as resistance: critical, indigenous and anti-oppressive approaches*, 2nd ed. Toronto: Canadian Scholars' Press.

Strobel, M. (1999) 'Becoming a historian, being an activist, and thinking archivally: documents and memory as sources', *Journal of Women's History*, 11(1), pp. 181–192.

Su, A. and Shyong, F. (2019) *The Chinese and non-Chinese internet are two worlds. Here's what it's like to use both*. Available at: www.latimes.com/world/la-fg-us-china-internet-split-20190603-story.html (Accessed: 27 October 2023).

Sud, N. and Sánchez-Ancochea, D. (2022) 'Southern discomfort: interrogating the category of the global South', *Development and Change*, 53(6), pp. 1123–1150.

Sugrue, T. (2020) P*re-existing conditions: what 2020 reveals about our urban future, Public Books*. Available at: www.publicbooks.org/preexisting-conditions-what-2020-reveals-about-our-urban-future/ (Accessed: 27 October 2023).

Sultana, F. (2007) 'Reflexivity, positionality and participatory ethics: negotiating fieldwork dilemmas in international research', *ACME: An International E-Journal for Critical Geographies*, 6(3), pp. 374–385.

Sultana, F. (2018a) 'An(Other) geographical critique of development and SDGs', *Dialogues in Human Geography*, 8(2), pp. 186–190.

Sultana, F. (2018b) 'The false equivalence of academic freedom and free speech: defending academic integrity in the age of white supremacy, colonial nostalgia, and anti-intellectualism', *ACME: An International Journal for Critical Geographies*, 17(2), pp. 228–257.

Surkis, A. and Read, K. (2015) 'Research data management', *JMLA: Journal of the Medical Library Association*, 103(3), pp. 154–156.

Swarr, A.L. and Nagar, R. (eds.). (2010) *Critical transnational feminist praxis*. Albany: State University of New York Press.

Sylvestre, P., Castleden, H., Martin, D. and McNally, M. (2017) ' "Thank you very much . . . you can leave our community now": geographies of responsibility, relational ethics, acts of refusal, and the conflicting requirements of academic localities in Indigenous research', *ACME: An International Journal for Critical Geographies*, 17(3), pp. 750–779.

Taiwo, A., Baird, S.L. and Sanders, J.E. (2022) 'Mutual aid on WhatsApp: reflections on an online support group for new and pre-tenured faculty', *Social Work with Groups*, 45(2), pp. 145–156.

Talbot, C.V. and Pownall, M. (2022) '"If your institution refuses to provide what you need, create it yourself": feminist praxis on #AcademicTwitter', *Feminism and Psychology*, 32(1), pp. 101–118.

Tamale, S. (2020) *Decolonization and Afro-feminism*. Wakefield: Daraja Press.

Tambe, A. and Thayer, M. (eds.) (2021) *Transnational feminist itineraries: situating theory and activist practice*. Durham: Duke University Press.

Tanyildiz, G., Peake, L., Koleth, E., Narayanareddy, R.S., Patrick, D. and Ruddick, S. (2021) 'A feminist urban theory for our time: rethinking social reproduction, the urban and its constitutive outside', in L. Peake, E. Koleth, G. Tanyildiz, R.S. Narayanareddy and D. Patrick (eds.) *A feminist urban theory for our time: rethinking social reproduction and the urban*. London: Antipode Book Series, Wiley Press, pp. 1–41.

The Counter-Cartographies Collective (3Cs). (n.d.) *Counter cartographies*. Available at: www.countercartographies.org/ (Accessed: 27 October 2023).

Thompson, C. (2023) *Why social media activism inevitably disadvantages Black people*. Available at: www.heliotropejournal.net/helio/social-media-activism-disadvantages-black-people (Accessed: 27 October 2023).

Thorat, D. (2019) 'Colonial topographies of Internet infrastructure: the sedimented and linked networks of the telegraph and submarine fiber optic Internet', *South Asian Review*, 40(3), pp. 252–267.

Till, K.E. (1996) *Place and the politics of memory: a geo-ethnography of museums and memorials in Berlin*, Doctoral Dissertation in Geography. Madison: University of Wisconsin.

Till, K.E. (2005) *The new Berlin: memory, politics, place*. Minneapolis: University of Minnesota Press.

Tissot, D. (2017) 'Transnational feminist solidarities and the ethics of translation', in O. Castro and E. Ergun (eds.) *Feminist translation studies: local and transnational perspectives*. London: Routledge, pp. 29–41.

Todd, Z. (2016) 'An Indigenous feminist's take on the ontological turn: "Ontology" is just another word for colonialism', *Journal of Historical Sociology*, 29(1), pp. 4–22.

Todd, Z. (2020) (*an answer*). Available at: https://anthrodendum.org/2020/01/27/an-answer/ (Accessed: 27 October 2023).

Tronto, J. (1993) *Moral boundaries: a political argument for an ethic of care*. New York: Routledge.

Trotz, A. (ed.) (2020) *The point is to change the world: selected writings of Andaiye*. London: Pluto Press.

Trovao, S.S., Ramalho, S.C.C. and David, M.I.P.T. (2015) 'Transnational families, religious participation and gender dynamics: Filipino, Sao Tomean and Indo-Mozambican immigrant women in Lisbon, Portugal', *Gender, Place, Culture*, 22(3), pp. 325–343.

True, J. (2003) 'Mainstreaming gender in global public policy', *International Feminist Journal of Politics*, 5(3), pp. 368–396.

Truth, S. (1851) *Ain't I a woman?* Speech delivered at Women's Convention in Akron, Ohio. Available at: https://thehermitage.com/wp-content/uploads/2016/02/Sojourner-Truth_Aint-I-a-Woman_1851.pdf (Accessed: 27 October 2023).

Tuck, E. and Yang, W.K. (2012) 'Decolonization is not a metaphor', *Decolonization: Indigeneity, Education, and Society*, 1(1), pp. 1–40.

Tummons, J. (2014) 'Using software for qualitative data analysis: research outside paradigmatic boundaries', in M. Hand and S. Hillyard (eds.) *Big data? Qualitative approaches to digital research*. Bingley: Emerald Publishing Limited, pp. 155–177.

Tuomi, I. (2000) 'Data is more than knowledge: implications of the reversed knowledge hierarchy for knowledge management and organizational memory', *Journal of Management Information Systems*, 16(3), pp. 103–117.

UK Data Service Guide. (2023) *Recommended formats*. Available at: https://ukdataservice.ac.uk/learning-hub/research-data-management/format-your-data/recommended-formats/ (Accessed: 27 October 2023).

UK Research and Innovation. (n.d.). *Research England*. Available at: https://re.ukri.org/research/ref-impact/ (Accessed: 27 October 2023).

UN Women. (2020) *The shadow pandemic: violence against women during Covid-19*. Available at: (Accessed: 27 October 2023).

UN Women and UN Department of Economic and Social Affairs. (2023) *Progress on the sustainable development goals: the gender snapshot 2023*. New York: United Nations.

US Department of Health and Human Services. (2018) *Read the Belmont report*. Available at: www.hhs.gov/ohrp/regulations-and-policy/belmont-report/read-the-belmont-report/index.html (Accessed: 27 October 2023).

United Nations, UN75. (n.d.) *Inequality – bridging the divide*. Available at: www.un.org/en/un75/inequality-bridging-divide (Accessed: 27 October 2023).

United Nations. (2015) *Transforming our world: the 2030 agenda for sustainable development (A/RES/70/1)*. New York. Available at https://sdgs.un.org/2030agenda (Accessed: 15 November 2023).

United Nations. (2018) '68% of the world population projected to live in urban areas by 2050, says UN', *United Nations Department of Economic and Social Affairs: News*. Available at: www.un.org/development/desa/en/news/population/2018-revision-of-world-urbanization-prospects.html (Accessed: 27 October 2023).

United Nations. (2019) *World urbanization prospects: the 2018 revision (ST/ESA/SER.A/420)*. New York: United Nations.

United Nations. (2023) 'The sustainable development goals report 2023: special edition: towards a rescue plan for people and planet.' *United Nations Statistics Division*. Available at: https://unstats.un.org/sdgs/report/2023/ (Accessed: 20 November 2023).

University of Kentucky. (n.d.) *Preparing future faculty*. Available at: https://gradschool.uky.edu/preparing-future-faculty (Accessed: 27 October 2023).

Underhill-Sem, Y.T.R.-R.-O.-T. (2017) 'Academic work as radical practice: getting in, creating a space, not giving up', *Geographical Research*, 55(3), pp. 332–337.

University of Lincoln. (2021) *Expansion of world's cities creating 'new ecological niches' for infectious diseases*. Available at: www.sciencedaily.com/releases/2020/04/200421112557.htm (Accessed: 27 October 2023).

University of Michigan. (2022) *Professional standards for faculty: standard practice guides*. Available at: https://spg.umich.edu/policy/201.96 (Accessed: 27 October 2023).

Uprichard, E. and Carrigan, M. (2015) 'Most Big Data is social data – the analytics need serious interrogation', *London School of Economics*. Available at: https://blogs.lse.ac.uk/impactofsocialsciences/2015/02/12/philosophy-of-data-science-emma-uprichard/ (Accessed: 23 October 2023).

Valencia, S.C., Simon, D., Croese, S., Nordqvist, J., Oloko, M., Sharma, T., Taylor, N., Versace, B. and Versace, I. (2019) 'Adapting the sustainable development goals and the new urban agenda to the city level: initial reflections from a comparative research project', *International Journal of Urban Sustainable Development*, 11(1), pp. 4–23.

Valentine, G. (2005) 'Tell me about . . .: using interviews as a research methodology', in R. Flowerdew and D.M. Martin (eds.) *Methods in human geography: a guide for students doing a research project*, 2nd ed. Essex: Addison Wesley Longman Ltd, pp. 110–127.

Valero-Garcés, C. (2015) 'Transcription and translation', in H. Mikkelson and R. Jourdenais (eds.) *Routledge handbook of interpreting*. Abington: Routledge, pp. 131–145.

Van den Hoonaard, W. (2017) 'About "othering" ourselves in a system with discrepant values: the research ethics review process today', in R. Iphofen (ed.) *Finding common ground: consensus in research ethics across the social sciences*. Bingley: Emerald Publishing, pp. 61–75.

Vanderbeck, R.M. (2005) 'Masculinities and fieldwork: widening the discussion', *Gender, Place and Culture*, 12(4), pp. 387–402.

Varley, A. (2015) 'At home in the city? Gender and urban poverty', in A. Coles, L. Gray and J. Momsen (eds.) *The Routledge handbook of gender and development*. London: Routledge, pp. 204–214.

Vasudevan, P. (2021) 'An intimate inventory of race and waste', *Antipode*, 53(3), pp. 770–790.

Vasudevan, P., Ramírez, M.M., Mendoza, Y.G. and Daigle, M. (2023) 'A manifesto for who we may become: storytelling earth and body', *Annals of the American Association of Geographers*, 113(7), pp. 1728–1744.

Vélez, T.I. (2019) 'So you're "*Latina* Latina": positionality and reflexivity in developing scholar-activist praxis', *Journal of Latin American Geography*, 18(3), pp. 215–220.

Velez, E.D. and Tuana, N. (2020) 'Toward decolonial feminisms: tracing the lineages of decolonial thinking through Latin American/Latinx feminist philosophy', *Hypatia*, 35(3), pp. 366–372.

Vickery, J.R. and Everbach, T. (eds.) (2018) *Mediating misogyny: gender, technology, and harassment*. Cham: Palgrave Macmillan.

Vrana, H. (2017) *This city belongs to you: a history of student activism in Guatemala, 1944–1996*. Berkeley: University of California Press.

Wainwright, J. and Bryan, J. (2009) 'Cartography, territory, property: postcolonial reflections on indigenous counter-mapping in Nicaragua and Belize', *Cultural Geographies*, 16(2), pp. 153–178.

Wakefield, S. and Koerppen, D. (2017) *Applying feminist principles to program monitoring, evaluation, accountability and learning*. Available at: https://doi.org/10.21201/2017.9965 (Accessed: 27 October 2023).

Wang, Q. (2020) *The Chinese Internet: the online public sphere, power relations and political communication*. Milton: Taylor and Francis.

Wa Ngugi, M. (2012) *Rethinking the global South*. Available at: www.globalsouthproject.cornell.edu/rethinking-the-global-south.html (Accessed: 27 October 2023).

Ward, K. (2010) 'Towards a relational comparative approach to the study of cities', *Progress in Human Geography*, 34(4), pp. 471–487.

Warren, G.C., Katz, C. and Heynen, N. (2019) 'Myths, cults, memories, and revisions in radical geographic history: revisiting the Detroit Geographical Expedition and Institute', in T.J. Barnes and E. Sheppard (eds.) *Spatial histories of radical geography: North America and beyond*. Oxford: John Wiley and Sons Ltd., pp. 59–86.

Watson, S. (1992) 'Femocratic feminisms', *The Sociological Review*, 39(1), pp. 186–204.

Weatherall, R. (2020) 'Even when those struggles are not our own: storytelling and solidarity in a feminist social justice organization', *Gender, Work and Organization*, 27(4), pp. 471–486.

Weir, J.K., Woelfle-Erskine, C. and Fuller, S. (2019) 'Investigating best practice: doctoral fieldwork experiences with and without Indigenous communities in settler-colonial societies', *ACME: An International Journal for Critical Geographies*, 18(6), pp. 1300–1320.

Wekerle, G. (2013) 'Interrogating gendered silences in urban policy: regionalism and alternative visions of a caring region', in L. Peake and M. Rieker (eds.) *Rethinking feminist interventions into the urban*. London: Routledge, pp. 1–22.

Welsh, E. (2002) 'Dealing with data: using NVivo in the qualitative data analysis process', *Forum: Qualitative Social Research*, 3(2), Art. 26. Available at: https://doi.org/10.17169/fqs-3.2.865 (Accessed: 27 October 2023).

Wenham, C., Smith, J. and Morgan, R. (2020) 'COVID-19: the gendered impacts of the outbreak', *The Lancet*, 395(10227), pp. 848–848.

Werner, M., Strauss, K., Parker, B., Orzeck, R., Derickson, K. and Bonds, A. (2017) 'Feminist political economy in geography: why now, what is different, and what for?' *Geoforum*, 79, pp. 1–4.

Westerman, A., Benk, R. and Green, D. (2020) *In 2020, protests spread across the globe with a similar message: Black Lives Matter*. Available at: www.npr.org/2020/12/30/950053607/in-2020-protests-spread-across-the-globe-with-a-similar-message-black-lives-matter (Accessed: 27 October 2023).

Whitacre, R., Oni-Orisan, A., Gaber, N., Martinez, C., Buchbinder, L., Herd, D. and Holmes, S. (2021) 'COVID-19 and the political geography of racialisation: ethnographic cases in San Francisco, Los Angeles and Detroit', *Global Public Health*, 16(8–9), pp. 1396–1410.

Whose Knowledge? (2019) *Decolonizing the internet: what is it about?* Available at: https://whose-knowledge.org/initiatives/decolonizing-the-internet (Accessed: 27 October 2023).

Wilder, C.S. (2013) *Ebony and ivy: race, slavery, and the troubled history of America's universities*. New York: Bloomsbury Press.

Wilkinson, M., Dumontier, M., Aalbersberg, I., Appleton, G., Axton, M., Baak, A., Blomberg, N., et al. (2016) 'The FAIR guiding principles for scientific data management and stewardship', *Scientific Data*, 3(160018). Available at: doi.org/10.1038/sdata.2016.18 (Accessed: 29 October 2023).

Williams, G. (2015) 'Researching with impact in the global South? Impact-evaluation practices and the reproduction of 'development knowledge', in J. Bannister and I. Hardill (eds.) *Knowledge mobilisation and the social sciences*. London and New York: Routledge, pp. 57–70.

Williams, M. (2003) *Making sense of social research*. London: Sage Publications.

Williams, R. (1961) *The long revolution*. London: Chatto and Windus.

Willis, K.D. and Yeoh, B.S.A. (2000) 'Gender and transnational household strategies: Singaporean migration to China', *Regional Studies*, 34(3), pp. 253–264.

Willis, W.S. (1972) 'Skeletons in the anthropological closet', in D. Hymes (Ed.), *Reinventing anthropology*. New York: Pantheon Books, pp. 121–152.

Wilson, S. (2008) *Research is ceremony: Indigenous research methods*. Halifax and Winnipeg: Fernwood Publishing.

Wittmer, J. and Qureshi, M. (2023) 'Navigating the emotion-embodiment-language nexus in international research: stories from a foreign researcher and local interpreter', *Emotion, Space and Society*, 49, 100990.

Wolf, D.L. (1996) 'Situating feminist dilemmas in fieldwork', in D.L. Wolf (ed.) *Feminist dilemmas in fieldwork*. London: Routledge, pp. 1–55.

World Health Organization. (2020) *Archived: WHO timeline – COVID-19*. Available at: www.who.int/news/item/27-04-2020-who-timeline – covid-19 (Accessed: 27 October 2023).

World Health Organization. (2022) *COVID-19 pandemic triggers 25% increase in prevalence of anxiety and depression worldwide*. Available at: www.who.int/news/item/02-03-2022-covid-19-pandemic-triggers-25-increase-in-prevalence-of-anxiety-and-depression-worldwide (Accessed: 27 October 2023).

World Medical Association. (n.d.) *Declaration of Helsinki 1964*. Available at: www.wma.net/what-we-do/medical-ethics/declaration-of-helsinki/doh-jun1964/ (Accessed: 27 October 2023).

Wright, S., Lloyd, K., Suchet-Pearson, S., Burarrwanga, L., Tofa, M. and Country, B. (2012) 'Telling stories in, through and with country: engaging with Indigenous and more-than-human methodologies at Bawaka, NE Australia', *Journal of Cultural Geography*, 29(1), pp. 39–60.

Wright, T. (2020) *COVID-19 has greater impact on women advocates say*. Available at: www.nationalobserver.com/2020/04/10/news/COVID-19-has-greater-impact-women-advocates-say (Accessed: 27 October 2023).

Wynter, S. (2003) 'Unsettling the coloniality of being/power/truth/freedom: towards the human, after man, its overrepresentation – an argument', *CR: The New Centennial Review*, 3(3), pp. 257–333.

Yeager, A. (2020) *How the COVID-19 pandemic has affected field research*. Available at: www.the-scientist.com/news-opinion/how-the-COVID-19-pandemic-has-affected-field-research-67841 (Accessed: 27 October 2023).

Yengoyan, A. (2006) *Modes of comparison: theory and practice*. Ann Arbor, MI: University of Michigan Press.

Yong, A., Shamsulm, H., McDonagh, D., Woodcock, A., Nordin, N.A. and Faiz, K. (2020) 'Gender transport inequalities in Malaysia and Pakistan: barriers to female mobility', in J. Berry, T. Moore, N. Kalms and G. Bawden (eds.) *Contentious cities: design and the gendered production of space*. Abingdon: Routledge, pp. 71–83.

York University. (n.d.-a) *Knowledge mobilization*. Available at: www.yorku.ca/laps/research/support-services/knowledge-mobilization/ (Accessed: 27 October 2023).

York University. (n.d.-b) *Research ethics at York University*. Available at: www.yorku.ca/research/research-ethics/ (Accessed: 29 October 2023).

Younas, F., Naseem, M. and Mustafa, M. (2020) 'Patriarchy and social media: women only Facebook groups as safe spaces for support seeking in Pakistan', in *Proceedings of the 2020 International Conference on Information and Communication Technologies and Development*. Presented at the ICTD2020: Information and Communication Technologies and Development, ACM, Guayaquil Ecuador, pp. 1–11. Available at: https://doi.org/10.1145/3392561.3394639 (Accessed: 27 October 2023).

Young, I.M. (1997) *Intersecting voices: dilemmas of gender, political philosophy, and policy*. Princeton: Princeton University Press.

Zaragocin, S. (2019a) 'Gendered geographies of elimination: decolonial feminist geographies in Latin American settler contexts', *Antipode*, 51(1), pp. 373–392.

Zaragocin, S. (2019b) 'Feminist geography in Ecuador', *Gender, Place and Culture*, 26(7–9), pp. 1032–1038.

Zaragocin, S. (2021) 'Challenging Anglocentric feminist geography from Latin American feminist debates on territoriality', in B. Gökariksel, M. Hawkins, C. Neubert and S. Smith (eds.) *Feminist geography unbound: discomfort, bodies, and prefigured futures*. Morgantown, WV: West Virginia University Press, pp. 235–252.

Zaragocin, S. (2023) 'Geographies of the global South and the hemispheric scale', *Dialogues in Human Geography*. Available at: https://doi.org/10.1177/20438206231179227 (Accessed: 28 October 2023).

Zaragocin, S. and Caretta, M.A. (2021) 'Cuerpo-territorio: a decolonial feminist geographical method for the study of embodiment', *Annals of the American Association of Geographers*, 111(5), pp. 1503–1518.

Zaragocin, S., Ramírez, M., García, M.A. and Mendoza, Y.G. (2022) 'Intervention – "LatinX and Latin American geographies: a dialogue," *Antipode Online*. Available at: https://antipode-online.org/wp-content/uploads/2022/08/LatinX-and-Latin-American-Geographies_A-Dialogue.pdf (Accessed: 28 October 2023).

Zerai, A. (2002) 'Models for unity between scholarship and grassroots activism', *Critical Sociology*, 28(1–2), pp. 201–216.

Zimmerman, S. (2017) 'The Canadian experience: a response to "Developing standards for research practice: some issues for consideration by James Parry"', in R. Iphofen (ed.) *Finding common ground: consensus in research ethics across the Social Sciences*. Bingley: Emerald Publishing, pp. 103–110.

Index

[Note: Numbers in **bold** indicate a table. Numbers in *italics* indicate a figure on the corresponding page]